# ENVIRONMENTAL APPLICATIONS OF DIGITAL TERRAIN MODELING

## New Analytical Methods in Earth and Environmental Science

Introducing New Analytical Methods in Earth and Environmental Science, a new series providing accessible introductions to important new techniques, lab and field protocols, suggestions for data handling and interpretation, and useful case studies.

This series represents an invaluable and trusted source of information for researchers, advanced students, and applied earth scientists wishing to familiarize themselves with emerging techniques in their field.

All titles in this series are available in a variety of full-color, searchable e-book formats. Titles are also available in an enhanced e-book edition which may include additional features such as DOI linking and high-resolution graphics and video.

*Ground-Penetrating Radar for Geoarchaeology*
by Lawrence B. Conyers

*Rock Magnetic Cyclostratigraphy*
by Kenneth P. Kodama and Linda A. Hinnov

*Techniques for Virtual Palaeontology*
by Mark Sutton, Imran Rahman, and Russell Garwood

*Structure from Motion in the Geosciences*
by Jonathan L. Carrivick, Mark W. Smith, and Duncan J. Quincey

# ENVIRONMENTAL APPLICATIONS OF DIGITAL TERRAIN MODELING

JOHN P. WILSON

*Spatial Sciences Institute, University of Southern California and Institute of Geographical Sciences and Natural Resources Research, Chinese Academy of Sciences*

WILEY Blackwell

This edition first published 2018


*Registered Offices*
John Wiley & Sons, Inc., 111 River Street, Hoboken, NJ 07030, USA
John Wiley & Sons Ltd, The Atrium, Southern Gate, Chichester, West Sussex, PO19 8SQ, UK

*Editorial Office*
9600 Garsington Road, Oxford, OX4 2DQ, UK

For details of our global editorial offices, customer services, and more information about Wiley products visit us at www.wiley.com

Wiley also publishes its books in a variety of electronic formats and by print-on-demand. Some content that appears in standard print versions of this book may not be available in other formats.



*Library of Congress Cataloging-in-Publication Data*

Names: Wilson, John P. (John Peter), 1955– author.
Title: Environmental applications of digital terrain modeling / John P. Wilson.
Description: First edition. | Hoboken, NJ : Wiley-Blackwell, 2018. | Series: New analytical methods in earth and environmental science | Includes bibliographical references and index.
Identifiers: LCCN 2017048368 (print) | LCCN 2018001634 (ebook) | ISBN 9781118936207 (pdf) | ISBN 9781118938171 (epub) | ISBN 9781118936214 (hardback)
Subjects: LCSH: Digital elevation models | Three-dimensional imaging. | Digital mapping. | BISAC: SCIENCE / Earth Sciences / Geology.
Classification: LCC GA139 (ebook) | LCC GA139 .W55 2018 (print) | DDC 551.410285–dc23
LC record available at https://lccn.loc.gov/2017048368

Cover Design: Wiley
Cover Image: Photograph taken to the north of the main channel looking southward to the highest peak which marks the southeast corner of the Cottonwood Creek, MT catchment. Photograph courtesy of William K. Wyckoff.

Set in 10/12.5pt Minion by SPi Global, Pondicherry, India

10 9 8 7 6 5 4 3 2 1

*For Duncan, Ha and Vanessa who made a project like this all the more meaningful for me and to Richard Bedford, Pip Forer, Kenneth Hare, Bruce Leadley, Michael Hutchinson, Ian Moore, and John Gallant and the many others I have encountered along the way for helping to lead me to this place.*

# Contents

# List of Figures

# List of Tables

# Preface

I started writing this book in January 2015 and the journey that produced the book you see now proved to be both an exhilarating and humbling one. My primary goal from start to finish has been to write a book that describes the typical digital terrain modeling workflow that starts with data capture, continues with data preprocessing and DEM generation, and concludes with the calculation of land surface parameters and objects.

The book itself consists of seven chapters The first introduces digital elevation models, the role of scale in this work, the applications that have exploded in number and sophistication during the past 30–40 years, and a study site that is used throughout the remainder of the book to illustrate key concepts and outcomes. The second chapter describes some of the ways in which LiDAR and radar remote sensing technologies have transformed the sources and methods for capturing elevation data. It next discusses the need for and various methods that are currently used to preprocess DEMs along with some of the challenges that confront those who tackle these tasks. The third and largest of the seven chapters describes the subtleties involved in calculating the primary land surface parameters that are derived directly from DEMs without additional inputs and the two sets of secondary land surface parameters that are commonly used to model the energy and thermal regimes and accompanying interactions between the land surface and the atmosphere on the one hand and water flow and soil redistribution on the other hand. The fourth chapter examines how the primary and secondary land surface parameters have been adopted and used to extract and classify landforms and other kinds of land surface objects from digital elevation data. The role of error pops up in various guises in the second, third and fourth chapters and this state of affairs motivated Chapter 5, which explores the various errors that are embedded in DEMs, how these may be propagated and carried forward in calculating various land surface parameters and objects, and the consequences of this state of affairs for the modern terrain analyst. The sixth chapter introduces the software and services that can be used to implement and execute the digital terrain modeling workflows illustrated in the first five chapters. The seventh and final chapter reviews how terrain analysis got started, where things stand today, and what will likely happen to digital terrain modeling in the future.

This was an exciting and exhilarating project for me once I realized how much had changed since I had published my first journal article on the topographic factor in the Universal Soil Loss Equation (Wilson, 1986) and the terrain analysis book I had helped to write and co-edit with John Gallant in 2000 (Wilson & Gallant, 2000a). The methods and data have changed tremendously along with the numbers and kinds of scholars and practitioners working with terrain and the results have exceeded my wildest expectations if I compare where things stand nowadays with the status quo in the early 1980s (when I was a PhD student at the University of Toronto in Canada). This book took me two years to write as I worked simultaneously to familiarize myself with all that has been accomplished thus far, which made it both the exhilarating and humbling journey it was for me.

Given this state of play, I would be remiss if I did not thank all those scholars who have shared their knowledge and showed me the way forward over the past four decades. Some I have come to know personally because I have been afforded the opportunity and pleasure to work with them directly – this group includes John Gallant, Michael Hutchinson, Ian Moore and Tian-Xiang Yue, among others – but there are many more whose work I have come to know and appreciate from afar. You will see the works of some of these individuals listed in Table 7.1 towards the end of the book because I have taken the opportunity to list the 25 works that both guided and inspired the contents and layout of the book that you now see.

A group of institutions and people have helped me with the preparation of the book itself. I owe thanks to all those connected with the Spatial Sciences Institute at the University of Southern California and the Institute of Geographic Sciences and Natural Resource Research at the Chinese Academy of Sciences for giving me the time and freedom to devote the many months it took me to write this book. Three people, in particular, deserve special thanks. The first is Petter Pilesjö who graciously shared the code for his TFM algorithm that I used to construct Figure 3.24; the second is Beau MacDonald who helped me to prepare the many maps and diagrams you will find scattered throughout the book and who graciously read the manuscript from start to finish and helped to identify numerous omissions and errors; and the third is my partner and confidant, Ha Nguyen, without whom none of what I have accomplished here would have been possible.

This said, I hope you will find something of value as you read this book and that you will remember that any shortcomings, blunders and errors you find were completely of my own making.

# Abbreviations

| | |
|---|---|
| **3DEP** | (USGS) 3D Elevation Program |
| **ADK** | aspect-driven kinematic single-flow direction algorithm |
| **AGNPS** | agricultural non-point source pollution model |
| **AIC** | Akaike's information criterion |
| **ALOS** | Advanced Land Observing Satellite |
| **ALSM** | airborne laser swath mapping |
| **AML** | Arc Macro Language |
| **ANGD** | Australian National Gravity Database |
| **ANI** | anisotropy index |
| **ANUDEM** | Australian National University Digital Elevation Model spline interpolation method |
| **AOI** | area of interest |
| **API** | application program interface |
| **ASTER** | Advanced Spaceborne Thermal Emission and Reflection Radiometer |
| **BR** | Braunschweiger relief model |
| **cCVT** | curvature-based centroidal Voronoi tessellation |
| **CFS** | Climate Forecast System |
| **CGIAR** | (CIAT) Consultative Group for International Agricultural Research |
| **CIAT** | International Center for Tropical Agriculture |
| **CIT** | channel initiation threshold |
| **CPE** | compound point extraction |
| **CSI** | (CGIAR) Consortium for Spatial Information |
| **CTI** | compound topographic index (same as TWI) |
| **CVT** | centroidal Voronoi tessellation |
| **D4** | deterministic four-node single-flow direction algorithm |
| **D6** | deterministic six-node single-flow direction algorithm |
| **D8** | deterministic eight-node single-flow direction algorithm |
| **D8-LAD** | deterministic eight-node least angular deviation single-flow direction algorithm |
| **D8-LTD** | deterministic eight-node least transversal deviation single-flow direction algorithm |
| **D∞** | infinity single-flow direction algorithm |
| **D∞-LAD** | deterministic infinite least angular deviation single-flow direction algorithm |

**D∞-LTD** deterministic infinite least transversal deviation single-flow direction algorithm
**DCW** Digital Chart of the World
**DEM** digital elevation model
**DEMON** digital elevation model network extraction multiple-flow direction algorithm
**Dev** deviation from mean elevation
**Diff** difference between elevation at the center of a local neighborhood and the mean elevation in this neighborhood
**DGPS** Differential Global Positioning System
**DLG** digital line graph
**DOI** degree of importance
**DrEICH** Drainage Extraction by Identifying Channel Heads channel initiation method
**DSM** digital surface model
**DTED** (US NGA) Digital Terrain Elevation Data
**DTM** digital terrain model
**$D_{trig}$** Shelef and Hilley multiple-flow direction algorithm
**ECIT** expanded channel initiation threshold
**EDM** electronic distance measurement unit
**ESA** European Space Agency
**EVAAL** Erosion Vulnerability Assessment for Agricultural Land
**FCM** fuzzy c-means clustering method
**FMFD** Freeman multiple-flow direction algorithm
**FGDC** (US) Federal Geographic Data Committee
**GAM** general additive model
**GAT** (Whitebox) Geospatial Analysis Tools
**GCP** ground control point
**GDAL** Geospatial Data Abstraction Library
**GDEM** (ASTER) Global Digital Elevation Model
**GEOS** Geometry Engine – Open Source software suite
**GFS** Global Forecasting System
**GIS** geographic information system
**GLM** generalized linear model
**GLMM** generalized linear mixed model
**GLOBE** Global Land One-km Base Elevation Data
**GLWD** Global Lake and Wetland Dataset
**GMTED** Global Multi-resolution Terrain Elevation Dataset (with a horizontal spacing of 15 arcseconds)
**GPS** Global Positioning System
**GRASS** Geographic Resources Analysis Support System
**GTL** geomorphic transport law (-based landscape development models)
**GTOPO30** Global Digital Elevation Model (with a horizontal spacing of 30 arcseconds)
**GUI** graphical user interface

| | |
|---|---|
| **HBV** | Hydrologiska Byråns Vattenbalansavdelning model |
| **HHSM** | hierarchical hexagonal surface model |
| **HIP** | hexagonal imaging processing system levels |
| **HLI** | heat load index |
| **HRRR** | High Resolution Rapid Refresh weather forecasting system |
| **HRS** | hillslope–riparian–stream |
| **HYDRO1k** | global topographic dataset derived from GTOPO30 DEM |
| **HydroSHEDS** | hydrologic data and maps based on Shuttle Elevation Derivatives at Multiple Scales |
| **ICESat** | Ice, Cloud, and Land Elevation Satellite |
| **IDL** | Interactive Data Language |
| **IDW** | inverse distance weighted interpolation method |
| **ILWIS** | Integrated Land and Water Information System |
| **IfSAR** | interferometric synthetic aperture radar |
| **IMI** | integrated moisture index |
| **InSAR** | interferometric synthetic aperture radar |
| **JAXA** | Japanese Aerospace Exploration Agency |
| **LAPSUS** | Landscape Process Modeling at Multi-dimensions and Scales landscape evolution model |
| **LAS** | laser scanning standard data exchange file format |
| **LiDAR** | light detection and ranging point cloud (i.e. data) |
| **LoD** | level of detail |
| **LOS** | large-over-small ratio |
| **LPJ-DH** | Lund–Potsdam–Jena-distributed hydrology |
| **LPJ-GUESS** | Lund–Potsdam–Jena general ecosystem simulator |
| **LPJ-wsl** | Lund–Potsdam–Jena Wald Schnee und Landschaft version |
| **LS** | length–slope (or topographic) factor in USLE and RUSLE |
| **LSM** | land surface model |
| **LV** | local variance method |
| **MCS** | Monte Carlo simulation |
| **MDEMON** | Moore digital elevation model network extraction multiple-flow direction algorithm |
| **MD∞** | triangular multiple-flow direction algorithm |
| **MDTA** | maximum depth tracing algorithm |
| **METI** | (Japanese) Ministry of Economy, Trade, and Industry |
| **MF** | mass flux multiple-flow direction algorithm |
| **MFD** | multiple-flow direction algorithm |
| **MFD-md** | multiple-flow direction local maximum downslope gradient algorithm |
| **MMFD1** | Moore multiple-flow direction algorithm (variant 1) |
| **MMFD2** | Moore multiple-flow direction algorithm (variant 2) |
| **MoRAP** | Missouri Resource Assessment Partnership |
| **MPI** | message parsing interface |
| **MRDB** | multiple representation database |

| | |
|---|---|
| **MRMS** | Multi-Radar/Multi-Sensor weather forecasting system |
| **MRRTF** | multi-resolution ridgetop flatness index |
| **MRS** | multi-resolution segmentation algorithm |
| **MRVBF** | multi-resolution valley bottom flatness index |
| **MTD** | mass transport and deposition index |
| **NAD83** | North American Datum 1983 |
| **NAIP** | (US) National Agriculture Imagery Program |
| **NASA** | (US) National Aeronautics and Space Administration |
| **NAW** | neighborhood analysis window |
| **NCAR** | National Center for Atmospheric Research |
| **NCEP** | National Centers for Environmental Protection |
| **NED** | (US) National Elevation Dataset |
| **NEE** | net ecosystem exchange |
| **NetCDF** | set of software libraries and self-describing, machine-independent data formats |
| **NGA** | (US) National Geospatial-Intelligence Agency |
| **NHD** | (US) National Hydrography Dataset |
| **NHDPlus** | (US) National Hydrography Dataset (Enhanced) |
| **NIMA** | (US) National Imagery and Mapping Agency (which has been subsumed in the NGA) |
| **NLCD** | (US) National Land Cover Database |
| **NMAS** | (US) National Map Accuracy Standards |
| **NOAA** | (US) National Oceanic and Atmospheric Administration |
| **NOMADS** | NOAA Operational Model Archive and Distribution System |
| **NRCS** | (USDA) Natural Resources Conservation Service |
| **NumPy** | numerical Python library for scientific computing |
| **NWM** | (US) National Water Model |
| **ORI** | ortho-rectified image |
| **PaRGO** | parallel raster-based geocomputation operators |
| **PCTG** | measures the elevation of the point in the center of a local neighborhood as a percentage of the elevation range |
| **PCTL** | ranking of the point in the center of a local neighborhood relative to all points within this local neighborhood |
| **PDF** | probability distribution function |
| **Pe** | Péclet number |
| **PMFD** | Pilesjö form-based multiple-flow direction algorithm |
| **PRISM** | Panchromatic Remote-sensing Instrument for Stereo Mapping |
| **PROMETHEE** | Preference Ranking Organization Method for Enrichment Evaluations |
| **QGIS** | Quantum GIS |
| **QMFD1** | Quinn multiple-flow direction algorithm (variant 1) |
| **QMFD2** | Quinn multiple-flow direction algorithm (variant 2) |
| **OpenMP** | Open-Multi-Processing programming model |
| **Range** | full range of elevations reported in a local neighborhood |

| | |
|---|---|
| **RAP** | (short-range) Rapid Refresh weather forecasting system |
| **REST** | representational state transfer web services architecture |
| **Rho8** | randomized eight-node single-flow direction algorithm |
| **RMSE** | root mean square error |
| **ROC-LV** | rate of change of local variance |
| **Rotor** | third measure of curvature describing the curvature of flow lines |
| **RRMSE** | relative root mean square error |
| **RST** | Regularized Spline with Tension interpolation method |
| **RUGN** | ruggedness index |
| **RUSLE** | revised universal soil loss equation |
| **SAGA** | System for Automated Geoscientific Analyses |
| **SAR** | synthetic aperture radar |
| **SCA** | specific catchment area |
| **SCI** | shape complexity index |
| **SD** | standard deviation of elevation within a user-defined local neighborhood |
| **SDFAA** | spatially distributed flow-apportioning algorithm |
| **SEI** | site exposure index |
| **SFD** | single-flow direction algorithm |
| **SI** | semantic import model fuzzy classification approach |
| **SLP** | stream longitudinal profile |
| **SOF** | saturation overland flow |
| **SPI** | stream power index |
| **SPOT** | Satellite Pour l'Observation de la Terre |
| **SQL** | Structured Query Language |
| **SR** | similarity relation model fuzzy classification approach |
| **SRAD** | solar radiation program included in TAPES suite |
| **SRF** | surface roughness factor |
| **SRI** | surface roughness index |
| **SRTM** | Shuttle Radar Topographic Mission |
| **SSURGO** | (US State) Soil Survey Geographic Database |
| **STI** | sediment transport index |
| **SWAMPS** | Surface Water Microwave Product Series (satellite-based) |
| **SWAT-VSA** | Soil and Water Assessment Tool, variable source area model |
| **SWBD** | SRTM water body data |
| **TAPES** | Terrain Analysis Programs for the Environmental Sciences |
| **TAS** | Terrain Analysis System |
| **TauDEM** | Terrain Analysis Using Digital Elevation Models |
| **TDR** | time-domain reflectometry |
| **TFM** | triangular form-based multiple-flow direction algorithm |
| **TFN** | triangular facet network multiple-flow direction algorithm |
| **TI** | topographic index (same as TWI) |
| **TIN** | triangulated irregular network |
| **TOPOGRID** | variant of topo-to-raster interpolation algorithm that was part of the ArcGIS platform |

| | |
|---|---|
| **TOPMODEL** | topography-based hydrology model |
| **TPI** | topographic position index |
| **TUCL** | total upstream channel length |
| **TWHC** | total water-holding capacity |
| **TWI** | topographic wetness index (same as CTI) |
| **UK** | United Kingdom |
| **USA** | United States of America |
| **USDA** | US Department of Agriculture |
| **USGS** | US Geological Survey |
| **USLE** | universal soil loss equation |
| **UTM** | Universal Transverse Mercator coordinate system |
| **VBF** | valley bottom flatness (score) |
| **VNIR** | visible and near-infrared portion of electromagnetic spectrum |
| **VR** | valley recognition drainage network delineation method |
| **VRT** | virtual raster format |
| **VSLF** | Variable Source Loading Function model |
| **WBD** | Watershed Boundary Dataset |
| **WEPP** | Water Erosion Prediction Project |
| **WGS** | World Geodetic System |
| **WRF-Hydro** | Weather and Forecasting Hydrologic Model |

# 1 Introduction

The land surface plays a fundamental role in modulating several of the Earth's dynamic systems including a large number of atmospheric, geologic, geomorphic, hydrologic, and ecological processes. The topography or shape of this surface constrains the operational scale of surface processes, and partially governs both climate and tectonic forcing (e.g. Molnar & England, 1990; Bishop et al., 2010; Koons, Upton & Barker, 2012). The strength of the linkage between form and process can range from weak to strong, and may or may not be inherently visible on the landscape depending on the history and complexity of the topography. Nevertheless, moderate to strong linkages have been observed, such that an understanding of the character of the land surface can provide insights about the nature and magnitude of the aforementioned processes (e.g. Zhu et al., 1997; Hutchinson & Gallant, 2000; Bishop et al., 2012b). Consequently, there is growing interest in quantitatively characterizing the land surface and segmenting the topography into fundamental spatial units, as the topography inherently represents the results of the interplay between various systems, and records an imprint of landscape dynamics (over some varying but typically finite time).

Applications that exploit this knowledge usually rely on digital elevation models (DEMs) to represent the surface and a steadily increasing and sophisticated range of techniques for topographic analysis, modeling, and visualization. Many of these innovations have accompanied the rapid proliferation of geographic information technologies, which have provided new data, algorithms, analysis, and modeling techniques for characterizing the Earth's surface. These techniques and the accompanying digital data represent the evolution of the field of geomorphometry, which in its broadest sense refers to the science of quantitative land surface characterization (Pike, 1995, 2000) or digital terrain modeling. For more details regarding the history, definitions and terminology used in geomorphometry,

*Environmental Applications of Digital Terrain Modeling*, First Edition. John P. Wilson.

see Wilson and Gallant (2000a), Li, Zhu and Gold (2005), Peckham and Jordan (2007), Zhou, Lees and Tang (2008), Hengl and Reuter (2009), Wilson (2012), and Wilson and Bishop (2013).

Modern geomorphometry focuses on the extraction of land surface parameters and the segmentation of the landscape into spatial entities or features (i.e. land surface objects) from digital topography. This characterization relies on the general and specific modes of geomorphometric analysis that were first defined by Evans (1972). The general mode attempts to describe the continuous land surface and the specific mode describes discrete surface features (i.e., landforms). Pike, Evans and Hengl (2009) have since updated these definitions, such that a land surface parameter is a descriptive measure of surface form (e.g. slope, slope azimuth or aspect, or curvature) and a land surface object is a discrete surface feature (e.g. a watershed, cirque, alluvial fan, stream, or drainage network). Although this definition represents an improvement, it is worth noting that this is a somewhat arbitrary distinction and there are already examples of work that show these two views are closely linked to one another (e.g. Gallant & Dowling, 2003; Hengl, Gruber & Shrestha, 2003; Fisher, Wood & Cheng, 2004; Deng & Wilson, 2008) and that anticipating and representing these linkages will likely grow in importance in future applications.

Geomorphometry is simultaneously a rapidly evolving and yet complicated field. This is partly due to its multidisciplinary nature and the rapid growth of geographic information and remote sensing technologies during the past 30 years. Similar to the field of geographic information science, it draws key concepts and ideas from and provides a variety of inputs and insights to many related disciplines. It not only attempts to deal with theoretical issues involving representation and spatiotemporal variation, but also includes issues of data collection and analysis, numerical modeling, and the utilization of knowledge from other domains for conceptual and practical problem-solving (Wilson & Bishop, 2013). Technological advances have provided an increasing number of digital remote sensing data sources and have transformed the computing platforms used to calculate selected terrain attributes. However, there are many subtleties involved in creating DEMs from these new as well as traditional sources, and it is important to recognize the empirical nature of many forms of spatial analysis and modeling and the implications this has for the assumptions and validity of various approaches (Goodchild, 2011; Bishop et al., 2012b).

Many questions still remain, and both scientists and practitioners must be aware of the advantages and disadvantages associated with various representations and data structures, metrics and indices, spatial modeling approaches, and their utility for scientific investigations. Furthermore, investigators must be familiar with the role of scale and the mathematical underpinnings of geomorphometric analysis in order to adequately use information and interpret the results (e.g. Wilson & Burrough, 1999; Bishop & Shroder, 2004; Yue et al., 2007; Minár & Evans, 2008; Bishop et al., 2012a,b; Florinsky, 2012; Evans, 2013).

These subtleties point to a series of key questions that at the highest or most general level include the following.

1. How should the land surface be represented?
2. What is the preferred scale and why?
3. What elevation sources are available and which would work best for the opportunity and/or problem at hand?
4. What preprocessing is required to produce a usable DEM?
5. How will DEM error get propagated and how should this uncertainty be handled throughout subsequent analyses?
6. What methods are best for calculating specific land surface parameters?
7. What methods are best for delineating specific land surface objects?
8. Is there a need to develop new land surface parameters and objects to address particular problems?
9. What approaches and metrics or indices are best suited to a particular mapping application and do methods even exist?
10. Does an adequate model exist or do we need to develop or modify one for the opportunity and/or problem at hand?

Many of these questions can be attributed to the steady growth in the number of parameters and algorithms for processing DEMs and extracting both the descriptive measures (parameters) and surface features (objects). The values of these parameters and/or the characteristics of the objects will vary depending on a variety of factors, including the parameterization scheme, the measurement scale of the data, the mathematical model by which they are calculated, the size of the search window, and the grid resolution.

Two sets of issues – the role of DEMs and scale in terrain analysis, modeling, and visualization – are taken up next since the ways we conceptualize and handle this pair of issues will influence all that we do.

## 1.1 Role of DEMs

The DEM has three components, as the name implies (Liu, Hu & Hu, 2015). The "D" in DEM, for example, stands for digital and of course refers to the kinds of digital data, such as digital line graphs (DLGs), triangulated irregular networks (TINs), grids, and light detection and ranging (LiDAR) point clouds, used to represent the terrain surface. Similarly, the "E" normally refers to the bare-earth elevation void of vegetation and non-natural features and the elevation of the surfaces of water bodies, but the term may include the aforementioned features on the land surface and/or the bathymetry of water bodies. These first two components have been described in great detail from a variety of perspectives during the past few decades and are discussed in more detail in Chapter 2.

The “M” in DEM, on the other hand, has received much less attention. Liu et al. (2015) have argued that a DEM can be expected to (i) serve as a schematic description of the terrain; (ii) account for the known or inferred properties of the terrain; and (iii) be used to further our understanding of terrain characteristics. The first of these requirements is straightforward because every DEM is made up of a finite number of points whereas the terrain itself has an infinite number of points. The challenge, therefore, is to be able to construct DEMs that account for the known and/or inferred properties of the terrain surface.

Making matters worse, much of the work has focused on the DEMs themselves rather than the terrain properties thus far. Two notable exceptions – the work of Hutchinson and colleagues (e.g. Hutchinson, 1989; Hutchinson & Gallant, 2000; Hutchinson et al., 2013) and Liu and colleagues (e.g. Hu, Liu & Hu, 2009a,b; Liu et al., 2012, 2015) – focus on the terrain properties and their role in building DEMs. Hutchinson and colleagues have long stressed the importance of surface shape and drainage structure when evaluating DEMs. Liu et al. (2015), on the other hand, recently described why a DEM must take account of three known or inferred properties: (i) that each terrain point has a single, fixed elevation; (ii) that terrain points have an order and sequence that is determined by their elevations; and (iii) that the terrain has skeletons which can provide a schematic description of the terrain surface. The views of these authors complement one another because the three aforementioned properties would capture the terrain shape and drainage structure. The three terrain properties noted by Liu et al. (2015) are explored in more detail below.

The first property, that each terrain point has a single fixed although possibly unknown elevation, has two implications for DEM generation. The first is that we need a DEM generation function that produces one estimate of elevation and ensures a one-to-one relationship between the predicted and real-world elevation values. Liu et al. (2015) refer to such a generation function as a bijective function or bijection and in previous work showed how first-order interpolators, such as linear interpolation in one dimension, TINs, and bilinear interpolation in a rectangle satisfy this bijection requirement automatically (Hu et al., 2009a). However, some higher-order, piecemeal, polygonal interpolation methods (e.g. Kidner, 2003; Li, Taylor & Kidner, 2005; Shi & Tian, 2006) that divide the topographic surface into contiguous and non-overlapping pieces so that the interpolation can be conducted piece-by-piece cannot guarantee that this correspondence holds everywhere and therefore fail this test. The second implication concerns the vertical accuracy and methods used to evaluate and ensure that the vertical error is acceptable. The root mean square error (RMSE) used by the US National Standards for Spatial Data Accuracy (FGDC, 1998) has been heavily criticized because it will only be effective if the vertical errors are random, independent and identically distributed, which seldom occurs in real-world landscapes (Fisher & Tate, 2006; Shortridge, 2006; Höhle & Höhle, 2009; Liu et al., 2012).

Hu et al. (2009b) have demonstrated how approximation theory can be used to evaluate vertical accuracy. This approach asks whether the largest error at a point in the entire terrain is acceptable and, if so, assumes that the errors at any other point must be acceptable. Liu et al. (2012) have shown how the error bands produced with this approach could be used to assess the vertical accuracy, and thus control the quality, of a DEM created by linear interpolation. They have argued that approximation theory can be used to not only assess overall accuracy, but also to point to those areas where user-desired accuracy is not met and more effort might be expended to collect additional reference data and reduce these errors further.

The sequence created by ordering or ranking terrain points according to their elevation constitutes the second important property of terrain (Liu et al., 2015). This property is related to the concept of isomorphism in set theory and means that any property true for one dataset (i.e. the real-world terrain values) is true for the other (i.e. the DEM). This property speaks to the shape of the terrain surface and means, for example, that if the DEM suggests that the flow direction is from point *a* to point *b*, then, in theory, this must be true in the field. However, the ability to achieve this result may be compromised by the limitations of the flow-direction algorithm used and/or the discretization of the landscape (O'Neil & Shortridge, 2013), as will be explained in subsequent chapters.

Hu et al. (2009a) described two requirements that must be satisfied to create an isomorphic DEM. The first is the ability to divide the terrain into a set of contiguous, monotonic patches with no "bumps" or "dips" so that each value can be reasonably modeled as a smooth facet and the second is that the DEM function as a bijection (Liu et al., 2015).

The third and final terrain property follows from the fact that the various points describing the terrain surface do not play the same roles in delineating the key characteristics of the terrain (Liu et al., 2015). Peaks, pits, passes, ridges, and channels form the basic structure of the terrain and must be retained with high fidelity during DEM generation for the resultant DEM to support the kinds of geomorphologic, hydrologic, and ecological applications highlighted in Section 1.3 and in the subsequent chapters.

Liu et al. (2015) described the challenges of achieving this result as either a sampling or a generalization task. Modern satellite remote sensing data collection systems provide grid-based data values and bring both of these aforementioned tasks to the fore. Many algorithms have been proposed to extract terrain features and the challenge is how to rank and displace the elevation values in a way that preserves their elevation sequence and order. Liu et al. (2015), for example, distinguished three groups of terrain points based on their information content: (i) critical points, such as peaks, pits, and saddles; (ii) special points located on a ridge or in a valley; and (iii) ordinary points. This issue is taken up again in Section 1.2 and reverberates throughout the descriptions of the methods and source data that can be used for constructing DEMs in Chapter 2.

These three properties also have important implications for how we construct and use the increasingly popular LiDAR-derived DEMs. LiDAR provides large numbers of high-density mass points and if processed appropriately can provide bare-earth terrain models with high vertical accuracy and the preservation of the terrain structure (i.e. the shape). Liu et al. (2015) argued that the future evolution of the efficacy of DEMs created from this and other sources should focus on their ability to satisfy the requirement of bijection, minimal vertical error, isomorphism, and generalization. The new airborne and space-borne remote sensing data collection systems will need more sophisticated assessments of their strengths and weaknesses than was provided for the legacy cartographic datasets given the more or less singular focus on horizontal resolution (i.e. scale) and vertical accuracy in those early assessments.

Many of these early assessments also took up the issue of scale, and the fundamental roles of scale in terrain analysis and modeling specifically and in geomorphology more generally are taken up next in Section 1.2.

## 1.2 Role of Scale

Goodchild (2011) recently described the three common uses of the term "scale" in a review of the major scale issues that arise when geographic information and remote sensing technologies are used in support of research on the Earth's surface and near-surface, as is typically the case nowadays in geomorphometry.

The first use is common in cartography where scale normally refers to the representative fraction that defines the scaling of the Earth's surface to a flat sheet of paper. This ratio has traditionally been used to define a map's level of detail, its content, and its potential accuracy notwithstanding the inevitable variability that results from representing the curved surface of the Earth on a flat sheet of paper. Goodchild had earlier argued that the representative fraction is undefined for digital data (Goodchild & Proctor, 1997) and, perhaps not surprisingly, most of the geomorphometry articles and applications during the past five decades have tended to focus their attention on the second and/or third meanings of scale.

The second and third meanings of scale, both of which can be defined for digital data, refer to the spatial extent of a study area and the resolution, or degree of detail (Figure 1.1). Both refer primarily to the spatial and sometimes the temporal dimensions, as would happen, for example, if the goal was to simulate water flow across a farm field or meadow following many years of cultivation and/or ecological succession. Both can be expressed as linear or areal measures and Goodchild (2001) has shown that their dimensionless ratio, which he termed the "large over small" ratio (LOS), is remarkably constant across a wide range of data sources and applications within the range 103–104. That said, most of the literature on scale in both

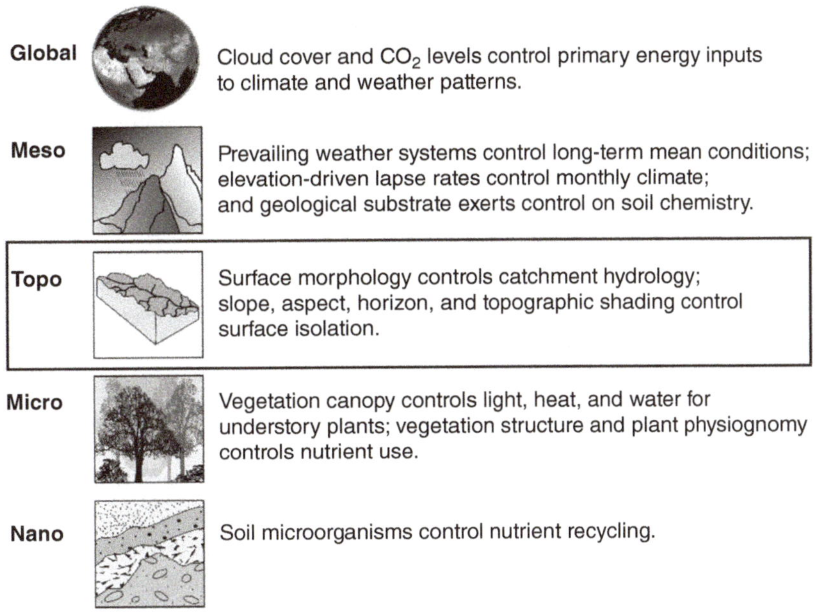


**Figure 1.1** Scales at which various biophysical processes dominate calculation of primary environmental regimes. *Source*: Mackey (1996). Reproduced with permission of NCGIA, Santa Barbara.

geomorphology and geomorphometry focuses on the "small" or resolution dimension of scale.

The choice of resolution used for digital terrain analysis and modeling should be guided by the spatial resolution of the processes and/or patterns of interest (see Figure 1.1). The resolution is typically constrained by our limited ability to sense, capture and handle massive volumes of data and, as a consequence, it is essential that the data used to model any given process include all of the important detail needed for accurate modeling of the said process.

However, few theories of process make spatial resolution explicit, and as Goodchild (2011, p. 6) has noted "the researcher whose model fits reality to a level that is less than perfect, as all models must, is left not knowing whether this misfit is due to the effects of spatial resolution, or to an imperfection of the model, or both." The results of analysis and modeling will clearly be misleading if the process of interest is significantly influenced by detail smaller than the spatial resolution of the data. The challenge, as we will see shortly, involves knowing when the level of detail will suffice to depict the surficial processes operating in one or more specific environments.

Most of the physical phenomena discussed in this book, from topography and precipitation to soil moisture content and land cover, are more often conceptualized using the general mode of Evans (1972) as continuous fields rather than as collections of discrete objects using the specific mode. These fields will be independent of scale in principle but their digital representations will in practice nearly always embody scale to some degree (Goodchild, 2011). In the raster case which is commonly used for DEMs (see Chapter 2 for additional details) the resolution is expressed by the size of the raster cells, which are almost always square in the two-dimensional case and cause numerous errors when the raster is laid over a curved surface (i.e. a mountainous area like the Himalaya with high relief).

This leads to the conclusion that nearly all of our work in geomorphometry is scale-specific. The length of drainage paths and stream networks and slope, for example, cannot be defined or measured independently of scale. Most slope calculations take the eight neighboring cells (i.e. the queen's case) and weight the eight neighbors depending on their distance from the central cell to obtain estimates of the two components of slope (e.g. Horn, 1981). In addition, the slopes and aspects calculated from 30-m DEMs (or any other resolution for that matter) will be representative of grid resolutions two or three times larger than the original DEM grid spacing (Hodgson, 1995). The problem then is that the resulting estimates will be dependent on the raster cell size. For example, larger cells will in general yield smaller estimates of slope (Goodchild, 2011). There is also the problem where topographic surfaces are often subject to breaks in slope and one or more of the topographic attributes or derivatives will be left undefined.

Goodchild (2011) attributed these problems, in part, to the lack of theory about scale and thus to the difficulty of formalizing it. He reviewed three frameworks – fractals, geostatistics, and spectral or Fourier analysis – that might support a formal discussion of scale.

In the case of fractals, Goodchild (1980) showed a long time ago that many geographic phenomena exhibit fractal behavior, which means that the rate of information loss or gain with scale changes is orderly and predictable through principles that have come to be known as scaling laws (Mandelbrot, 1977). Many authors have documented the fractal properties of topographic surfaces (e.g. Clarke & Schweizer, 1991; Klinkenberg, 1994) and river networks (Tarboton, Bras & Rodriguez-Iturbe, 1988; La Barbera & Rosso, 1989; Liu, 1992; Nikora, Sapozhnikov & Noever, 1993) during the past 30 years.

Geostatistics, on the other hand, proposes to refine the spatial resolution of a point dataset artificially by replacing a finite number of observations with potentially an infinite number by modeling the spatial autocorrelation among point measurements using the correlogram or variogram (Goovaerts, 1997). These two aforementioned graphs show the distances over which the values are strongly correlated and give this class of techniques, generally known as Kriging, their claim to theoretical rigor (Goodchild, 2011).

The third and final framework is provided by spectral or Fourier analysis, which decomposes any field variable into its harmonic components. Clarke (1988), for example, used these techniques to discard unnecessary detail in topographic surfaces and, more recently, Gallant and Hutchinson (1997) used a positive wavelet representation with one-dimensional topographic data (i.e. profiles) to identify changes in topographic structure with scale.

Notwithstanding these gains, the scientist will, more often than not, still be left with the dilemma that they cannot determine whether the failure of a model to fit perfectly is due to the model itself, or to the coarse spatial resolution of the data, or both (Goodchild, 2011). This state of affairs is exemplified by the multitude of studies that have examined the sensitivity of selected topographic attributes to some combination of data source, structure, and/or grid resolution (i.e. cell size) over the past 25 years. Some examples of these studies are described briefly below and the discussion closes with some works that have attempted to identify the best resolution supported by the digital elevation data at hand or, even better, to identify the best grid resolution to study the phenomena or problem at hand.

Two of the earliest studies, by Panuska, Moore and Kramer (1991) and Vieux and Needham (1993), examined the effects of data structure and cell size on Agricultural Non-Point Source Pollution (AGNPS) model inputs and showed how the computed flow-path lengths and upslope contributing areas varied with cell size. Issacson and Ripple (1991) compared 1° US Geological Survey (USGS) 3-arcsecond (~90 m) and 7.5-minute USGS 30-m DEMs and Lagacherie et al. (1993) examined the effect of DEM data source and sampling pattern on computed topographic attributes and the performance of a terrain-based hydrologic model. Vieux (1993) examined the sensitivity of a surface runoff model to the effect of cell size aggregation and smoothing using different-sized windows. Moore, Lewis and Gallant (1993c) examined the sensitivity of computed slope and steady-state topographic wetness index (TWI) values across 22 grid resolutions for three moderately large catchments (~100 km$^2$) in southeastern Australia. Chairat and Delleur (1993), Wolock and Price (1994), and Zhang and Montgomery (1994) examined the effects of DEM cell spacing on the TWI and TOPMODEL (topography-based hydrology model) watershed model predictions.

Garbrecht and Martz (1994) examined the impact of DEM resolution on extracted drainage properties for a 84 km$^2$ study site in Oklahoma using hypothetical drainage configurations and DEMs of increasing size. They derived various quantitative relationships and concluded that the grid resolution must be selected relative to the size of the smallest drainage features that are considered important for the work at hand. Gyasi-Agyai, Wilgoose and de Troch (1995) examined the effects of the vertical accuracy and map scale of DEMs on the topographic attributes used in hydrology and concluded that a ratio of the average drop per pixel and vertical resolution greater than unity could be used to define the minimum grid resolution for reliable channel network definition for any given vertical resolution. Schoorl, Sonneveld and Veldkamp (2000) constructed a simple process model and used several artificial DEMs in a series of experiments to show the effects of DEM resolution and the choice of flow routing method (both steepest descent and multiple flow directions) on modeled erosion and sedimentation rates.

Wilson, Repetto and Snyder (2000), Tang et al. (2002), Kienzle (2004), Wu, Li and Huang (2008), and Vaze, Teng and Spencer (2010) examined the effect of grid resolution on a variety of computed topographic attributes. Deng, Wilson and Bauer (2007) documented how scale dependencies in terrain analysis varied across a series of landform classes that were defined in a reproducible way using an unsupervised landform clustering procedure. Nguyen and Wilson (2010) examined the sensitivity of a quasi-dynamic

TWI to the choice of DEM resolution, flow routing algorithm, and the representation of soil variability. Drăguţ and Eisank (2011) calculated the homogeneity of a given land-surface parameter or a combination of several parameters across multiple scales (i.e. grid resolutions) and proposed using this method to delineate landform objects. Hasan, Pilesjö and Persson (2013b) examined how the slope, drainage area, and TWI varied with the resolution of six DEMs with spatial resolutions ranging from 0.5 to 90 m for a flat peatland area in Sweden. Mukherjee et al. (2013) examined the effect of grid resolution on the TWI using four small watersheds in the Himalaya with varying relief (i.e. roughness) and showed that the mean of the TWI surface increased and the standard deviation decreased with an increase in grid spacing, and that this effect was more pronounced in rough terrain.

The aforementioned studies provide some useful findings but stop short of connecting the DEM grid resolution with either the quality of the elevation source data and/or the scales at which the phenomena of interest occur or operate on the Earth's surface. Nearly 20 years ago, Bates, Anderson and Horrit (1998) showed how high-frequency information is lost at progressively larger grid resolutions. However, the reverse may not always be true and the likelihood that finer resolution grids will yield valuable hydrologic information, for example, will depend on both the phenomena of interest and the characteristics of the study site (e.g. Grayson et al., 1993; Wilson & Gallant, 2000b). That said, there have been several studies published during the past 10–15 years that have tackled the grid resolution as an optimization problem and several others that have tried to link the choice of grid resolution to the phenomena of interest in one or more specific landscape settings. Several examples from these two groups of studies are discussed in turn below.

In the first of two optimization studies, Hengl (2006) proposed a series of empirical and analytical rules to select a suitable grid resolution for output maps based on the inherent properties of the input data. In the second study, Albani et al. (2004) performed a series of numerical simulations with DEMs from British Columbia, Canada, and showed how increasing the grid resolution reduced the propagation of elevation errors in the derived topographic variables but at the expense of topographic detail. These authors then proposed a methodology to evaluate quantitatively the loss of topographic detail and an analytical model of error propagation to select the scale or range of scales at which to calculate topographic variables from a DEM.

Fortunately, the second group of four studies has gone further and endeavored to identify the level of detail needed to use topographic attributes to support the study of specific phenomena (i.e. soil moisture control on the composition and vigor of vegetation cover) in specific landscape settings.

In the first of these studies, Florinsky and Kuryakova (2000) proposed a statistical method to determine an adequate grid size ($w$) for DEMs used in landscape studies. Their method included four steps as follows: (i) the derivation of a DEM set using a series of $w_i$; (ii) performing a correlation analysis of data on a landscape property and a topographic attribute estimated with various $w_i$; (iii) plotting of the correlation coefficients obtained versus $w$; and (iv) the determination of smoothed plot portions indicating intervals of an adequate $w$. The authors next employed this method to study the influence of topography on the spatial resolution of soil moisture ($M$) at the microscale (i.e. using DEMs with grid resolutions ranging from 1 to 7 m). Specifically, they measured the dependence of $M$ on four topographic attributes (slope gradient, horizontal, vertical, and mean curvatures) across 13 values of $w_i$ and showed how this method allowed the estimation of an adequate area of landform (i.e. square grid cells measuring 2.25–3.25 m on a side) to measure the topographic control of soil moisture distribution for a study site on the East European Plain near the city of Pushchino in Russia.

The second study by Pain (2005) examined the character and scale of slopes near Picton in New South Wales, Australia and concluded that only an original ground survey and, to a lesser extent, a 25-m DEM gave any indication of the shape of the land surface. The 50-m and 100-m DEMs barely resembled the land surface and values from this latter pair of DEMs did not reflect the original slope form and gave no indication of the slope values and the processes operating in this landscape. The results showed that a grid resolution of 5 m was adequate to capture the scale of the surface processes and therefore the likely variation of related phenomena such as soil attributes. Pain (2005) also noted that the resolution required may be as small as 1 m or as large as 100 m in other areas and that this can only be determined by geomorphic analysis of landform shape and process. Pain (2005) thought this would require ground survey in most instances and that the slope values derived from most available DEMs will be of limited value as descriptions of real-world landscapes and processes unless the data are at a resolution that equals or exceeds the scale of the slope and regolith processes.

The third study by Martinez et al. (2010) examined the effects of DEM grid resolution for accurate representation of the land surface and drainage network in a small agricultural catchment in the Upper Hunter Valley region of New South Wales, Australia. These authors used a hierarchical scaling approach to investigate the effect of increasing DEM grid size on a number of geomorphic and hydrologic descriptors and the results showed that average slope gradients decreased and the drainage network became increasingly simplified as DEM grid size increased. A comparison of long-term field soil moisture data with TWIs derived from DEMs showed that a 5-m (or better) DEM was required for the current study site in order to accurately capture catchment geomorphology and hydrology and to model the spatial distribution of soil moisture.

The fourth, and in many ways most impressive work in this second group is that of Leempoel et al. (2015). These authors set out to explore how very high resolution DEM-derived variables could be used to characterize mosaic habitats along a 2-km long alpine ridge encompassing the subalpine–alpine ecotone in the western Swiss Alps. The topography was assumed to be a major driver of air temperature, humidity and soil moisture and therefore the distribution of plants, and the study set out to assess the goodness-of-fit and significance of a series of models connecting several DEM-derived variables (i.e. primary and secondary land surface parameters) and climatic variables measured in the field based on DEM-derived variables computed at resolutions of 0.5, 1, 2, and 4 m. The DEM-derived variables at 1, 2, and 4 m spatial resolutions were generated from the original 0.5 m resolution of the processed LiDAR source data using a Gaussian pyramid. Their associations with local climatic factors and ecological indicators derived from species compositions were assessed with multivariate generalized linear models (GLMs) and generalized linear mixed models (GLMMs). The results showed that the slope, aspect, curvature and less popular topographic wetness and ruggedness indices showed significant associations with measured ambient humidity and soil moisture and that the spatial resolution of the DEM-derived variables had a significant influence on the model's strength, with coefficients of determination decreasing with coarser resolutions and showing a local optimum at about 2 m resolution, depending on the variable considered. Leempoel et al. (2015, p. 2) noted how "most studies in ecology use variables at a single resolution with no consideration of scale representativeness" and that the results of this and similar studies (e.g. Lassueur, Joost & Randin, 2006; Kalbermatten et al., 2012; Cavazzi et al., 2013) support the use of multiscale topographic attributes to provide surrogates for important climatic variables and suitable surrogates for direct measurements in evolutionary ecology studies at local scales.

Given the findings from the large number of studies which have tackled the effect of grid resolution (i.e. scale) on computed topographic attributes and their meaning, coupled with the more or less continuous and for the most part, implicit analysis of the benefits of detail versus cost, we are fortunate that the past quarter century has been marked by dramatic increases in the availability of very high resolution elevation data fueled by technological innovation (as will be described in Chapter 2).

With this commentary as background, we turn our attention next to a review of some of the ways in which terrain modeling has been used to support environmental knowledge discovery and problem-solving during the past 50 years.

## 1.3 Survey of Applications

The land surface parameters and objects described in this book have been adopted and used in a large number and variety of environmental applications and settings during the past five decades. The examples introduced in this section give a sense of the breadth and depth of their deployment across a series of environmental science domains (i.e. tectonics; climatology; soil genesis and mapping; land cover/land use; water flow, drainage and floods; slope hazards; and non-point source pollution).

For tectonics, land surface parameters have been used to assess the role of surface processes in relief production and mountain evolution (e.g. Burbank et al. 1996; Bishop, Shroder & Colby, 2003; Bishop & Shroder, 2004), the characterization of tectonic–geomorphic features (e.g. Z. Lin et al., 2013), the delineation of active faulting zones (e.g. Chan et al., 2007; Begg & Mouslopoulou, 2010), the surface perturbations following volcanic activity and the susceptibility of the land surface to lava flows (e.g. Baldi et al., 2000, 2002; Csatho et al., 2008; Neri et al., 2008; Fornaciai et al., 2010; Kereszturi et al., 2012), rheology and pyroclastic flows (e.g. Jessop et al., 2012) and last but not least, the characterization and monitoring of alpine glaciers and their role in shaping the land surface (e.g. Evans, Hengl & Gorsevski, 2009; Bishop et al., 2012a).

The work connecting the height and shape of the Earth's surface with climatology has transcended multiple scales. At one end, there is a substantial body of work that has used existing weather stations along with altitude and aspect to interpolate monthly and annual climate surfaces as well as anomalies over large areas to support science and management (e.g. Dubayah & van Katwijk, 1992; Hetrick et al., 1993a; Hetrick, Rich & Weiss, 1993b; Miklánek, 1993; Dubayah & Rich, 1995; Hutchinson, 1995; Corbett & Carter, 1996; Hofierka, 1997; Kumar, Skidmore & Knowles, 1997; Thornton, Running & White 1997; Thornton & Running, 1999; Thornton, Hasenauer & White, 2000; Jarvis & Stuart, 2001; Daly et al., 2002; Fu & Rich, 2002; Šúri & Hofierka, 2004; Reuter, Kersebaum & Wendroth, 2005; Böhner & Antonić, 2009; Sheng, Wilson & Lee, 2009). At the other end, there is an increasing number of studies that have used land surface parameters to predict and/or explain the presence or absence of specific plants over very short distances such as a ridgeline or hillslope forest stand (e.g. Coughlan & Running, 1996; Leempoel et al., 2015). The indices in this second group focus on the toposcale (i.e. hillslopes and small to medium-sized catchments) and use either simple indices (e.g. Balice et al., 2000; McCune & Keon, 2002) or more complicated topographic radiation and temperature indices (e.g. Wilson & Gallant, 2000c; Šúri & Hofierka, 2004) to capture the variability of the energy and thermal regimes across these land surfaces (e.g. Moore, Norton & Williams, 1993d; Davies, Bates & Miller, 2007; Davies et al., 2010; Austin, Gallant & Van Niel, 2013).

Turning next to soil genesis and mapping, many land surface parameters have been used to help drive a series of automated soil landscape modeling workflows (e.g. Moore et al., 1993a; Gessler et al., 1995, 2000; Zhu et al., 1997, 2001; McKenzie et al., 2000; Thompson, Bell & Butler, 2001; Hengl, Gruber & Shrestha, 2004; Bishop & Minasny, 2005; Dobos & Hengl, 2009), estimate water and/or carbon budgets (e.g. Band, 1993; Bell, Grigal & Bates, 2000) and map specific soil properties such as solum depth (Gessler, McKenzie & Hutchinson, 1996) and soil organic matter (e.g. Qin et al., 2012).

Similar strategies have been deployed for land use/land cover modeling. Many different land surface parameters have been used to predict the landscape structure (e.g. Austin, Cunningham & Fleming, 1984; Antonić,

Pernar & Jelaska, 2003; Hoechstetter, Thinh & Walz, 2006; Hoechstetter et al., 2008) as well as the patterns and performance of specific plants and plant communities in specific settings (e.g. Tajchman & Lacey, 1986; Bolstad & Lilliesand, 1992; Moore et al., 1993b; Franklin, 1995; Burrough et al., 2001; Van Niel, Laffan & Lees, 2004; Persson, Pilesjö & Eklundh, 2005; Kopecký & Čížková, 2010; Leempoel et al., 2015), soil–water–vegetation dynamics (e.g. Coughlan & Running, 1996; Tang et al., 2014, 2015), and wildfire propagation (e.g. Hernández Encinas et al., 2007).

There is also, not surprisingly, a large and ever-increasing body of work that has employed land surface parameters to simulate hydrologic processes and outcomes, including soil wetness, runoff production, stream discharge, and floods. Some of these studies have focused their attention on soil moisture (i.e. wetness), surface saturation zones and the topographic controls of channel initiation and headwater streams (e.g. Burt & Butcher, 1985; Jones, 1986, 1987; O'Loughlin, 1986; Moore, Burch & MacKenzie, 1988a; Moore, O'Loughlin & Burch, 1988b; Montgomery & Dietrich, 1989, 1992; Phillips, 1990; Barling, Moore & Grayson, 1994; Pilesjö, Persson & Harrie, 2006; Sørensen, Zinko & Seibert, 2006; James, Watson & Hansen, 2007; Brooks & Colburn, 2011; Jencso & McGlynn, 2011). Many others have examined flow routing, the delineation of upslope contributing areas, and the magnitude and timing of runoff production from hillslopes (e.g. Onstad & Brakensiek, 1968; Sivapalan, Beven & Wood, 1987; Jenson & Domingue, 1988; Martz & de Jong, 1988; Morris & Heerdegen, 1988; Fairfield & Leymarie, 1991; Freeman, 1991; Moore & Grayson, 1991; Rieger, 1992; Montgomery & Foufoula-Georgiou, 1993; Costa-Cabral & Burges, 1994; Holmgren, 1994; Wolock & McGabe, 1995; Garbrecht & Martz, 1997; Tarboton, 1997; Wood, Sivapalan & Robinson, 1997; Pilesjö, Zhou & Harrie, 1998; Liang & Mackay, 2000; Endreny & Wood, 2003; Orlandini et al., 2003; Fisher & Welter, 2005; Bogaart & Troch, 2006; Hancock & Evans, 2006; Jones et al., 2007; Qin et al., 2007; Seibert & McGlynn, 2007; Pilesjö, 2008; Jencso et al., 2009; Orlandini & Moretti, 2009; Stanislawski, 2009; Gironás et al., 2010; Stanislawski & Buttenfield, 2011; Zhou, Pilesjö & Chen, 2011; Buchanan et al., 2012; Persson et al., 2012; Peckham, 2013; Shelef & Hilley, 2013; Pilesjö & Hasan, 2014; Wright, Moore & Leonard, 2014; Lindsay & Dhun, 2015).

One of the most important and enduring outcomes from this last group of studies has been the development of a series of methods for routing the flow of water on and below the Earth's surface to the channel network. These methods, which are examined in more detail in Chapter 3, play critical roles in many of the terrain modeling workflows used today.

Finally, some other hydrologic studies have employed one or more land surface parameters to examine the location, form, and functioning of riparian areas (e.g. Weller, Jordan & Correll, 1998; Weller, Baker & Jordan, 2011; Fried et al., 2000; McGlynn & Seifert, 2003; Baker, Weller & Jordan, 2006a,b, 2007; Hollenhorst, Host & Johnson, 2008) and stream networks (e.g. Smith, Zhan & Gao, 1990; Tarboton, Bras & Rodriguez-Iturbe, 1991; Soille & Gratin, 1994; Passalacqua et al., 2010; Pirotti & Tarolli, 2010), various forms of hydrologic modeling across a variety of spatiotemporal scales (e.g. Beven & Kirkby, 1979; Band, 1989; Quinn et al., 1991; Quinn, Beven & Lamb, 1995; Band, Vertessey & Lammers, 1995; Peckham, 1998; Olsson & Pilesjö, 2002; Lindsay & Evans, 2006; Notebaert et al., 2009; Marthews et al., 2015), and the morphology and flow dynamics of river systems as well as flood and streambank erosion hazards (e.g. Cobby, Mason & Davenport, 2001; Lohani & Mason, 2001; Thoma et al., 2005; Jones et al., 2007; Cavalli et al., 2008; Heritage & Milan, 2009; McKean et al., 2009; De Rose & Basher, 2011; Legleiter, 2012). The use of these methods to create and launch a real-time national water model for the USA in April, 2016 is described in Chapter 5 to illustrate the kinds of applications that are possible nowadays.

Many studies have also deployed land surface parameters and objects to assess the likelihood of one or more forms of slope hazard. For example, Moore and Burch (1986a–c), Moore and Wilson (1992, 1994), Desmet and Govers (1996a), Bishop and Shroder (2000), Van Remortel, Hamilton and Hickey (2001), Van Remortel, Maichle and Hickey (2004), Lewis, Verstraeten and Zhu (2005), and H. Zhang et al. (2013a,b) have proposed and/or used various combinations of slope gradient and length to estimate sheet erosion rates. Mitášová et al. (1995) went one step further and used a variety of land surface parameters to model both the erosion and deposition processes that contribute to soil redistribution. Moore et al. (1988a), Desmet and Govers (1996b), James et al. (2007), Kheir, Wilson and Deng (2007), and Zakerinejad and Maerker (2015) used a variety of land surface parameters to evaluate the susceptibility of the land surface to gully erosion. Montgomery and Dietrich (1994), Montgomery, Sullivan and Greenberg (1998), Duan and Grant (2000), Guzzetti et al. (2005), Glenn et al. (2006), Ardizzone et al. (2007), Verdin et al. (2007), Gruber, Huggel and Pike (2009), Kasai et al. (2009), Wieczorek and Snyder (2009), Burns et al. (2010), Ventura et al. (2011), Band et al. (2012), Jaboyedoff et al. (2012), Qin et al. (2013b), and Tseng et al. (2013) have deployed a variety of land surface parameters to assess the susceptibility of landscapes to landslides and to depict the changes to the land surface following landslide activity.

And finally, some studies have examined the topographic controls of nutrient dynamics (e.g. Gold et al., 2001), phosphorus runoff (e.g. Buchanan et al., 2013), and pollution loadings and water quality outcomes (e.g. Panuska et al., 1991; Saunders & Maidment, 1996; Velleux, England & Julien, 2008; Chinh et al., 2013).

This last group of applications, like all those noted earlier, span several decades and must therefore be interpreted carefully given the tremendous changes that have occurred in terms of the terrain analysis, modeling and visualization techniques and data that have been used over the same period. Numerous examples illustrating these changes will be introduced throughout Chapters 2–6 and the significance of the aforementioned observation about the breadth and magnitude of these changes will be taken up again in more detail in Chapter 7.

## 1.4 Study Site and Software Tools

The north fork of Cottonwood Creek, located immediately west of the Madison River and to the east of State Highway 287 on the Montana State University Red Bluff Research Ranch near Norris, Montana, is used throughout this book to illustrate numerous terrain modeling tasks and products (Figure 1.2). The north fork of Cottonwood Creek is used for this purpose because it contains a nice mix of flat areas and moderate to steep slopes. The watershed, which is shaped like a pear, runs twice as long north–south as it does east–west, and covers a surface area of approximately 209 ha.

The elevation ranges from 1628 m at the catchment outlet to 1977 m in the southeast corner of the watershed (Figure 1.3). A series of small peaks helps to define the southern, western, and northern boundaries of the watershed but the elevation gradually declines moving north such that a 1820 m peak marks the northern limit of the watershed. The eastern boundary of the watershed to the south of the channel outlet is dominated by a prominent ridge. The local relief from the stream channels to the ridgetops ranges from 75 to 145 m throughout most of the study catchment (Aspie, 1989). The catchment itself contains moderate to steep slopes with a variety of aspects. The southeastern quadrant is dominated by steep slopes that face north and west and the northern and southern ends of the watershed are split by higher ground that runs east–west in an easterly direction from the western boundary of the watershed. The main channel is spring-fed and runs year round. It is flanked by small, intermittent seeps that feed water laterally into this channel and several ephemeral channels. The main channel is incised approximately 0.2–1.5 m along most of its length and flows across a series of small alluvial fans that occur in draws along the footslopes of the stream valley (Aspie, 1989). The National Hydrography Dataset Enhanced (NHD-Plus) included three first-order streams but missed several others (based on the shapes of the contours shown in various parts of the watershed in Figure 1.3).

The soils are deep, moderately well-drained, and formed in colluvium and material derived from gneiss, schist, and granite. Most of the soils have gravelly loam, loamy sand, or coarse sandy loam surface textures (Boast & Shelito, 1989). The climate is cool (8.3°C mean annual temperature) and semi-arid (25–70 cm annual precipitation), and the vegetation is strongly correlated with landscape position and aspect (Aspie, 1989; Jersey, 1993). Grasses occupy south-facing slopes and ridgetops and cover approximately 60% of the catchment. Maple, aspen, willow, and snowberry occupy north-facing slopes and lower stream bottoms and cover about 10% of the catchment. Sagebrush interspersed with small stands of conifers dominates the remainder of the catchment. The north and south forks of Cottonwood Creek join together a few kilometers northeast of the study catchment and join the Madison River in Beartrap Canyon approximately 16 km southeast of Norris, MT.

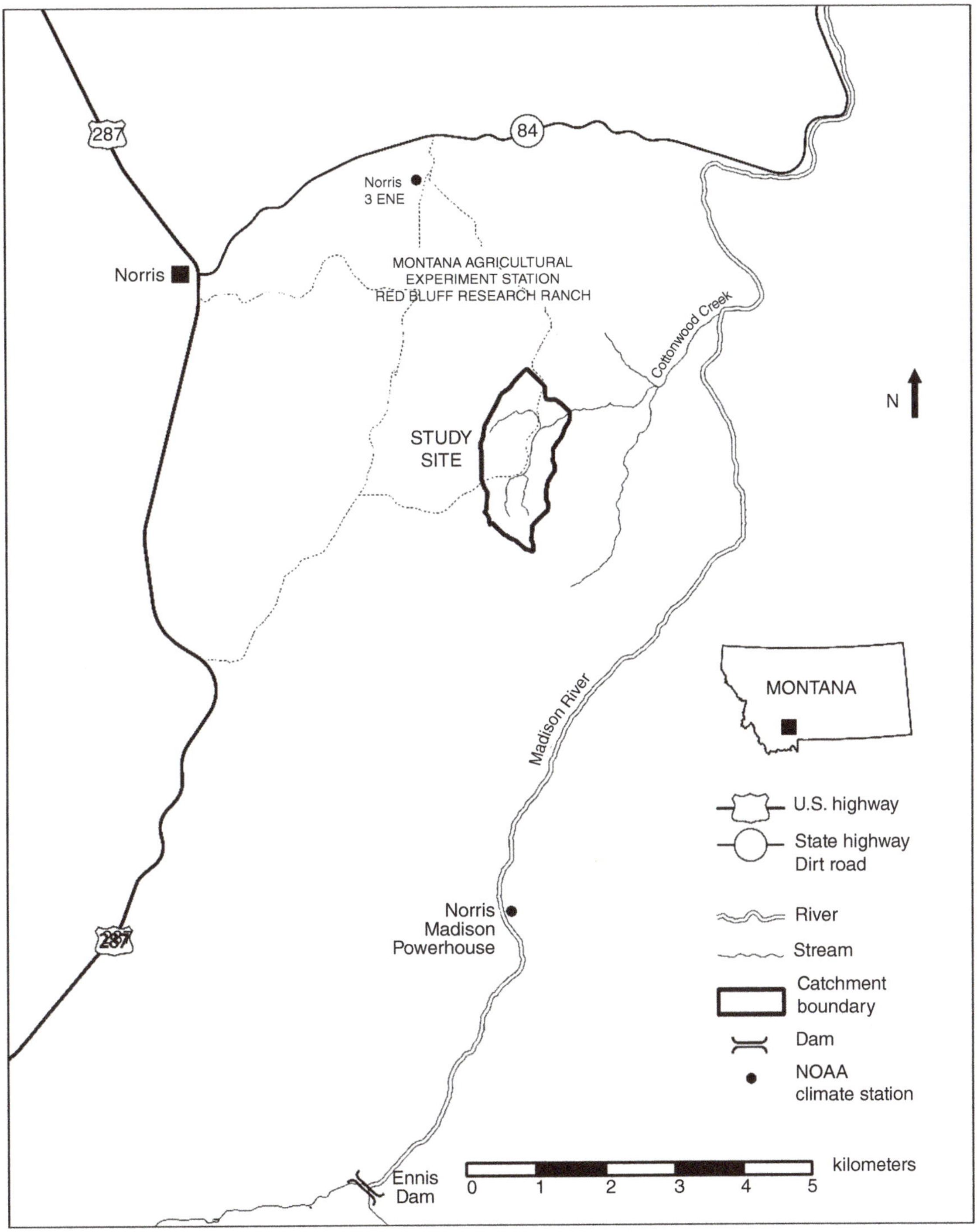


Figure 1.2 Map of Cottonwood Creek, MT study site.

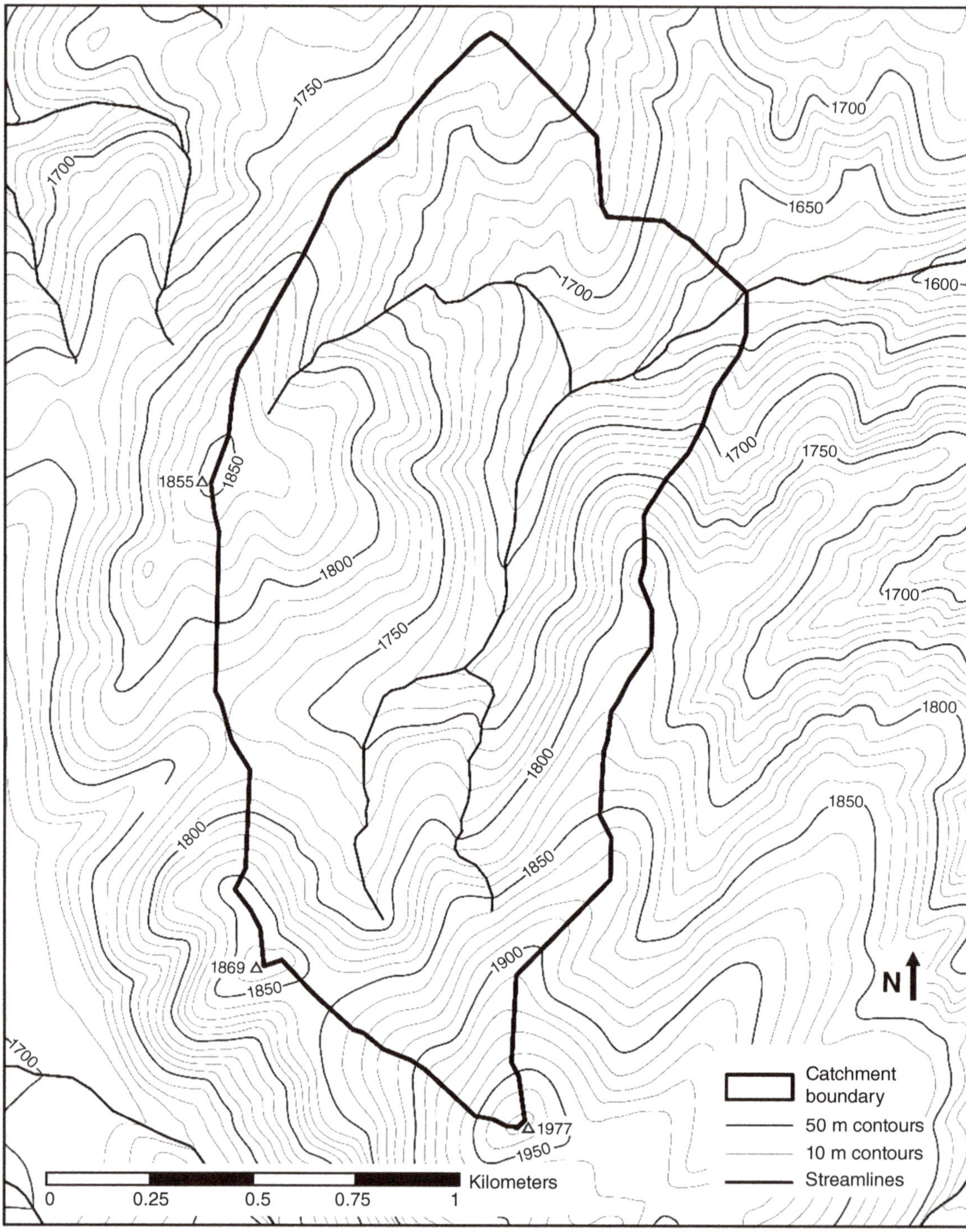


**Figure 1.3** NED 10-m contour and NHD-Plus streamline data for the Cottonwood Creek, MT study site, with the catchment boundary overlaid.

This is the same watershed that was used with different elevation input data and terrain analysis and modeling tools to illustrate selected terrain modeling tasks and products by Hutchinson and Gallant (2000), Gallant and Wilson (2000) and Wilson and Gallant (2000b,c) in Chapters 2–4 of Wilson and Gallant (2000a). The terrain modeling work undertaken for this book used the 10-m DEM from the US National Elevation Dataset (NED), the NHD-Plus, and the ArcGIS terrain modeling tools noted below whereas the Wilson and Gallant (2000a) book chapters used 20-foot contours and streamlines extracted from a USGS 1:24,000-scale topographic map quadrangle and the TAPES-G terrain analysis tools (Gallant & Wilson, 1996, 2000; Wilson & Gallant, 1996, 2000b,c).

ArcGIS was first used to project the 10-m NED DEM into the NAD83 (North American Datum 83)/UTM (Universal Transverse Mercator) Zone 12 N coordinate system using the bilinear resampling tool. This process generated a square-grid DEM measuring 8.68 m on a side and this grid is used as the base for the primary land surface parameters displayed in Figures 3.4–3.9 and 3.20–3.22 for example. The catchment reproduced in Figure 1.3 was then delineated with the watershed tool using a user-designated pour point as input along with the DEM and the deterministic eight-node (D8) flow-direction algorithm (O'Callaghan & Mark, 1984). Figure 1.3 also shows the stream network that was downloaded from the high-resolution NHD-Plus.

The first series of five maps reproduced in Section 3.1.2 (Figures 3.4–3.9) show several terrain parameters related to slope, aspect, and curvature that were calculated with the standard tools in ArcGIS, but for the fact that the positive and negative profile and plan curvatures reported in Figures 3.8 and 3.9 refer to convex and concave curvatures, respectively. This is the usual practice and opposite of the assignments adopted by ArcGIS.

The next two pairs of maps reproduced in Section 3.1.3 (Figures 3.12–3.13 and 3.23–3.24) show the uplope contributing areas calculated with two single-flow and two multiple-flow direction algorithms. Figures 3.12 and 3.13 show the upslope contributing areas calculated with the D8 (O'Callaghan & Mark, 1984) and D∞ (Tarboton, 1997) flow routing algorithms, respectively, and Figures 3.23 and 3.24 show the upslope contributing areas calculated with the MD∞ (Seibert & McGlynn, 2007) and TFM (Pilesjö & Hasan, 2014) flow routing algorithms, respectively.

Two of the three maps that comprise the third series of maps reproduced in Section 3.1.4 – the difference from mean elevation (Figure 3.26) and standard deviation of elevation (Figure 3.28) – were calculated using the Focal Statistics tools in ArcGIS. The elevation percentile values shown in the second map (Figure 3.27) were calculated with the Whitebox Geospatial Analysis Tools (Lindsay, 2014, 2016c), because ArcGIS still does not support "<" and ">" scripting within user-specified neighborhoods.

The fourteenth Cottonwood Creek map reproduced in Section 3.2.1.1 shows the steady-state TWI (Figure 3.31) and the fifteenth and final Cottonwood Creek map reproduced in Section 4.2 shows the D8

(O'Callaghan & Mark, 1984) flow-direction grid. This final map would have served as an intermediate output in the workflow leading to the D8 upslope contributing area map noted earlier and reproduced in Figure 3.12, but it was included in Chapter 4 because this map would have also served as an intermediate output in a workflow using one or more upslope contributing area thresholds to prune a flow-direction grid like this one to predict the channel network present in the Cottonwood Creek catchment.

## 1.5 Structure of Book

This book examines the methods and data sources that have been used to generate DEMs and calculate land surface parameters and objects during the past five decades. The primary goal is to describe the state-of-the-art for a typical digital terrain modeling workflow that starts with data capture, continues with data preprocessing and DEM generation, and concludes with the calculation and deployment of one or more primary and secondary land surface parameters for adoption and use in landform classification and a variety of environmental modeling applications. The remainder of the book is organized as follows.

The next chapter describes the first three tasks that must be completed to construct DEMs: (i) the choice of one or more elevation data networks; (ii) the sampling of the land surface; and (iii) the creation of a model from the sampled heights. The strengths and weaknesses of the three principal ways of structuring a network of elevation data – contour-based methods, grid-based methods, and triangulated irregular networks – are described. The ways in which the data sources and processing methods for sampling the land surface and generating DEMs have evolved rapidly over the past few decades – from ground surveying and topographic map conversion to passive methods of remote sensing and more recently to active sensing with LiDAR and radar interferometry are described next. The strengths and weaknesses of the principal data sources, including ground surveys, GPS surveys, topographic maps, photogrammetry, airborne laser scanning and interferometric synthetic aperture radar (InSAR), are described in turn. Three global datasets – the Shuttle Radar Topographic Mission (SRTM), Advanced Spaceborne Thermal Emission and Reflectance Radiometer (ASTER), and World DEMs – are given special attention. The kinds of decisions that need to be made when using elevation data to construct DEMs are noted and the most popular methods for completing these tasks are described in some detail. The chapter concludes with a brief description of the continuously evolving multi-source, multi-resolution and seamless elevation data products provided as part of the US NED and the use of web portals like the National Map to organize and distribute these elevation products.

Chapter 3 describes nine groups of primary land surface parameters that are derived directly from DEMs without additional inputs and the two sets of secondary land surface parameters that are commonly used to model water flow and soil redistribution on the one hand and the energy and heat regimes operating at the Earth's surface on the other hand. This is the most substantial chapter in the whole book. The first two-thirds of this chapter describe the calculation and significance of 68 primary land surface parameters organized into nine groups: (i) elevation and area; (ii) slope, aspect, and curvature; (iii) slope direction and width; (iv) flow accumulation; (v) elevation residuals; (vi) statistical parameters; (vii) upslope parameters; (viii) downslope parameters; and (ix) visibility and visual exposure. The rationale and guiding principles along with the similarities and differences between 24 different single- and multiple-flow direction algorithms are described as well. This topic is important because many new algorithms have been proposed during the past 15 years and flow directions are required to estimate flow accumulation. This last parameter serves, in turn, as a key input for many of the flow-based secondary land surface parameters. The secondary land surface parameters are described in the final part of the chapter in which the various forms of the TWI, stream power index (SPI), and various sediment transport, radiation and temperature indices take center stage. A series of maps of selected primary and secondary land surface parameters for the Cottonwood Creek study site are used to illustrate the selected land surface parameters and to help set the stage for the delineation of land surface objects and landforms in Chapter 4.

Chapter 4 marks the shift from general to specific geomorphometry, where general geomorphometry describes the continuous surface (and the land surface parameters that were the subject of Chapter 3) and specific geomorphometry describes a series of land surface objects and landforms, such as alluvial fans, drumlins, eskers, mountains, and valleys. The approaches used to extract and classify specific landform elements, land surface objects based on flow variables, specific (fuzzy) landforms such as mountains and valleys, and repeating landform types are described separately using a series of exemplary applications which culminated with the release of a new global Hammond landform classification by Karagulle et al. (2017) in the fourth quarter of 2016. The key role played by fuzzy concepts and fuzzy classification methods in many of the approaches used to extract and classify these land surface objects and landforms is highlighted as well. The last section of this chapter considers whether discrete geomorphometry, with its focus on coupling multiscale pattern recognition and object delineation, can provide a more robust transition from grid cells (i.e. general geomorphometry) to landforms and land surface objects (i.e. specific geomorphometry).

Chapter 5 discusses the various kinds of errors that are embedded in DEMs and how this error may be propagated and carried forward with the calculation of various land surface parameters and objects. The discussion starts with a review of the various ways in which researchers have identified and treated error and uncertainty in digital terrain modeling. The concepts of error and uncertainty are defined and the most important papers focused on both concepts are reviewed. The discussion of error is organized around the typical terrain modeling workflow (see Figure 2.1) such that (i) the error associated with the choice of data model; (ii) the discretization of space, choice of elevation source data, and the interpolation and/or filtering approach that was used; (iii) the preprocessing of the source data to remove unwanted depressions, resolve flow directions in flat terrain, and merge the DEM with streamlines acquired from some other vector hydrography source; and (iv) the methods used to calculate the many primary and secondary land surface parameters which have been proposed, including the flow-based parameters, that must process activity in the correct order are discussed in separate subsections. The discussion of uncertainty is organized around a series of papers describing key methods and exemplary case studies (i.e. applications). The next two sections of this chapter also rely on a series of exemplary studies, this time to explore some of the ways in which error and uncertainty intersect with two concepts – fitness-for-use and scale – that were first introduced in this chapter and which pervade nearly all digital terrain analysis work. Chapter 4 concludes with a description of the US National Water Model that was recently launched and shows how useful products and services can be constructed notwithstanding the presence of error and uncertainty.

Chapter 6 describes the evolution of terrain modeling tools over time. The shift from specialist stand-alone software programs to full-service GIS over the past 30 years and the pervasive changes to data capture and computing systems that are currently underway, and how these are likely to affect the ways in which scientists and practitioners will perform digital terrain modeling in the future, are described first. The focus then shifts to the kinds of terrain modeling tools that are currently included in the ArcGIS ecosystem and the roles and characteristics of three third-party add-ons – the ArcGIS Geomorphometry, Arc-Geomorphometry and Gradient Metrics, and ArcGeomorphometry toolboxes – that have been developed to extend the native capabilities of the ArcGIS ecosystem itself. The roles and characteristics of nine other proprietary and free- and open-source terrain analysis modeling products – the Geographic Resources Analysis Support System (GRASS), the Integrated Land and Water Information System (ILWIS), LandSerf, MicroDEM, Quantum GIS (QGIS), RiverTools, the System for Automated Geoscientific Analyses (SAGA), the Terrain Analysis Using Digital Elevation Models (TauDEM) toolbox, and the Whitebox Geospatial Analysis Tools (GAT) project – are described next. Several of these systems are still under development but all have attracted sizable user communities. The chapter concludes with some final thoughts for how most of us are likely to conduct terrain analysis and modeling work during the next 10–20 years.

Chapter 7 concludes the book by summarizing the major accomplishments of the past 10–15 years and the current state of the art, describing some immediate needs and opportunities, and a brief call to action that specifies the kinds of collaborations that will be needed to push digital terrain analsysis and modeling to even greater heights in the years ahead.

# 2 Constructing Digital Elevation Models

The construction of DEMs typically incorporates four interrelated tasks: (i) the choice of one or more elevation data networks or models; (ii) the sampling of the land surface (i.e. the gathering of height measurements or estimates); (iii) the creation of a surface model from the sampled heights; and (iv) the correction of the errors and artifacts in the surface model (Hengl & Reuter, 2009). Each phase in the production of a DEM is critical for determining its utility, and in assessing the amount of error that will propagate through the analysis, interpretation, and visualization phases. The general workflow is depicted in Figure 2.1 but this needs to be interpreted carefully because the various tasks may be reordered and combined in various ways to produce a more or less unique workflow that serves the specific use or application at hand (e.g. Vivoni et al., 2005). This workflow will be relevant to those who start with an existing DEM as well, because they will want to check the documentation to understand what choices and decisions were made in developing this DEM and whether or not this particular DEM is suitable for their own ongoing work.

That said, here we take up the first three of the four aforementioned tasks and leave the fourth to subsequent chapters because we are more often than not interested in how these errors and artifacts influence our analysis as opposed to elevation per se.

## 2.1 Elevation Data Networks

The first task is to choose a surface representation knowing that the ideal structure will vary with the purpose at hand and perhaps the kinds of elevation data sources that are available for the area of interest.

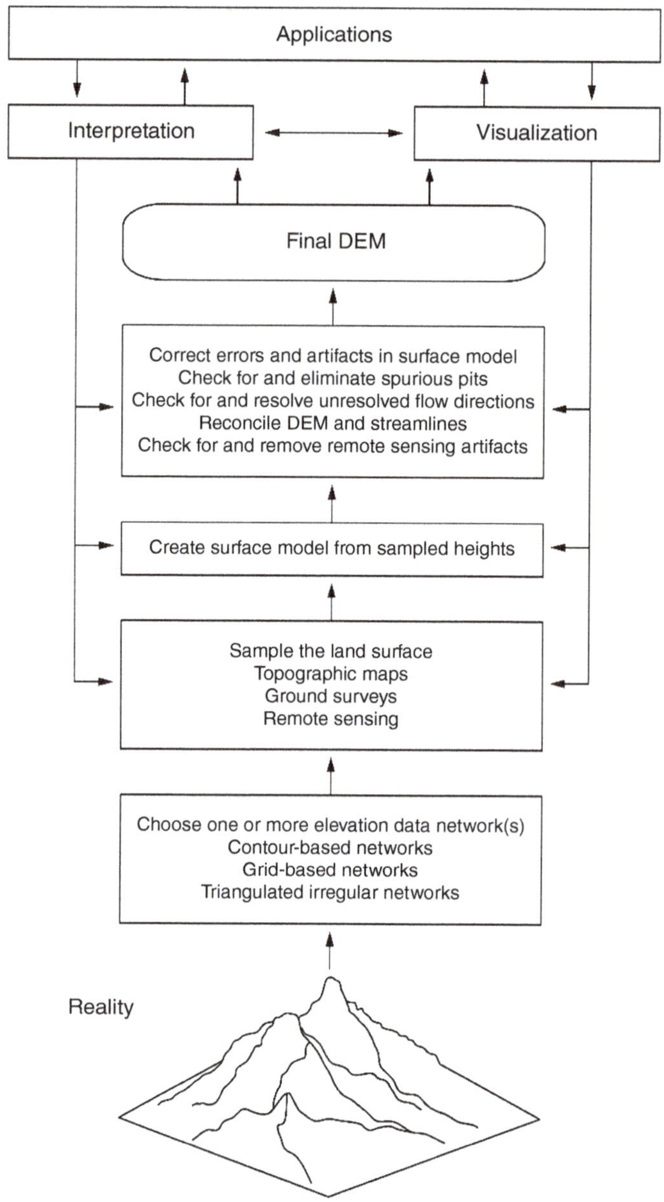


**Figure 2.1** The main tasks associated with digital terrain modeling. *Source*: Hutchinson & Gallant (2000, p. 30). Reproduced with permission of John Wiley & Sons, Inc.

There are three principal ways of structuring a network of elevation data consisting of contour-based methods, grid-based methods, and triangulated irregular networks, as illustrated in Figure 2.2. However, there are numerous examples of work that combine two or all three of these structures. Yang, Shi and Li (2005), for example, proposed an integrated TIN and grid method for constructing and rendering multi-resolution digital terrain models in a three-dimensional GIS and Pilesjö (2008) and Zhou, Pilesjö and Chen

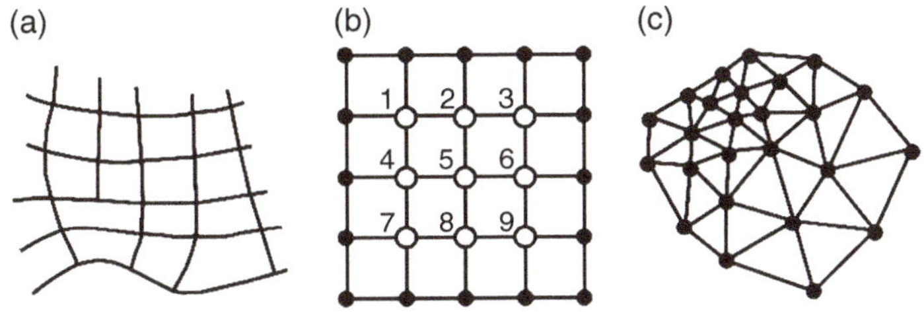


Figure 2.2 The three principal methods of structuring an elevation data network: (a) a contour-based network; (b) a square-grid network showing a 3 × 3 moving window; and (c) a triangulated irregular network (TIN). *Source*: I.D. Moore et al. (1991, p. 4). Reproduced with permission of John Wiley & Sons, Inc.

(2011) have proposed creating a triangular facet network (TFN) on top of a grid-based DEM to estimate surface flow paths more accurately.

The contour-based methods consist of digitized contour lines that are typically stored as digital line graphs (DLGs) in the form of *x,y* coordinate pairs bounded by adjacent contour lines and streamlines (Moore, Grayson & Ladson, 1991), based on the stream path analogy first proposed by Onstad and Brakensiek (1968). Visually, contour lines are useful because they provide opportunities to interpret the land surface at multiple scales and they support the search for both regions with specific characteristics (i.e. steep or gentle slopes based on the density of contour lines in different parts of a map) and specific land surface forms (i.e. a series of closed, more or less circular contours signifying hilltops or closed depressions depending on the elevation of the contour representing the center of the circle; see Figure 1.3).

The grid-based methods may use a regularly spaced triangular, square or rectangular mesh or a regular angular grid, such as the 1- and 3-arcsecond spacing used by the SRTM datasets. The choice of grid-based method depends primarily on the extent of the area of interest. The data can also be stored in a variety of ways but the most common is as *z* coordinates corresponding to sequential points along a profile with the starting point and grid spacing specified. The elemental area is the cell bounded by three or four adjacent grid points for the triangulated and rectangular grid networks, respectively (I.D. Moore et al., 1991).

The square-grid network is the most widely used of the various options because of its simplicity and computational efficiency (Collins & Moon, 1981). However, they suffer from several disadvantages, including the following.

1 The re-interpolation of DEMs that use geographic coordinates, including the USGS NED (Gesch et al., 2002), the National Geospatial-Intelligence Agency's Digital Terrain Elevation Data (NGA DTED; National Imagery and Mapping Agency, 2000) and the SRTM datasets (Slater et al., 2006), can introduce unwanted artifacts in the DEM, take unnecessary processing time, and require unnecessary

storage space. No matter how it is accomplished, re-interpolation changes elevations and cannot produce a better elevation surface (Guth, 2010).

2 They cannot easily handle sharp or abrupt changes in elevation as sometimes occurs in mountainous terrain or when cliffs, escarpments, and landslide scars are present.
3 The size of the grid mesh affects the results obtained and the computational efficiency (Panuska et al., 1991; Zhang & Montgomery, 1994; Wilson et al., 2000; Thompson et al., 2001; Albani et al., 2004; Kienzle, 2004; Hengl, 2006; Deng et al., 2007; Vaze et al., 2010).
4 The computed upslope flow paths tend to zigzag and therefore are somewhat unrealistic, notwithstanding the many new flow routing algorithms that have been proposed during the past 25 years. See Wilson et al. (2007, 2008) for comparisons of several of these models.
5 Some precision is lacking in the definition of specific catchment areas (SCAs), although Gallant and Hutchinson (2011) and Hutchinson et al. (2013) have recently proposed solutions for this problem.
6 The redundancy can be substantial in sections with smooth terrain assuming the spacing chosen for the regular grid was adjusted to the roughest terrain (Peucker et al., 1978).

A TIN, on the other hand, samples surface-specific points – peaks, ridges, and breaks in slope – and forms an irregular network of points stored as a set of $x$, $y$, and $z$ coordinates together with pointers to neighbors in the net (Mark, 1975; Peucker et al., 1978). The elemental area or facet is the plane joining three adjacent points and these networks are more efficient and flexible than regular grid structures in representing complex terrain surfaces. However, many of the digital elevation data sources described in Section 2.2 start out as grid-based structures of one kind or another, and so the initial challenge for those wanting to work with TINs is the need to transform the grid-based surface elevation data (which may miss some or all of the important surface-specific points noted above) to form a TIN.

Notwithstanding these kinds of challenges, a large number of methods have been proposed for constructing TINs from grid-based surface elevation data over the years. Lee (1991), Kumler (1994), Heckbert and Garland (1997), and Zhou and Chen (2011), among others, have reviewed and classified these algorithms, and the five categories included in the following list are taken from Zhou and Chen (2011, p. 39).

1 The *3D line generation method* is based on the Douglas–Peucker algorithm (Douglas & Peucker 1973) and uses perpendicular distance as a criterion to find the farthest point from each line segment, so that the critical points can be detected along a curve. See Fowler and Little (1979), Saalfeld (1999), and Fei and He (2009) for additional details.

2 The *filtering method* employs a moving window and evaluates the importance of each central point by calculating how well the point can be estimated by its neighbors. See Chen and Guevara (1987) and Weibel (1992) for additional details.
3 The *point additive method* starts with a minimum initial approximation and with each iteration, one or more points, which have the maximum variation to the surface defined by the TIN generated by the points selected in the previous iterations, are added to form a new TIN, until some desired error threshold level is reached. See de Floriani, Falcidieno & Penovi (1984), DeHaemer and Zyda (1991), Scarlatos and Pavlidis (1992), and Garland and Heckbert (1995) for additional details.
4 The *point subtractive method* reverses the point-additive process by starting with a triangulation of all the points and iteratively dropping points from the triangulation until some predefined threshold of simplification is reached. See Lee and Schachter (1980) and Hughes, Lastra & Saxe (1996) for additional details.
5 The *feature point method* selects a set of important terrain features such as peaks, pits, ridges, valleys and saddles, and uses them as the vertex set for triangulation. See Southard (1991), Zakšed and Podobnikar (2005), and Palomar-Vázquez & Pardo-Pascual (2008) for additional details.

The choice of method to construct TINs from digital grid-based elevation sources is an active and ongoing area of research. Vivoni et al. (2004), for example, suggested that TINs should be generated based on hydrologic similarity and Vivoni et al. (2005) showed how three different steady-state landscape indices – the TWI (Beven & Kirkby, 1979; O'Loughlin, 1986), a sediment erosion index that is analogous to the length–slope (LS) factor in the revised universal soil loss equation (RUSLE; Renard et al., 1991) adapted to three-dimensional terrain (Moore & Wilson, 1992, 1994), and the critical steady-state rainfall required to initiate a landslide using the assumptions of infinite slope stability and cohesionless soil (Montgomery & Dietrich, 1994; Montgomery et al., 1998) – might be used to generate triangulated meshes that reflect the spatial variability of some locally significant landscape process. More recently, Zhou and Chen (2011) proposed a compound method by integrating the aforementioned *point-additive* and *feature-point methods* for extracting critical points from a square-grid DEM to generate a drainage-constrained TIN that would realize some of the same benefits that Vivoni et al. (2004, 2005) were aiming for.

Notwithstanding these new approaches and the subtleties embedded in nearly all of them, the most important observation for the moment is that topographic attributes, such as slope, SCA, aspect, plan and profile curvature, can be calculated from all three types of elevation data networks. The most efficient DEM structure for calculating these attributes is the grid-based method (I.D. Moore et al., 1991). Contour-based methods require an order of magnitude more data storage and do not provide any computational advantages (I.D. Moore et al., 1991; Carara et al., 1997). The irregularity of the TIN makes the computation of some land surface parameters easier (i.e. slope and aspect) and others more difficult than for grid-based methods because there may be difficulties in determining the upslope connections of one or more facets (Gruber & Peckham, 2009). Further details of how to calculate land surface parameters are provided in Chapter 3.

However, the use of grid structures for partitioning topographic data will likely not work so well for dynamic hydrologic modeling and many other more sophisticated hydrologic, geomorphic, and ecological applications (Mark, 1978; Olsson & Pilesjö, 2002). I.D. Moore et al. (1991), for example, noted how hydrologic models simulate the flow of water across a surface and that the elemental areas of the DEMs should reflect this requirement. Much of the preprocessing work discussed in Section 2.3 involves working with square-grid DEMs in one way or another in an attempt to satisfy this requirement, but with varying levels of success. Moore and Grayson (1991) showed that the contour-based methods offer important advantages in this regard because the structure of their elemental areas is based on the way water flows on the land surface. A variety of attempts has also been made over the years to simulate the flow of water across the surface using TIN-based flow routing algorithms (e.g. Jones et al., 1990; Tachikawa, Shiba & Takasao, 1994; Tucker et al., 2001). However, many of the TIN-based algorithms delineate drainage networks by defining flow routing along the edges of triangles and, as a consequence, they are seldom used to model the flow on terrain surfaces before channels are formed (Zhou et al., 2011).

Pilesjö and colleagues (Pilesjö, 2008; Zhou et al., 2011; Pilesjö & Hasan, 2014) have recently proposed a method in which the grid cells in a square-grid DEM were further partitioned by creating triangular facets between the cell centers according to their surface form and then combining all the grid cells to form a TFN to eliminate gaps between adjacent cells and facets. This approach means that the vector flow lines or surface flow paths from each facet can be uniquely tracked and the catchment area at any given contour line segment can be estimated by counting the number of flow paths passing that segment. Zhou et al. (2011) showed that this method performed favorably compared to several grid-based flow routing algorithms, but noted that it required substantial computing resources and that further computational optimization and algorithm parallelization would be needed to apply this method over large DEMs.

Last but not least, the way we conceptualize the surface is becoming more and more important, notwithstanding the continued popularity of square-grid DEMs (Hengl & Evans, 2009). There is not only the problems that arise because of the loss of many locally significant terrain features (i.e. ridgelines, stream bottoms) and the scale dependency of many of the descriptors we use to describe the topographic surface when we are limited to coarse DEMs (see Zhang & Montgomery, 1994; Kienzle, 2004; Raaflaub & Collins, 2006 for descriptions of these kinds of problems), but there is also a need for better algorithms to filter out vegetation, buildings, and other non-natural structures in the new and more accurate DEMs that can be generated from remote sensing systems (i.e. InSAR, and LiDAR point clouds), and what exactly constitutes the surface then becomes more problematic. Indeed, this same ambiguity is true of stream channels given that the initiation of, and paths followed by, these features may vary from storm to storm in upland areas (Montgomery & Dietrich, 1988, 1989, 1992; Sheng et al., 2007). Finally, the generation of "bare earth" DEMs or digital terrain models (DTMs) as opposed to digital surface models (DSMs) that is often held up as the preferred result would seem less than optimal for a number of applications, such as non-point source pollutant applications that seek to trace pollutants from their sources to the ocean across the engineered landscapes that dominate large metropolitan areas and many of the world's most productive agricultural regions.

Once the preferred elevation data network is chosen, we can turn our attention to the rapidly expanding choices we have in terms of elevation data sources and some of the advantages and disadvantages associated with each of these sources and some of the DEMs made from them.

## 2.2 Elevation Data Sources

The data sources and processing methods for sampling the land surface and generating DEMs have evolved rapidly over the past 20–30 years, from ground surveying and topographic map conversion to passive methods of remote sensing and more recently to active sensing with LiDAR and radar interferometry. Nelson, Reuter and Gessler (2009) distinguished three general classes of DEM data – those collected from ground survey techniques, including electronic theodolites, total stations, electronic distance measurement (EDM) and Global Positioning System (GPS) units; those extracted from existing topographic maps (i.e. contours, streams, lakes and spot heights); and those acquired by way of remote sensing (both airborne and satellite photogrammetric/stereo methods, airborne laser systems, and airborne and satellite radar using interferometry) – and offered a succinct summary of the major features of each of these options (see Table 2.1 for additional details).

The rapid growth in sources of mass-produced DEMs during the past two decades, such as the SRTM, ASTER and LiDAR surveys, has seen DEM resolution improve considerably although the current state-of-the-art and range of applications that can be supported are still, more often than not, a function of the geographic extent of the area of coverage and/or interest. However, some of the work with DEMs that was only possible in small watersheds 20–30 years ago can now be performed for all of the Earth's surface (compare the landslide assessments of Duan & Grant (2000) and Verdin et al. (2007), for example).

**Table 2.1** List of key characteristics of elevation data sources described in this chapter.

| Source | Resolution (m) | Accuracy | Footprint (km²) | Post-processing requirements |
|---|---|---|---|---|
| Ground survey | Varies, but usually <5 | Very high vertical and horizontal | Varies, but usually small | Low |
| GPS | Varies, but usually <5 | Medium vertical and horizontal | Varies, but usually small | Low |
| Table digitizing | Varies with map scale and contour interval of source map | Medium vertical and horizontal | Varies with map footprint | Medium |
| On-screen digitizing | Varies with map scale and contour interval of source map | Medium vertical and horizontal | Varies with map footprint | Medium |
| Scanned topo-map | Varies with map scale and contour interval of source map | Medium vertical and horizontal | Varies with map footprint | High |
| Ortho-photography | <1 | Very high vertical and horizontal | Varies with method of acquisition | High |
| SPOT | 30 | 10 m vertical, 15 m horizontal | 72,000 per swath | Medium |
| LiDAR | 1–3 | 0.15–1 m vertical, 1 m horizontal | 30–50/hour | High |
| InSAR/IfSAR | 2.5–5 | 1–2 m vertical, 2.5–10 m horizontal | Varies with method of acquisition | High |
| SRTM, Band C | 30–90 | 16 m vertical, 20 m horizontal | 60° N to 58° S | High |
| SRTM, Band X | 30 | 16 m vertical, 6 m horizontal | 60° N to 58° S available | High |
| ASTER | 30 | 7–50 m vertical, 7–50 m horizontal | 3600 | Medium |
| WorldDEM | 12–24 | 2–4 m vertical, 2–10 m horizontal | Global | High |

*Source*: Several of the metrics reported here were taken from a similar table presented in Nelson et al. (2009, pp. 83–84, reproduced with permission of Elsevier) as well as other sources.

These new DEMs also signify a shift in the distribution of work. Thirty years ago the leading scientists and practitioners probably constructed their own DEMs and therefore could choose the elevation data network and preprocessing methods that were best suited to the tasks and/or applications at hand. Today's scientists and practitioners can still do this when working with small areas of a few square kilometers (e.g. Podobnikar, 2005), but those working with larger regions or watersheds will more than likely choose among a series of national and global digital elevation datasets. These offer tremendous time savings (see Gesch et al., 2002 for an excellent example illustrating how multi-source, multi-resolution elevation data can be acquired from the NED for a project on the Upper Colorado River headwaters region) but carry with them the need to learn about and anticipate the strengths and weaknesses of each potential candidate dataset before settling on the preferred dataset.

The discussion that follows starts with those elevation datasets that cover a few square kilometers and are typically collected by individuals or small groups (various forms of ground survey, including GPS surveys), before moving to some that might be collected by one or more individuals working alone or by large teams using airborne- and satellite-based surveys that cover thousands of square kilometers (topographic maps, ortho-photography, laser scanning, and synthetic aperture radar or SAR), and finishing up with three examples of very large continental and global datasets (SRTM, ASTER, and TanDEM-X) that were collected using SAR and processed by large multinational teams of scientists and engineers. The US NED is described in detail in Section 2.5 (following brief discussions of DEM preprocessing and fitness-for-use) because this multi-source, multi-resolution suite of elevation datasets and the accompanying terrain attributes has been built on top of a very elaborate infrastructure and workflow, and the elevation datasets themselves will be substantially revised and improved as the USGS's 3D Elevation Program (3DEP) initiative (Stoker, Harding & Parrish, 2008; Snyder, 2012; Sugarbaker et al., 2014) gathers momentum during the next 10 years.

### 2.2.1 Ground Surveys

We start this description of elevation data sources with ground surveys, which have been used in one way or another to describe selected parts of the Earth's surface for thousands of years, with one notable example being the topographic assessments that must have preceded the construction of the gravity-fed water distribution systems build across southern Europe in the first century AD. The growth and dispersion of the world's population along with the emergence of cities and the need to support steadily increasing human populations in many parts of the world has led to ever larger numbers and types of ground surveys, including small area mapping projects, detailed topographic surveys of specific areas, measurement of cross-sections and longitudinal profiles, and stratigraphic analysis of active gully systems, among others. All of these surveys combine field measurement and interpolation and of course the ease and precision with which such surveys can be conducted has increased enormously during the past few decades due to the development and widespread deployment of electronic theodolites, total stations, and other electronic distance measurement devices.

Ground surveys might be used in one or more of three ways nowadays. The first and perhaps most obvious is the opportunity to use ground surveys to prepare new, detailed topographic surface models of areas of a few square kilometers in extent (Table 2.1). The second is the opportunity to use repetitive surveys to describe the geomorphic processes at work and spatially specific rates of erosion and deposition. James et al. (2012), for example, illustrated how detailed topographic surveys completed more than 70 years ago were used to construct DEMs and compute difference maps to estimate geomorphic change in Cox Gully in the upper Piedmont of South Carolina, the Feather River at Shanghai Bend and the Lower Yuba River channels (both in California). This particular study used modern DEMs to estimate the present-day topographic surface but Trimble (1983, 1999, 2009) and others have used various kinds of repetitive ground surveys to achieve similar results. The third and final approach involves the collection and use of ground survey data with topographic maps and other elevation data sources to improve the final DEMs (see Podobnikar, 2005 in Section 2.2.3 for a nicely executed example of this type of application).

### 2.2.2 Kinematic GPS Surveys

High-resolution DEMs can also be generated relatively quickly with kinematic GPS surveys (Baldi et al., 2000, 2002; Abd Aziz, 2008; Abd Aziz et al., 2009). These systems have also been used to validate DEMs generated in other ways, such as LiDAR (e.g. Webster & Diaz, 2006) and InSAR (e.g. Lee, Chang & Ge, 2005), which are discussed in more detail later.

Nico et al. (2005b), for example, generated a high-resolution DTM of the Torre di Satriano archaeological site in the Basilicata Region of Italy by kinematic GPS survey and showed that the accuracy of the DTM depends on both the operational precision of the device and protocols used to collect the data (10 cm in this instance) and the density of the GPS samples. These authors used Delaunay triangulation and its corresponding Voronoi tessellation to connect the nodes (i.e. GPS sample points) whose associated Voronoi polygons shared a common edge and a simple linear interpolation was then performed to fit the terrain surface. The results showed that the accuracy of the topographic surface was <1 m in areas where the density of the samples was as low as one sample per 100 $m^2$ and approximately 30 cm in areas with at least one sample per 10 $m^2$.

The aforementioned application illustrates how these surveys can provide fast and accurate mapping of terrain surfaces of a few square kilometers in extent (Table 2.1) and, as such, how they are well suited to the production of DEMs for specific areas, such as archaeological sites (e.g. Nico et al., 2005b), volcanic and landslide-prone areas (e.g. Baldi et al., 2000, 2002), and small agricultural fields (e.g., Abd Aziz, 2008; Abd Aziz et al., 2009).

We turn our attention next to the sources that might have been available, and which in many instances have been used, to generate relatively high-resolution DEMs for much larger areas than is practical using ground and kinematic GPS surveys.

### 2.2.3 Topographic Maps

We will start here with published topographic maps and then move to the various kinds of ortho-photography that were used to create many of the topographic maps published in the second half of the twentieth century and close with reviews of several passive and active airborne and satellite-based remote sensing systems that have been deployed and used to generate high-resolution DEMs for large areas of Earth during the past 20–30 years in subsequent subsections.

A large number and variety of topographic maps have been published during the past two centuries by national mapping agencies around the world and these can be used to construct DEMs if no suitable ground survey or remote sensing data is available. The USGS, for example, published 190,000 topographic quadrangle maps at a variety of scales from the inception of the US topographic mapping program in 1884 until the last topographic map using lithographic printing technology was published in 2006 (Allord et al., 2014). Many of the 1:24,000- and 1:100,000-scale topographic map sheets were digitized or scanned and used for hydrologic, geomorphic, and ecological analysis prior to the large-scale production of digital DEMs and hydrography datasets from photogrammetric and other kinds of remotely sensed imagery in the 1980s and 1990s. The topographic maps may still be of value today for those with limited budgets and these data at hand (e.g. Jobin, Prasannakumar & Vineetha, 2015), and especially those interested in hydrologic, geomorphic and/or ecological change detection (e.g., Stein et al., 2007; James et al., 2012; Beattie, 2014).

The USGS's topographic map series, for example, extends back more than 150 years in some instances and therefore offers opportunities for change analysis that cannot be matched by remote sensing sources, because the topographic maps extend back further in time and/or they cover heavily vegetated areas where traditional remote sensing methods cannot penetrate the canopy (James et al., 2012). However, these historical maps must be used carefully because the source materials and processing methods used for the construction of topographic maps have changed dramatically over time (Hodgson & Alexander, 1990). Many of the early maps produced in the USA prior to the 1940s, for example, were artistically drawn with few field observations between field measurements, whereas the more recent maps are remote sensing-based and used a comparatively dense set of observations (i.e. every few meters horizontally) for contour construction (James et al., 2012). National cartographic accuracy standards were also established in the USA in the 1940s and similar standards were adopted in many other

Table 2.2 Horizontal National Map Accuracy Standards (NMAS) used in the USA since 1947.

| | On map[a] | | | On ground[a] | |
|---|---|---|---|---|---|
| **Scale factor** | **(inch)** | **(inch)** | **(mm)** | **(m)** | **(feet)** |
| >20,000 | 1/50 | 0.0200 | 0.508 | – | – |
| <20,000 | 1/30 | 0.0333 | 0.847 | – | – |
| **Examples** | | | | | |
| 250,000 | 1/50 | 0.0200 | 0.508 | 127.0 | 417 |
| 100,000 | 1/50 | 0.0200 | 0.508 | 50.8 | 167 |
| 62,500 | 1/50 | 0.0200 | 0.508 | 31.8 | 104 |
| 24,000 | 1/50 | 0.0200 | 0.508 | 12.2 | 40.0 |
| 12,000 | 1/30 | 0.0333 | 0.847 | 10.2 | 33.3 |
| 5,000 | 1/30 | 0.0333 | 0.847 | 4.23 | 13.9 |

[a] No more than 10% of well-defined points can be beyond these error limits. These values may be interpreted as the 90% confidence limit with the understanding the NMAS do not assume a specific probability distribution.
*Sources*: James et al. (2012, p. 186); US Geological Survey (1999).

countries around the same time (Marsden, 1960). The US National Map Accuracy Standards (NMAS) used an observed error threshold of 90% for all maps that met the NMAS such that 90% of the tested points on the maps satisfying this standard were within some specified distance of the true position (Table 2.2) and 90% of the tested point elevations differed by no more than half the contour interval (James et al., 2012). Both the horizontal and vertical thresholds show how the map accuracy standards varied with map scale and why combining topographic maps drawn at varying scales may cause numerous challenges.

The challenges and/or potential errors that are encountered using published topographic maps and/or the products derived from such maps may or may not stop there. Walsh, Lightfoot and Butler (1987) and Walsh (1989), for example, referred to the aforementioned kinds of errors on the source maps as "inherent" errors and explained how various "operational" errors produced from the processing approaches used to digitize or scan the topographic map sheets might cause additional problems (i.e. errors). These operational errors might be caused by misregistration during coordinate transformations, digitizing blunders, or when interpolating from contour or point data.

More recently, Podobnikar (2005) has described a complex, multi-step iterative process for generating DEMs from a variety of data sources and showed for a part of Slovenia that the result was 50–80% better than the 1:25,000-scale topographic map that he started with. The final workflow utilized geodetic network points, land cadastre boundary points, building information models, spot elevations, highway construction and hydrologic network measurements along with the contours on a 1:25,000-scale topographic map.

These last three studies by Walsh et al. (1987), Walsh (1989) and Podobnikar (2005) suggest the need to look carefully at the lineage of the topographic maps – which may cover the source materials, period of time used for preparation, date of publication, processing methods, and the various decisions made to produce the maps – before deciding whether or not to use them for a specific purpose or application.

### *2.2.4 Photogrammetry Datasets*

Many of the more recent DEMs produced in various parts of the world have been produced with photogrammetric methods. For example, many of the oldest DEMs in the USGS NED corresponding to the 7.5-minute topographic map quadrangles were produced with photogrammetric methods (Kelly, McConnell & Mildenberger, 1978). Many of these DEMs included striping artifacts when they were first released and various approaches were proposed to filter the DEMs in order to reduce or remove these artifacts (e.g. Brown & Bara, 1994; Garbrecht & Starks, 1995; Oimoen, 2000). The "mean profile filter" (Oimoen, 2000) has been applied to isolate the high-frequency artifacts and subtract them from the USGS's DEMs that suffered from these problems, and Gesch et al. (2002) showed how this post-processing produced significant improvements in the derivation of elevation derivatives notwithstanding the small magnitude of many of the artifacts (i.e. typically <1 m).

These and other DEMs show how the stereoscopic interpretation of aerial photography can deliver elevations to meter and sub-meter accuracy (Lemmens, 1978). These methods and sources have been used to generate DEMs with planimetric and vertical accuracies of about 1 m, although the successful processing of these images will vary with the sun illumination, cloud conditions, and ground cover (Toutin & Cheng, 2000). These same stereoscopic methods have also been applied to SPOT (Satellite Pour l'Observation de la Terre) and IKONOS imagery (e.g. Konecny et al., 1987; Day & Miller, 1988; Toutin & Cheng, 2000).

There may well be a resurgence in the popularity of DEMs produced from photogrammetric methods in the next 5–10 years for two reasons. The first is the continued advances in the collection of satellite-based ortho-rectified images (ORIs) and the opportunities these provide for repetitive mapping (i.e. change detection). The Japan Aerospace Exploration Agency (JAXA), for example, has generated a global DEM/DSM and ORI using archived data from the Panchromatic Remote-sensing Instrument for Stereo Mapping (PRISM) onboard the Advanced Land Observing Satellite (ALOS), which collected data from 2006 to 2011. The mission utilized a 25-m spatial resolution to collect precise elevation measurements and 3 million stereoscopic images with less than 30% cloud cover (Tadono et al., 2014). Preliminary results show that DEMs with a spatial resolution of 5–30 m and a height accuracy of ≤5 m (RMSE) can be produced from these same data (Tadono et al., 2009). The second is that balloons, kites, and unmanned aerial vehicle mapping kits (or drones as they are commonly called) along with free and open source image processing software (e.g. MapKnitter 2.0 distributed by the Public Laboratory for Open Technology and Science) allow the opportunity to collect and process custom datasets. These products have found widespread use in community and school projects all around the world.

### *2.2.5 Airborne Laser Scanning Datasets*

Airborne laser scanning can also provide high-resolution topographic data (Wehr & Lohr, 1999; Wack & Wimmer, 2002). LiDAR surveys have grown quickly in popularity and this source now dominates local and regional projects nearly everywhere: small countries such as Belgium and the Netherlands, for example, produced national LiDAR DSMs at resolutions of 2–5 m several years ago and much finer resolution DEMs have been produced for many smaller areas as well (Nelson et al., 2009). Stoker et al. (2008) led calls to compile a national LiDAR dataset for the USA and the USGS recently launched the 3DEP initiative to respond to the growing need for high-quality topographic data and for a wide range of other three-dimensional representations of the nation's natural and engineered features (Snyder, 2012). The primary goal of 3DEP is to systematically collect enhanced elevation data in the form of high-quality LiDAR data over the conterminous USA, Hawaii, and the US territories during the period 2014–2022 plus InSAR data over Alaska, where cloud cover and remote locations preclude the use of LiDAR over much of the state (Sugarbaker et al., 2014).

The creation of a standard data exchange format (LAS) for working with and interpolating laser scanning data (Graham, 2005) and the recently launched US 3DEP initiative both illustrate the rapid advancements in LiDAR data quality and processing capabilities that have been made over the past 25 years (e.g. Zhang et al., 2003; Sithole & Vosselman, 2004). The advantages of using LiDAR include the high density of sampling, high vertical accuracy, and the opportunity to derive two or more sets of surface models given that some laser scanning systems can model both the vegetation canopy (first returns) and the ground surface (last returns). This helps, of course, with the modeling of the land surface in heavily vegetated areas (i.e. forests) and engineered environments (i.e. urban areas). Tarolli (2014) has described how the high-resolution topographic data obtained from airborne and/or terrestrial LiDAR have provided new opportunities for the analysis of landslides (e.g. Glenn et al., 2006; Ardizzone et al., 2007; Kasai et al., 2009; Burns et al., 2010; Ventura et al., 2011; Jaboyedoff et al., 2012; Tseng et al., 2013), hillslope and channelization processes (e.g. Cobby et al., 2001; Jones et al., 2007; Passalacqua et al., 2010; Pirotti & Tarolli, 2010; De Rose & Basher, 2011), river morphology (e.g. Lohani & Mason, 2001; Thoma et al., 2005; Cavalli et al., 2008; Heritage & Milan, 2009; McKean et al., 2009; Notebaert et al., 2009; Legleiter, 2012), active tectonics (e.g. Chan et al., 2007; Begg & Mouslopoulou, 2010; Lin, Z. et al., 2013), volcanic landforms (e.g. Csatho et al., 2008; Neri et al., 2008; Fornaciai et al., 2010; Jessop et al., 2012; Kereszturi et al., 2012), and anthropogenic signatures on the Earth's surface (e.g. Bailly et al., 2008; Lindsay & Dhun, 2015).

However, the small footprint and measurement challenges encountered in areas with tall buildings, dense vegetation canopies and water surfaces make this a relatively expensive option and numerous studies have noted how the accuracy of LiDAR elevation data may vary in specific kinds of hydrologic and geomorphic settings (e.g. Raber et al., 2002, 2007; Hodgson et al., 2003, 2005; Hodgson & Bresnahan, 2004; James et al., 2007; Aguilar & Mills, 2008; Aguilar et al., 2010; Wheaton et al., 2010; Estomell et al., 2011). Simply increasing the mean point density does not guarantee significantly improved vertical accuracy in these kinds of applications and settings because the scanners do not penetrate well through thickly vegetated (or other surface cover) environments, and it is now well accepted that errors will vary with land cover class (i.e. greater errors in forested vs. non-forested land cover for example). A variety of filters have been proposed to remove non-ground measurements (e.g. Zhang et al., 2003; Sithole & Vosselman, 2004) and Lloyd and Atkinson (2006) have shown how geostatistical methods (i.e. ordinary kriging, kriging with a trend model) might be used to "fill-in" areas (i.e. predict ground surface elevations) after the removal of above-surface features.

Taken as a whole, the high-resolution topographic data offered by airborne and terrestrial LiDAR provide numerous new and exciting opportunities to help us better understand the hydrologic, geomorphic and ecological processes operating at the Earth's surface, but we still need to gather, analyze, and interpret these data with care to successfully link these processes to topographic signatures.

### 2.2.6 Interferometric Synthetic Aperture Radar Datasets

InSAR (or IfSAR) is another active remote sensing source that can be used to construct medium- to high-resolution DEMs. This geodetic method uses the stereoscopic effect caused by two or more SAR images taken from slightly different positions to generate DEMs using differences in the phase of the waves returning to the satellite or aircraft. There is a rapidly expanding suite of space-borne, airborne, and ground-based systems that can be used to collect and process SAR images.

The initial experimentation with satellite-based InSAR dates back to the 1980s, but expanded in the 1990s with the launch of the ERS-1 (1991), JERS-1 (1992), RADARSAT-1 and ERS-2 (1995) satellite platforms. The 11-day NASA STS-99 mission in February 2000 used a SAR antenna mounted on the space shuttle *Endeavor* to gather data for the SRTM products discussed in more detail in Section 2.2.7. The launch of new instruments has continued during the past 15 years and the recent launch of the Sentinel-1A and 1B satellites by the European Space Agency (ESA) will provide InSAR coverage on a global scale on a 6-day repeat cycle. The latest platforms have expanded the available sensors to include the L- and X-bands in addition to the C-band used by most of the early satellite platforms (which can offer high-resolution datasets that cover the entire globe).

The airborne data acquisition systems built by companies like Fugro Geospatial, Intermap Technologies, and Orbisat Indústria also offer higher resolution and multiple bands as well. There are now also ground-based techniques that are typically used for displacement monitoring of slopes, rock scarps, volcanoes, and landslides, among others, and that generally provide high-resolution images with spatial and vertical accuracy of a few meters for areas of a few square kilometers (e.g. Nico et al., 2004, 2005a).

### 2.2.7 Shuttle Radar Topographic Mission DEMs

The 1- and 3-arcsecond SRTM DEMs illustrate the enormous gains that have been at the continental and global extents during the past two decades (Farr & Kobrick, 2000; Rabus et al., 2003; Farr et al., 2007). The 3-arcsecond SRTM DEM (SRTM-3) covers a large fraction of the globe (from 60° N to 58° S) and has quickly become one of the most consistent, complete and popular environmental datasets in the world (Nelson et al., 2009; Zandbergen, 2008). The 3-arcsecond (~90 m) grid spacing is much better than the 1-km spacing of the worldwide GTOPO30 DEM and an accuracy assessment using kinematic GPS data showed good absolute height accuracy, with 90% of the errors less than 5 m (Rodriguez, Morris & Belz, 2006). The release of the 1-arcsecond SRTM DEM (SRTM-1), which was initially limited to the USA, provided a nominal 30-m DEM with vertical accuracy of 15 m and discretization to the nearest meter. A 1-arcsecond, customized SRTM for Australia was released in 2011 and the USA released the generic 1-arcsecond SRTM dataset for the whole globe from 60° N to 58° S in 2014.

Both the SRTM-1 and SRTM-3 DEMs offer unprecedented coverage and look impressive at first sight; however, they must be used carefully because of the following.

1. They provide a DSM rather than a bare-earth DEM or DTM.
2. The surface characteristics (i.e. mountains, terrain roughness, presence of sand, proximity to water, and presence and/or density of vegetation cover) may affect accuracy (e.g. Gallant & Read, 2016).
3. Voids often occur at land–water margins as well as in other locations. The finished-grade version of SRTM (also referred to as Version 2; Slater et al., 2006) still contained data voids covering some 836,000 km$^2$ for example (Reuter, Nelson, & Jarvis, 2007).
4. Problems may occur in desert and mountain areas due to foreshadowing and shadowing effects (Rodriguez et al., 2005).

5 The resolution of one or both of these DEMs may not be fine enough for the mapping of soils, vegetation and similar phenomena (Gessler et al., 2009).
6 A series of recent studies has shown a positive relationship between elevation error and the height of the canopy (e.g. Carabajal & Harding, 2006; Hofton et al., 2006; Shortridge, 2006; Berry, Garlick & Smith, 2007; Bhang, Schwartz & Braun, 2007; Gallant, Read & Dowling, 2012) and one study in particular showed how low-lying riparian areas may be represented as substantially higher than the surrounding agricultural areas (leading to an inverted terrain model; LaLonde, Shortridge & Messina, 2010), highlighting the need to assess fitness-for-use before deploying one or more of these datasets for a specific application and/or study area.

The SRTM-1 and SRTM-3 DEMs have been continuously improved since their initial release but the latest versions may or may not have solved all of the aforementioned problems (Slater et al. 2006). For example, the Consultative Group for International Agricultural Research Consortium for Geospatial Information (CGIAR-CSI) provides the best global resolution (SRTM-3) and has released numerous versions during the past decade (see Table 2.3 for additional details). The aforementioned voids, for example, have been filled using a range of interpolation algorithms in conjunction with other sources of elevation data. Reuter et al. (2007) reviewed a series of void-filling interpolation methods for SRTM data and recommended the best algorithms for filling each of the 3,339,913 voids in the SRTM-3 data based on their size and a typology of their surrounding terrain. Their recommendations are worth restating here because the preprocessing challenges noted in the next section will focus on interpolation algorithms as well.

1 Kriging or inverse distance weighting interpolation were recommended for small- and medium-sized voids in relatively flat, low-lying areas.
2 Spline interpolation was recommended for small- and medium-sized voids in high-altitude and dissected terrain.
3 TINs and inverse distance weighting were recommended for large voids in very flat areas.
4 An advanced spline method (ANUDEM), which is described in more detail in Section 2.4, was recommended for large voids in other terrain settings (Reuter et al., 2007).

These recommendations were later adopted and used to prepare the CGIAR-CSI SRTM Version 4.1 product (Table 2.3).

Numerous attempts have been made to solve the aforementioned problems. Selige, Böhner and Ringeler (2006), for example, developed a procedure to preprocess the phase data instead of the more popular approaches that rely on the filtering and smoothing of the DEM itself and

**Table 2.3** SRTM-3 versions produced and distributed by CGIAR-CSI.

**Change from Version 3 to Version 4**
- Version 4 used a number of interpolations techniques, as described by Reuter et al. (2007)
- Version 4 used some additional auxiliary DEMs to fill the voids and SRTM-1 for large voids
- Version 4 differs from Version 3 with a half-grid pixel shift to eliminate the confusion caused by the earlier shift (see below)

**Change from Version 2 to Version 3**
- Version 3 included "finished" grade SRTM data
- Version 3 used the SRTM water body data (SWBD) to clip the coastlines
- Version 3 used auxiliary DEMs to fill the voids
- Version 3 differs from Version 2 with a half-grid pixel shift

**Changes from Version 1 to Version 2**
- Version 2 included DEM data for Australasia and small islands in the Atlantic, Indian and Pacific Oceans
- Version 2 had the shorelines clipped
- Version 2 had no "cliffs" on the tile joins, which were caused by insufficient overlap in interpolation in Version 1

**Known issues and future improvements**
- CGIAR-CSI plans to continue improving the data as and when high-resolution auxiliary datasets become available. Updates are planned that will use high-resolution ASTER DEMs for filling holes in particularly troublesome areas (the Sahara Desert, for example)

*Source*: modified from the CGIAR-CSI website http://www.cgiar-csi.org/data/srtm-90m-digital-elevation-database-v4-1#version_history

produced a significantly improved SRTM DEM for geomorphic applications in a landscape near Landshut, Bavaria. Gallant (2011), on the other hand, described an adaptive smoothing technique to remove random noise, and two approaches that can be used with tree canopy (e.g. Hansen et al., 2013; Chen et al., 2015) and the HH polarization data distributed with the ALOS PALSAR L-band SAR dataset to remove the offsets due to tree cover (Gallant et al., 2012; Gallant & Read, 2016). These papers illustrate the large number of studies that have endeavored to assess the fitness-for-use of either SRTM-1 or SRTM-3 for one or more specific applications (i.e. catchment hydrology, flood inundation, soil and habitat mapping) by comparing the new DEMs to spot heights, other DEMs, and related geospatial datasets.

### 2.2.8 Advanced Spaceborne Thermal Emission and Reflectance Radiometer DEMs

The release of the ASTER Global Digital Elevation Model (GDEM) in 2009 (Abrams et al., 2010) provided another "global" DEM that has been compared with the aforementioned SRTM DEMs. This new ASTER dataset was created by compiling 1.3 million images taken from the visible and near-infrared (VNIR) portion of the electromagnetic spectrum using single-pass stereoscopic correlation techniques. This new product initially offered better resolution (1 vs. 3 arcseconds, but this advantage was nullified with the full release of the STRM-1 dataset in 2014) and better spatial coverage (83° N to 83° S vs. 60° N to 58° S) as well as comparable vertical and horizontal accuracy (7–20 m vertically and 30 m horizontally compared to 16 and 20 m, respectively, for SRTM) (Hirano, Welch & Lang, 2003; Eckert, Kellenberger & Itten, 2005; Nelson et al., 2009; Slater et al., 2009). In addition, the missing data problems with the SRTM datasets caused by clouds were expected to be easier to fill because of the open-ended acquisition schedule but the 30-m resolution may still not be sufficient to support the mapping of soils, vegetation and similar phenomena in some landscapes. A second and improved version of the ASTER GDEM (Version 2) was released in 2011 (Krieger, Curtis & Haase, 2010; Gesch et al., 2011; Tachikawa et al., 2011a,b).

Several evaluations comparing the quality of the ASTER GDEM to the SRTM DEMs and other products as well as various sources and forms of reference data have also been published. Hirt, Filmer and Featherstone (2010), for example, compared the quality of the ASTER-GDEM Version 1 dataset from the US National Aeronautics and Space Administration (NASA) and the Japanese Ministry of Economy, Trade, and Industry (METI) (measured in terms of elevation accuracy and systematic errors) with two other publicly available DEM datasets for Australia: the 9-arcsecond GEODATA DEM-9S Version 4.1 dataset from Geoscience Australia and the Australian National University and the 3-arcsecond SRTM Version 4.1 dataset from CGIAR-CSI. The results showed that (i) ASTER contains numerous artifacts related to residual cloud patterns and striping effects; and (ii) the vertical accuracy of DEM-9S is ~9 m, SRTM ~6 m, and ASTER ~15 m based on a set of geodetic ground control points (GCPs) over Western Australia. These results must be interpreted carefully because the superior performance of the SRTM product, for example, may be due in part to the fact that the GCPs were located in sparsely vegetated areas (SRTM is a first return system and often returns canopy heights in areas with dense vegetation) and the fact that the relatively coarse 9-arcsecond resolution of the DEM-9S dataset meant that this product produced large errors of up to 100 m in areas with rugged terrain.

Guth (2010) compared the ASTER GDEM with SRTM 3-arcsecond data for 46 locations in Europe using the NextMap IfSAR DEM and DTM pairs for absolute calibration and six locations in North America using the US NED, Canadian cartographic DEMs, and the SRTM 1-arcsecond data as well as the two aforementioned datasets for calibration. The results showed that the GDEM, at its best, is equivalent to the SRTM 3-arcsecond data and that, like the SRTM 1-arcsecond data, the 30-m spacing claimed for GDEM does not match the 30-m or 1-arcsecond cartographic DEMs but closely approximates 3-arcsecond cartographic DEMs at roughly nine times the storage cost. And like Hirt et al. (2010), Guth (2010) found that much of

GDEM, perhaps up to 20% of the data tiles (based on looking at 52 locations spread across Europe and North America), contained data anomalies that would compromise its usefulness for many hydrologic, geomorphic and ecological applications.

Finally, Rexer and Hirt (2014) compared ASTER GDEM2 and two DEMs based on the SRTM released by the USGS (SRTM-3 Version 2.1) and by the CGIAR-CSI (Version 4.1) over the Australian continent. The results showed the following.

1. There was good agreement between both SRTM DEMs (1.2 m RMSE) and the voids in the USGS model over steep terrain had been filled in the CGIAR-CSI model.
2. The ASTER GDEM2 has a northeast to southwest-aligned striping error at the 10-m level and shows an average height bias of ~5 m relative to the two SRTM models.
3. The RMSE obtained from the differences between the ASTER GDEM2 and SRTM heights over Australia was ~9.5 m.
4. The vertical accuracy of ASTER GDEM2 is ~8.5 m, USGS SRTM-3 is ~6 m, and CGIAR-CSI SRTM is ~4.5 m over Australia based on 228,000 “accurate” station heights from the Australian National Gravity Database (ANGD).
5. This last result shows that the ASTER GDEM2 has improved significantly over its predecessors as the comparisons of ASTER GDEM1 with leveling and GPS/leveling heights carried out by Hirt et al. (2010) revealed RMSEs of 13.1 and 15.7 m, respectively.
6. The ASTER GDEM2 is shifted by −0.007/−0.042 arcseconds in the north–south direction and by −0.100/−0.136 arcseconds in the east–west direction relative to both the SRTM DEMs.
7. Analyzing the height differences between the DEMs and 775,437 ANGD stations as a function of three land cover groups (generalized from ESA’s GLOBCOVER 2009 map; Bontemps et al., 2011) revealed a true height offset (over bare ground) of −4.2 m for the ASTER GDEM2 and +2.7 m for both SRTM DEMs.
8. Analyzing the height differences between the DEMs as a function of terrain roughness revealed a high correlation between terrain roughness and accuracy: ASTER GDEM did relatively better (RMSE 11.3 m) than the CGIAR-CSI SRTM Version 4.1 and USGS SRTM-3 (RMSE 15.1 m), which might be linked to the higher spatial resolution of ASTER GDEM2, but relatively worse than SRTM over all other (i.e. less rough) terrain types.

These three studies, taken as a whole, paint both the first and second versions of the ASTER GDEM as “global” products with persistent and serious artifacts, a lower true resolution than the nominal 30-m resolution claimed at the outset, and as products that in many ways are not as good as the leading CGIAR-CSI SRTM-3 Version 4.1 "global" dataset. However, the next product, the first truly global WorldDEM, is likely to supplant the CGIAR-CSI SRTM-3 DEM as the most accurate global product in the next few years.

### 2.2.9 WorldDEM Datasets

The Airbus Defense and Space WorldDEM™ products rely on data gathered by TanDEM-X, another space-borne SAR mission. These datasets promise to set new standards not only in terms of their truly global coverage but also their accuracy (predicted vertical accuracy: 2 m relative, 10 m absolute) (Rexer & Hirt, 2014). The first validation results reported by Gruber et al. (2013) show that with a block adjustment approach and GCPs as ties, an absolute vertical accuracy of 1–2 m is possible. This is a private venture so some customers will need to pay for the data products they acquire and use. However, one of the derivative products, the WorldDEM4Ortho that features vertical accuracy of 4 m in a 24-m square grid, is currently distributed as one of several elevation datasets in Esri's ArcGIS Online platform and is thereby available for use in a variety of ArcGIS-based terrain modeling applications (see Chapter 6 for additional details).

## 2.3 Fitness-For-Use

Three observations stand out from this review of digital elevation sources. The first is that each source brings specific strengths and weaknesses and so the initial task of every potential user is to judge the fitness-for-use of the product or source for the task and/or application at hand. The second is that great care is needed to successfully combine elevation data drawn from two or more sources and to generate a fused or composite DEM that is better than the source DEMs (see Leitão, Prodanović and Maksimović, 2016 for comparisons of three conventional grid-based DEM merging methods that are provided by commercial GIS software and a new method that provided significant improvements when compared to the conventional methods in terms of elevation, slope, and aspect). And the third is that the presence and propagation of errors is a consistent source of concern that pervades the entire digital terrain modeling workflow (see Figure 2.1).

Not surprisingly, the choice of the data source and/or data product is usually just the first step because some preprocessing will typically be required no matter what the source or intended application. These tasks and some of the challenges and issues that accompany them are taken up in Section 2.4 and the ongoing efforts by USGS to build and support the NED are described in Section 2.5.

## 2.4 Data Preprocessing and DEM Construction

The preparation of elevation data for geomorphometric analysis is tricky because elevation, in isolation, is typically not the attribute of interest and this means that the true geomorphic accuracy can only be assessed by measuring surface parameters and objects such as drainage lines, landforms or viewsheds in the field and then comparing their shapes, distributions, and location with the values obtained by geomorphometric analysis (e.g. Fisher, 1998, Wilson et al., 2008; Leitão et al., 2016).

Reuter et al. (2009, p. 90) suggested that the true applicability of DEMs for geomorphometric analysis can only be assessed by providing answers to the following questions.

1. How accurately is the surface roughness represented?
2. How accurately is the shape of the land surface represented (i.e. concave and convex shapes, erosion and deposition, water convergence or divergence, but see Hutchinson & Gallant, 2000 for additional ways to evaluate this aspect of data quality)?
3. How accurately are the "real" world ridgelines and streamlines detected?
4. How consistently are elevations measured over the whole area of interest?

The answers to these and similar questions are interrelated, and there will almost certainly still be errors present in the available and/or preferred DEM(s) notwithstanding the answers to these important questions. The frequency and magnitude of errors will depend on the technology and methods used to collect the source data, the preprocessing algorithms that are applied, and the characteristics of the land surface itself.

Not surprisingly, the horizontal and vertical resolution of the elevation data used to portray a terrain surface will have a significant influence on (i) the level of detail; (ii) the accuracy of the portrayal of surface features; and (iii) the values of the land surface parameters that are computed from a DEM (MacMillan & Shary, 2009). Numerous authors have documented the effects of grid spacing on the value and accuracy of land surface parameters and landform objects (e.g. Zhang & Montgomery, 1994; Florinsky, 1998; Jones, 1998; Wilson et al., 2000; Thompson et al., 2001; Shary, Sharaya & Mitusov, 2002; Tang et al., 2002; Kienzle, 2004; Warren et al., 2004; Zhou & Liu 2004a; Raaflaub & Collins, 2006; Clarke & Lee, 2007). The increasing interest in various forms of multiscale analysis (see Gallant and Dowling, 2003 and Sulebak and Hjelle, 2003 for two notable examples) and the enduring need to be able to move seamlessly across scales suggests that more work is needed on these scale relationships and effects. However, the most immediate and fundamental challenge for those wanting to build and interpret terrain models is typically the need to choose an operational scale that can be supported by the source data and that suits the work at hand (as noted in Section 1.2).

Beyond the aforementioned kinds of relationships and the ensuing impacts, the decisions made about unwanted depressions (i.e. spurious pits or sinks) will have a large impact on the subsequent analysis and interpretation of the results of geomorphometric analysis. Two approaches have found frequent use: one relies on progressively filling the sinks by increasing their elevation values until the elevation of their lowest outflow point is reached (e.g. Jenson & Domingue, 1988; Martz & de Jong, 1988; Soille & Gratin, 1994; Planchon & Darboux, 2001; Wang & Liu, 2006; Jiang & Tang, 2015), whereas the second method creates a descending path from the bottom of the sink by carving the terrain along this path until the nearest point is reached which has an elevation lower than the bottom of the sink (e.g. Morris & Heerdegen, 1988; Rieger, 1992; Martz & Garbrecht, 1999; Soille, Vogt & Colombo, 2003; Soille, 2004).

However, Reuter et al. (2009) recently used both of these approaches along with one that combined sink filling and carving such that the sum of the differences in elevation between the input DEMs and the output DEMs that did not have sinks were minimized. This combined approach produced superior results for the Baranja Hill study catchment in Croatia. Grimaldi et al. (2007) proposed an alternative physically based approach and Lindsay and Creed (2005a,b, 2006) combined elements from several of the aforementioned approaches and used them to distinguish real and spurious (i.e. artifact) depressions in digital elevation data and to propose a minimum impact approach for removing the artifact depressions in relatively flat landscapes like those occurring on the Canadian Shield.

The motivation for all of these depression filling techniques is the desire to fill the depressions to ensure the continuity of flow. Some authors have proposed methods to achieve this outcome without filling the depressions first. Chow et al. (2004), for example, applied the PROMETHEE (Preference Ranking Organization Method for Enrichment Evaluations) technique to determine flow directions using unfilled DEMs. This method reverses the flow directions of the cells within the depressions and extracts profiles along the network but does not calculate the upslope contributing areas. Arnold (2010) also proposed a "fill and overflow" method to calculate flow accumulation without disruption across unfilled DEMs but this algorithm does not calculate suitable flow paths to simulate the flow of water through the depressions. Byun and Seong (2015) adopted these two aforementioned methods and used them to generate more useful data for extracting flow paths, flow accumulation within each of the depressions, and stream longitudinal profiles. Their algorithm calculates flow accumulation sequentially heading downstream by first constructing the topological network for the depressions during the process of reversing the flow directions of the cells following the line of maximum depth in the depression and, second, calculating flow accumulation based on the ordered depression data from the highest to the lowest elevations. This approach ensures the continuity of flow across the unfilled DEM and allows the calculation of flow accumulation within depressions, thereby enabling the extraction of more accurate stream profiles and other parameters of interest (Byun & Seong, 2015).

There are at least two arguments that might be used to pick one of the last three algorithms. The first is the unintended consequences and bias that will be introduced when elevations are altered. Filling the depressions, for example, will change the gradient of the channel and reduce slopes on both sides of the valley and change the various terrain attributes that incorporate slope as an input (Wechsler, 2007). The second is related to the changing character of the source data and the increasing likelihood that the depressions nowadays represent small depressions that were captured with LiDAR and other remote sensing platforms, as opposed to errors embedded in the 10- and 30-m DEMs that represented the state of the art in the 1980s and 1990s (Byun & Seong, 2015).

There are at least two other related challenges that may need to be addressed as well. The first concerns the problem of unresolved flow directions on flat terrain because the assignment of flow directions relies on the presence of elevation differences between adjacent cells to drive the flow. The presence of lakes and reservoirs and reliance on the first of the aforementioned approaches for filling sinks may exacerbate this challenge by creating artificial flat regions as well. Whatever the cause, one of two approaches is typically used to remove or minimize these kinds of problems. The first relies on an iterative procedure to assign a single flow direction to a neighboring cell without alteration of the elevation values (Jenson & Domingue, 1988), whereas the second method makes small alterations to the elevation of the flat cell(s) in order to create a small artificial gradient (Garbrecht & Martz, 1997). The solutions obtained with these approaches will vary slightly from one another and an intimate knowledge of the field conditions will usually be required to know whether one of these approaches produces superior results in most landscape settings.

The second challenge is the need to reconcile the DEM and drainage lines acquired from some other source (dataset) (Lindsay, Rothwell & Davies, 2008). One approach relies on stream "burning" where the local topography is altered to provide consistency with some existing vector hydrography dataset (Saunders & Maidment, 1996) and the second method utilizes the stream network as part of the surface fitting approach used to generate square-grid DEMs (Hutchinson, 1989; Hutchinson et al., 2013). We have used the latter approach almost exclusively in our own work, in part because it is now embedded in two separate modeling environments (ANUDEM Version 5.3 and the Topo-to-Raster tool in ArcGIS Desktop Version 10.4) and it tackles the last three problems (i.e. unwanted depressions, unresolved flow directions in flat terrain, and reconciliation of elevation and hydrography datasets) simultaneously.

The early versions of ANUDEM (Hutchinson, 1989, 2000, 2008) worked with contours and spot heights and used an iterative finite difference interpolation technique to build one or more square-grid DEMs. The method relied on data smoothing and local grid cell conditions to produce high-quality interpolated DEMs. It basically started with a coarse grid and then enforced the drainage conditions (using one or more depictions of the streamlines), increased the spatial resolution, enforced the drainage conditions again, and so on, until the desired resolution was reached. This proved to be a very popular technique because it produced a hydrologically correct land surface model (i.e. one in which the ridges were retained, streams enforced, and spurious sinks removed) although like all interpolators, this approach may not produce optimal results if poor input parameters are selected (Wise, 2000b).

That said, the choice of the preferred interpolation technique may depend in part on the source of the data (precise height measurements might lead one to choose an exact interpolator whereas noisy data would direct our attention to an approximate interpolation technique) and the characteristics of the application. See Pain (2005) for an extended discussion of these issues. A large number and variety of interpolators have been used to date, including inverse distance weighting (IDW), nearest neighbor, linear and non-linear triangulation-based methods, various forms of kriging, and splines; see Mitášová and Hofierka (1993), Mitášová and Mitas (1993), Mitas and Mitášová (1999), Meyer (2004), Aguilar et al. (2005), and Arun (2013) for examples of how some of the other interpolation methods have been deployed. However, the recommendations by Reuter et al. (2007) suggesting the best ways to address the various errors and voids in the SRTM dataset show how the choice of the best interpolator might be site- and/or application-specific (i.e. it likely depends on the type of terrain, the types of problems with the source data that need to be addressed, and the ways in which the final data will be used).

Another positive feature of the ANUDEM approach is that it has undergone more or less continuous improvement since its launch in 1989 (see Hutchinson, 1988, 1989, 1996, 2000, 2008 for additional details). For example, the latest version of ANUDEM (Hutchinson, 2011; Hutchinson et al., 2013) includes two major upgrades – the first addresses the impacts of significant noise in dense remotely sensed elevation data and the second, the impacts of spurious intersections in streamline data when these are incorporated in regular grids.

Several of the airborne and space-borne sensors described in Section 2.2 generate high-resolution DEMs over large areas, but suffer from two generic limitations. The first is that they cannot measure ground elevations underneath dense vegetation cover, water or non-natural structures, leading to significant errors and the second limitation is that all remote sensing data have significant random errors that depend on the inherent limitations of the observing instrument as well as the surface slope and roughness (Harding, Bufton & Frawley, 1994; Farr et al., 2007). However, the growing availability of remotely sensed data sources affords new opportunities for characterizing and using the local grid cell conditions for DEM preprocessing.

For example, the vertical standard error of the input data was set to zero when traditional contour-based elevation datasets were used as the primary inputs and the discretization errors were used to weight the data residuals in early versions of ANUDEM (Hutchinson, 1989). A new and improved approach is possible nowadays because there are estimates of the vertical standard error for remotely sensed elevation datasets, but they will often vary in terms of their reliability (Farr et al., 2007). The SRTM dataset has an added problem because it often shows local correlation in data errors for neighboring cells. The latest version of ANUDEM (Hutchinson, 2011) can accommodate both of these changes to the local grid cell conditions. For example, the multigrid algorithm that lies at the heart of the ANUDEM approach can detect the local correlation and provides a simple method for taking it into account. Hutchinson (2011) also describes how the estimates of the standard error with varying reliability can be calibrated within ANUDEM by experimenting with different levels of user-supplied vertical standard error.

The streamline procedure used in the early versions of ANUDEM has been revised as well to take and use the initial streamline heights from the preceding (coarser) grid in the multigrid interpolation process instead of the error-prone actual data point heights that lie on the streamlines themselves. The elevations were initialized at each grid resolution by linearly interpolating along streamlines between grid points to which elevations had been assigned, and the elevation data points on streamlines that were higher than the upstream data points were usually removed in the early versions of ANUDEM. This approach is problematic when working with noisy remotely sensed elevation data because it can remove accurate data points and produce streamlines that are systematically too low relative to the neighboring landscape along extended sections of rivers and streams (Hutchinson et al., 2009, 2013).

The latest version of ANUDEM also includes new approaches to remove sharp corners from gridded stream segments and spurious gridded stream junctions and disjunctions (Figure 2.3). These new approaches maximize the fidelity of the shifted gridded streamlines to the original (source) streamline network and are likely to yield major improvements in the characterization of the land surface and connected drainage structure in large catchments with extensive flat areas.

The rapid evolution of the ANUDEM package shows how the growth in sources of mass-produced, remotely sensed DEMs during the past two decades has demanded new forms of DEM preprocessing. The problems with these new data sources have spawned many proposals for ways to identify and eliminate errors and artifacts. Webster and Dias (2006) and Reuter et al. (2009), for example, describe various approaches and opportunities for ortho-rectifying DEMs, reducing local outliers and noise, filtering water surfaces, filtering pure noise, filtering forests in SRTM DEMs, reducing padi terraces (i.e. areas with closed contours where all the surrounding pixels show the same value), filling voids and sinks, mosaicking adjacent DEMs,

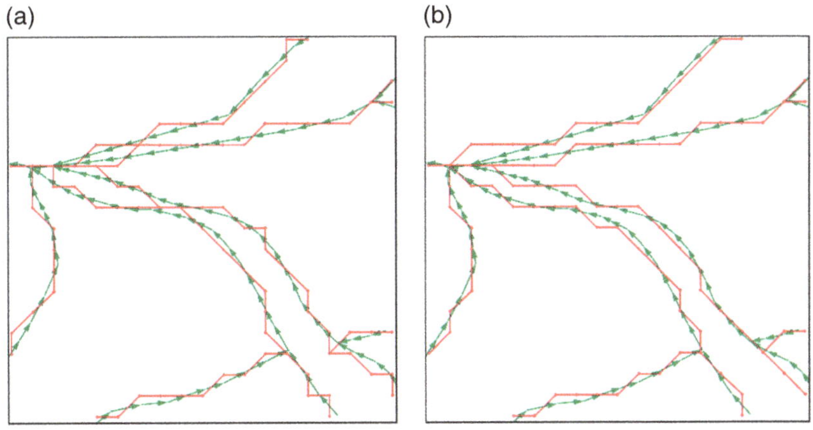


**Figure 2.3** Streamline data in green and (a) initial gridded streamlines at 1-second resolution in red and (b) adjusted gridded streamlines at 1-second resolution in red. *Source*: Hutchinson et al. (2013, p. K-1-3), with permission from geomorphometry.org and the authors.

and filtering LiDAR DEMs. Some problems are more difficult to fix than others and numerous authors have noted the presence of systematic and random errors that are not so easy to detect and correct in LiDAR datasets for example (i.e. Filin, 2003; Katzenbeisser, 2003). The detection of distance errors, of varying deflection errors and of time delays between measurements are especially difficult and may be specific to the LiDAR sensor system that is deployed and the large number of parameters that were assigned when this sensor system was manufactured.

Reuter et al. (2009) also noted two additional trends that have emerged in recent years. The first is the integration of topographic and auxiliary information (such that the location of lakes, streams, ridges, and/or breaks will be identified from satellites and incorporated in the DEM processing chain similar to what happens with ANUDEM) and the second is the increased use of fully data-driven simulation methods that reduce some or all of the aforementioned errors by calculating the average value of the land surface parameter from multiple equiprobable realizations of the DEM (e.g. Burrough, van Gaans & MacMillan, 2000; Hengl et al., 2004; Raaflaub & Collins, 2006).

A third trend can be added here, since the TIN has mounted a comeback of sorts during the past decade as well. Several authors over the years have advocated using TINs in place of square-grid DEMs because of the advantages they offered in terms of data storage and/or the calculation of one or more terrain attributes (e.g. Jones et al., 1990; Lee 1991; Lea 1992; Nelson, Jones & Miller, 1994; Nelson, Jones & Berrett, 1999; Tachikawa et al. 1994; Tucker et al. 2001; Vivoni et al. 2004). The recent work over the past decade has focused on the advantages TINs offer in terms of generalization and the accuracy with which various terrain attributes can be calculated.

Zhou and Chen (2011), for example, have proposed a method of restructuring square-grid elevation data to form a TIN that is optimized to keep the important terrain features and slope morphology with the minimum number of sample points. Their compound method integrates the traditional point-additive and feature-point methods to extract the critical points of the terrain surface from the square-grid DEM based on their significance, measured not only by their local relief, but also by their importance in identifying geomorphologic and drainage features in the DEM. The drainage-enforced TINs constructed with this new method outperformed those constructed with filtering as well as the point-additive and feature-point algorithms by keeping the most significant terrain features, maintaining the RMSE at an acceptable level, and reducing the elevation points by more than 99%. Their new method was also better at retaining drainage features at the same level of RMSE.

Zhou et al. (2011) used this compound method to create a TFN over which the surface flow paths were determined. They argued, as others had done earlier, that the estimation of flow direction and divergence/convergence is less complicated compared to traditional raster-based solutions because each facet in the network has a constant slope and aspect (see Figure 2.2). The performance of this new approach was compared with that of seven algorithms reported in the literature by estimating the SCA over a series of mathematical surfaces, as well as a single real-world DEM. The algorithms selected for the comparison included three single-flow direction algorithms – the deterministic eight-node (D8; O'Callaghan & Mark, 1994), deterministic eight-node least transversal deviation (D8-LTD; Orlandini et al., 2003; Orlandini & Moretti, 2009), and deterministic infinite-node (D∞; Tarboton, 1997) algorithms – and four multiple-flow direction algorithms – the triangulated multiple-flow direction (MD∞; Seibert & McGlynn, 2007), Freeman multiple-flow direction (FMFD; Freeman, 1991), Quinn multiple-flow direction (QMFD1; Quinn et al., 1991), and digital elevation model network extraction (DEMON; Costa-Cabral & Burges, 1994) algorithms – with software implementations that are widely available. Each of these flow routing algorithms is described in more detail in Section 3.1.3. Zhou and Chen (2011) showed that the TFN algorithm produced the closest outcomes to the theoretical values of the SCA compared with other algorithms, deriving more consistent outcomes and being less influenced by surface shapes. The real-world DEM test showed that the TFN was capable of modeling flow distributions without noticeable "artifacts" and that its ability for tracking flow paths makes it a viable platform for dynamic surface flow simulation.

## 2.5 US National Elevation Dataset

The USGS NED (Gesch et al., 2002; Gesch, 2007) covers the USA, Canada and Mexico, and offers a suite of continuously evolving multi-source, multi-resolution and seamless elevation data products using consistent datums, elevation units, and coordinate reference systems (Table 2.4). The impetus for these datasets extends back more than 20 years and the first US dataset, based on the 10-m, 30-m, 2-arcsecond, and 3-arcsecond source data, was

**Table 2.4** Elevation data sources included in the US National Elevation Dataset (NED) as of August, 2015.

- High-resolution data (3 m or better) derived from LiDAR or digital photogrammetry and often with edited water bodies
- Moderate-resolution raster data (4–10 m) derived from digital photogrammetry and IfSAR
- 10-m DEMs derived from 1:24,000-scale cartographic contours and hydrography: these data were produced by the USGS as a standard elevation product for many years and although they currently represent the predominant source in the NED, they continue to be replaced as better source data are acquired
- 30-m DEMs from 1:24,000-scale cartographic contours: these are similar in many respects but usually of lower overall quality than their 10-m counterparts and they are currently only used as a source for Puerto Rico and as void fill around the conterminous USA
- 30-m photogrammetrically derived DEMs: these are the oldest DEMs in the 7.5-minute series and were derived directly from stereo photography, either by a human operator or by an early form of electronic image correlation. They are badly marred by production artifacts that have been addressed to the greatest practical extent by digital filtering within the NED production process. These DEMs are the source for most of Canada and all of Mexico
- 2-arcsecond DEMs are standard USGS product that was derived from cartographic contours at a scale of 1:63,360 over the state of Alaska
- 1-arcsecond SRTM data are the NED source over the Aleutian Islands
- 3-arcsecond DEMs are another standard USGS product and are currently only used with the NED as a source of fill values over large water bodies and small slivers at the borders of the USA and Canada/Mexico
- Bathymetric data for some coastal areas are collected using LiDAR, sonar, and other methods of subsurface terrain measurement, transformed to a common datum, and resampled to only 0.11 (i.e. 1/9) arcsecond resolution within the NED.

completed in 1997, and the first 10- and 30-m DEMs derived from 7.5-minute DEM source data and covering the whole of the continental USA were produced in 1999 (Gesch et al., 2002). The seamless NED layers available as of August 15, 2015 included the following horizontal resolutions: (i) 0.33-arcsecond resolution (approximately 10 m) for the conterminous USA and portions of Alaska; (ii) 1-arcsecond resolution (approximately 30 m) for the conterminous USA, Hawaii, Puerto Rico, the US Territorial Islands, Mexico, and Canada; and (iii) 2-arcsecond resolution (approximately 60 m) over all of Alaska (US Geological Survey, 2015).

The overall vertical accuracy of the NED within the conterminous USA expressed as a RMSE of 25,310 reference points is 1.55 m, although the vertical accuracy varies significantly across the USA due to source quality, terrain relief, land cover, and other factors (as is to be expected). That said, the NED data are collected at much higher resolutions and have much greater inherent accuracies than the SRTM and ASTER datasets for example (Gesch, Oimoen & Evans, 2014).

This coverage will change quickly over the coming months and years. For example, NED data are available at 0.11-arcsecond (approximately 3 m) resolution across one-third of the USA and the first 1-m datasets collected using LiDAR source were released in 2015. The 1-m LiDAR-based DEMs will eventually cover the conterminous USA through the 3DEP and the coverage for Alaska at one-third and 1-arcsecond resolutions will increase significantly through 2016 as 5-m InSAR source data are acquired under the Alaska Statewide Digital Mapping Initiative (US Geological Survey, 2015).

The current elevation data sources and products listed in Table 2.4 are also supported by spatially referenced metadata that capture all of the metadata from the source DEMs and NED processing stages on a quadrangle basis. The NED was assembled from approximately 57,000 files of quadrangle-based source DEMs (Gesch et al., 2002) and the various layers are distributed through the National Map (http://nationalmap.gov/). This includes the various collections noted above as well as the metadata layer and, starting in 2014, the USGS started distributing newly acquired DEMs and LiDAR point cloud data in their original projection and resolution through the National Map as well (US Geological Survey, 2015).

The same two innovations reported here – the launch of new elevation data collection initiatives under the auspices of the NED and the development and use of web portals like the National Map to organize and distribute elevation products – are being replicated in many countries and will provide many new and important DEMs for those interested in terrain modeling in the next few decades.

The next pair of chapters take these DEMs and illustrate how they can be used to calculate suites of land surface parameters and/or land surface objects.

# 3 Calculating Land Surface Parameters

The typical digital terrain modeling workflow, once there is a suitable DEM at hand, will focus on either the extraction of measures (i.e. land surface parameters) and/or spatial features (i.e. landforms and other kinds of surface objects) (see Figure 2.1). This chapter focuses on the calculation of land surface parameters and their significance and the chapter which follows focuses on the extraction and classification of landforms and other kinds of land surface objects.

Most land surface parameters are calculated from projected (planar) square-grid DEMs. However, the rise of global environmental threats coupled with the increasing availability of global elevation sources has led to a steady growth in the number of studies that have calculated land surface parameters from grids using geographic coordinates during the past decade (e.g. Guth, 2010; Marthews et al., 2015). And finally, some of the land surface parameters described in this chapter can be calculated from TINs and contour-based data structures as well. TINs, in particular, have experienced something of a renaissance in recent years and some progress has been made with the development of new methods to calculate hydrology-related land surface parameters from TINs (e.g. Zhou et al., 2011; Pilesjö & Hasan, 2014).

That said, the remainder of this chapter focuses on the calculation of more than 100 land surface parameters from projected, square-grid DEMs. The result is a new grid representing the new land surface parameters with the same dimensions as the source DEM. Two groups of land surface parameters – primary and secondary – are usually distinguished and the same distinction is used in the discussion which follows here. The meaning, significance, and interpretation of the most common land surface parameters included in each of the aforementioned groups are described in turn.

*Environmental Applications of Digital Terrain Modeling*, First Edition. John P. Wilson.

## 3.1 Primary Land Surface Parameters

The primary land surface parameters are derived directly from the DEMs without additional inputs. A variety of terms has been used to describe these parameters. Olaya (2009), for example, referred to them as "basic" land surface parameters and noted that they can be calculated directly from the DEM without further knowledge of the area represented. He then distinguished local and regional parameters because the latter consider additional parts of the DEM apart from the exact area for which parameters are to be calculated. Florinsky (1998) also distinguished local primary parameters that are calculated as a function of their surroundings and non-local primary parameters that require the analysis of a larger, non-local land surface area from a computational perspective. Wilson and Burrough (1999) later explained the distinction between local versus non-local terrain parameters in terms of the existence of local interactions between neighboring points and "action-at-a-distance" forces (see Figure 3.1 for details). Typical examples of local primary land surface parameters include slope, aspect, profile and tangential curvatures; non-local primary land surface parameters include flow path length, proximity to nearest ridgeline or channel, and upslope contributing area.

Table 3.1 lists 68 of the most frequently used primary land surface parameters. These individual land surface parameters are assigned to nine groups as a way to organize the discussion which follows and references are given for parameter descriptions where appropriate. Some of these parameters can be calculated using a variety of software but others can only be calculated using one or more specialist software packages (see Chapter 6 for additional details).

### *3.1.1 Elevation and Surface Area*

The two land surface parameters – elevation and surface area – that comprise the first group in Table 3.1 are local site-specific parameters that are extracted and/or calculated for single grid cells. The elevation, more often than not, is extracted directly from the source DEM. The suitability of this elevation estimate for a specific project will depend on the grid resolution, vertical accuracy, and characteristics of the study area. Gruber and Peckham (2009), for example, showed how a DEM with a vertical resolution of 1 cm and a grid spacing of 10 m has a minimum resolvable slope of $1/1000 = 0.001$. This lower bound means that slopes on hillsides can usually be calculated with small error, but slopes calculated for channels crossing floodplains and other flat areas would be less satisfactory because the slopes in channels in these landscapes will be much smaller than the lower bound reported in this example. This example shows how grid resolution and vertical accuracy might be a concern, depending on the character of the study area and the application at hand.

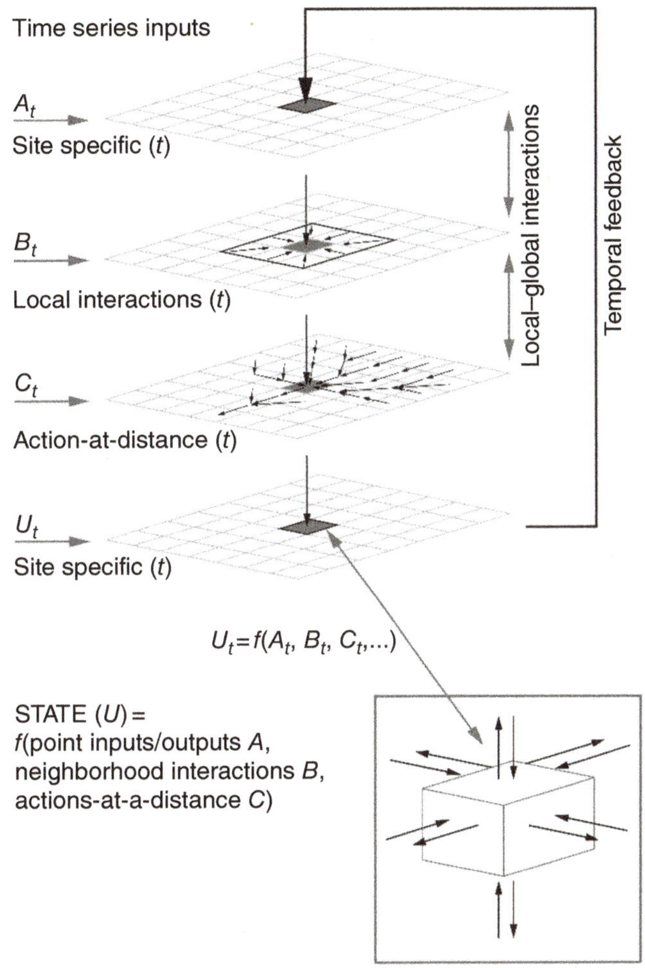


**Figure 3.1** Schematic showing site-specific, local, and regional interactions as a function of time. *Source:* Wilson & Burrough (1999, p. 739). Reproduced with permission of Taylor & Francis.

The surface area is fundamental and is frequently used in the evaluation of land and biological resources, including the calculation of upslope contributing areas for channels, carbon budgets and how storage varies across the landscape and the home ranges of one or more species of interest, among others. The traditional approach treated the surface as two-dimensional and used projected area as a surrogate for the surface area. This approach works well over small areas with flat terrain but grows more and more problematic in areas with steep and variable terrain and when land and biological resource assessments are conducted at regional to global scales (Jenness, 2004; Hoechstetter et al., 2006). Horizontally projected areas will have different surface areas in topographically complex areas such as China, where mountains and hills account for 65% of the total land area (Zhao, 1995).

Ying et al. (2014) hypothesized that the difference between the surface and projected areas is influenced by the topographic complexity (i.e. slope

Table 3.1 List of primary land surface parameters and their significance.

| Land surface parameter | Type | Description |
|---|---|---|
| **Group 1: Elevation and area** | | |
| Elevation | Site-specific | Height above sea level |
| Surface area | Site-specific | Surface area of grid cell, TIN facet, or flow element |
| **Group 2: Slope, aspect, and curvature** | | |
| Slope | Local | Slope gradient (Zevenbergen & Thorne, 1987) |
| Maximum downward slope | Local | Slope to the lowest neighbor in a moving 3 × 3 cell window. Sometimes referred to as the D8 slope and/or the primary flow direction (Gallant & Wilson, 2000) |
| Aspect | Local | Direction of maximum downward gradient (Zevenbergen & Thorne, 1987) |
| Slope northness | Local | Cosine of the aspect (Olaya, 2009) |
| Slope eastness | Local | Sine of the aspect (Olaya, 2009) |
| Profile curvature | Local | Downslope curvature (Zevenbergen & Thorne, 1987) |
| Tangential curvature | Local | Curvature in an inclined plane (Mitášová & Mitas, 1993; Mitášová & Hofierka, 1993) |
| Plan curvature | Local | Along-slope curvature (Zevenbergen & Thorne, 1987) |
| Total curvature | Local | Curvature of the surface itself (Gallant & Wilson, 2000) |
| Surface curvature index | Local | Calculation and use of total curvature to indicate whether or not the center cell in a 3 × 3 moving window has a predominantly convex shape (positive values) or concave shape (negative values) or is a flat surface or one in which the convex and concave curvatures cancel each other out, as might occur on a saddle (zero values) (Blasczynski, 1997) |
| Mean curvature | Local | Average of the profile and tangential curvatures |
| Unsphericity curvature | Local | How the surface differs from a sphere (Shary, 1995) |
| Difference curvature | Local | Half the difference between the vertical and horizontal curvatures (Shary, 1995) |
| Minimal curvature | Local | Difference between the mean and unsphericity curvatures (Shary, 1995) |
| Maximal curvature | Local | Sum of the mean and unsphericity curvatures (Shary, 1995) |
| Horizontal excess curvature | Local | Difference between the unsphericity and difference curvatures (Shary, 1995) |
| Vertical excess curvature | Local | Sum of the unsphericity and difference curvatures (Shary, 1995) |
| Total Gaussian curvature | Local | Square of the mean curvature minus the square of the unsphericity curvature (Shary, 1995) |
| Total accumulation curvature | Local | Square of the mean curvature minus the square of the difference curvature (Shary, 1995) |
| Total ring curvature | Local | Square of the unsphericity curvature minus the square of the difference curvature (Shary, 1995) |
| Rotor curvature | Local | Curvature of the flow lines which are perpendicular to the contours (which is a measure of the twisting of the flow lines) (Florinsky, 1998) |
| Surface curvature index | Local | Total curvature within a group of grid cells (Blaszczynski, 1997) |
| **Group 3: Slope direction and width** | | |
| Flow direction | Local | Direction of flow calculated using one of 24 flow-routing algorithms (see Table 3.2 for additional details) |
| Flow width | Local | Width associated with flow leaving a cell |

Table 3.1 *(Continued)*

| Land surface parameter | Type | Description |
|---|---|---|
| **Group 4: Flow accumulation** | | |
| Upslope contributing area | Regional | The region from which water from precipitation is collected. Sometimes referred to as contributing area or catchment area (Gruber & Peckham, 2009) |
| Specific catchment area | Regional | The contributing area per unit contour width. Sometimes referred to as specific contributing area |
| Rate of change of specific catchment area | Regional | The specific catchment area leaving a cell less the average of the specific catchment area entering a cell, divided by the flow length across the cell (Gallant & Wilson, 2000) |
| **Group 5: Elevation residuals** | | |
| Mean elevation | Local | Average elevation in a user-defined angular, circular, or square window (Gallant & Wilson, 2000) |
| Difference from mean elevation | Local | Difference between the elevation in the center of a user-defined window and the mean elevation in the window (Gallant & Wilson, 2000) |
| Standard deviation of elevation | Local | Standard deviation of elevations in a user-defined window (Gallant & Wilson, 2000). Sometimes referred to as surface roughness |
| Elevation range | Local | Difference between maximum and minimum elevations in a user-defined window (Gallant & Wilson, 2000) |
| Deviation from mean elevation | Local | Difference from the mean elevation divided by the standard deviation of the elevation in a user-defined window (Gallant & Wilson, 2000) |
| Local elevation percentile | Local | Ranking of the point in the center of a user-defined window relative to all the points in the window (Gallant & Wilson, 2000) |
| Local elevation percentage | Local | Ranking of the point in the center of a user-defined window based on the percentage of the elevation range in the window (Gallant & Wilson, 2000) |
| **Group 6: Statistical parameters** | | |
| Skewness coefficient | Local | Measure of the asymmetry of the probability distribution of elevation about its mean in a user-defined angular, circular, or square window (Olaya, 2009) |
| Kurtosis coefficient | Local | Measure of the "tailedness" of the probability distribution of elevation in a user-defined window (Olaya, 2009) |
| Ruggedness index | Local | The difference between the highest and lowest elevation divided by the square root of the area covered by the local region of interest (Melton, 1965) |
| Surface roughness factor | Local | The *x*, *y*, and *z* components of the unit vector normal to the land surface at each cell in a square or circular user-specified window divided by the number of cells (Hobson, 1972) |
| Topographic ruggedness index | Local | The average elevation change between the center cell in a 3 × 3 cell moving wndow and the eight cells immediately surrounding it (Riley et al., 1999) |
| Surface roughness index | Local | Ratio between the fitted nugget variation and the local variance for densely measured elevations (i.e. LiDAR data) (Olaya, 2009) |
| Anisotropy index | Local | Ratio between the minimum and maximum range parameter of spatial dependence, fitted for various directions (Olaya, 2009) |
| Anisotropic coefficient of variation | Local | Difference of the first derivative in four directions divided by the average value (Olaya, 2009) |
| Fractal dimension | Local | Measure of terrain roughness since surfaces with higher values of fractal dimension are more complex than those with lower values (Olaya, 2009) |
| Shape complexity index | Local | Measure of how compact (or oval) a polygon feature or DEM slices are (Hengl et al., 2003) |

*(Continued)*

**Table 3.1** *(Continued)*

| Land surface parameter | Type | Description |
|---|---|---|
| **Group 7: Upslope parameters** | | |
| Maximum upslope flow-path length | Regional | The longest flow path from a peak or ridge marking the catchment divide to the cell of interest |
| Minimum upslope flow-path length | Regional | The shortest flow path from a peak or ridge marking the catchment divide to the cell of interest |
| Mean upslope flow-path length | Regional | The mean length of all the flow paths from a peak or a ridge on the catchment boundary to the cell of interest |
| Number of upslope neighbors | Local | Number of neighbors in a $3 \times 3$ window that are of a higher elevation |
| Number of inflowing cells | Local | Number of neighbors in a $3 \times 3$ window that flow into the center cell |
| Elevation relative to peaks and divides | Regional | Elevation relative to nearest peak and/or divide (Lindsay, 2009) |
| Average upslope elevation | Regional | Mean height of the upslope contributing area which provides a measure of potential energy |
| Average upslope slope | Regional | Mean slope of the upslope contributing area which provides a measure of runoff velocity |
| Average upslope curvature | Regional | Mean profile, plan, or tangential curvature of the upslope contributing area which provides a measure of the flow convergence/divergence and runoff velocity |
| Area–distance function | Regional | The fraction of the upslope contributing area that is at some user-specified distance from the cell of interest (one of several hypsometric curves used for morphometric analysis) |
| Cumulative area function | Regional | The fraction of the upslope contributing area that has a contributing area greater than some user-specified threshold (one of several hypsometric curves used for morphometric analysis) |
| **Group 8: Downslope parameters** | | |
| Number of downslope neighbors | Local | Number of neighbors in a $3 \times 3$ window that are of a lower elevation |
| Maximum downslope elevation change | Local | Maximum elevation drop to a neighbor in a $3 \times 3$ window |
| Average downslope elevation change | Local | Average elevation drop to a neighbor in a $3 \times 3$ window |
| Elevation relative to channels and pits | Regional | Elevation relative to nearest channel and/or downslope depression cell (Lindsay, 2009) |
| Elevation above target cell | Regional | Elevation above nearest downslope target cell (Lindsay, 2009) |
| Downslope flow-path length | Regional | Downslope distance along flow path to outlet (Lindsay, 2009) |
| Average downslope slope | Regional | Mean slope of dispersal area |
| Specific dispersal area | Regional | The area downslope of a unit length of contour to a channel and/or a downslope depression into which water can drain (Speight, 1980) |
| **Group 9: Visibility and visual exposure** | | |
| Viewshed | Regional | Area that can be seen from a point (Franklin & Ray, 1994) |
| Visibility index | Regional | The number of cells that can be seen from each cell or, alternatively, the number of viewsheds to which a cell belongs (Olaya, 2009) |
| Openness | Regional | Maximum angle of vision from a point on the land surface within a user-defined maximum radial distance (Yokoyama et al., 2002) |

*Source*: modified from Wilson & Gallant (2000b, p. 7), Olaya (2009, p. 142), and Lindsay (2009, pp. 384–386).

steepness and surface roughness), the algorithm used to calculate surface area, and the spatial resolution of the underlying DEM that is used. They calculated incremental area coefficients (Hoechstetter et al., 2006, 2008) based on two slope algorithms and nine DEMs derived from the ASTER GDEM with resolutions varying from 30 to 20,000 m for mainland China. The first algorithm relied on the slope calculation method in Esri's ArcGIS Spatial Analyst toolbox, which assumes the local surface is an inclined plane, and the second used the method proposed by Jenness (2004), which breaks each cell into eight triangles in three-dimensional space to reflect the topographic relief more accurately. The surface area increments calculated with the two slope algorithms were 4.31 and 4.89% over China using the 30-m resolution DEM and the largest increment for a 50 × 50 km cell was over 45%. Using multivariate regression and hierarchical variation partitioning, Ying et al. (2014) also showed how the variation in the regional surface area increment was influenced by the average slope (29.3%), slope variability (21.7%), and regional elevation range or relief (17.5%). The DEM resolution independently contributed another 10% to this variation.

One solution that would avoid some, if not all, of these kinds of discrepancies would be to use geographic (latitude, longitude) grids to calculate land surface parameters, although the level of success achieved using this approach would likely vary with the geographic extent of the study area and strategies used to estimate the diminishing width of grid cells with increasing latitude (Florinsky, 2017).

### 3.1.2 Slope, Aspect, and Curvature

The second group of land surface parameters listed in Table 3.1 are concerned with slope, aspect and curvature, and are best handled by moving a 3 × 3 window across a grid and calculating land surface parameters for the target grid cell (i.e. the central grid cell in the 3 × 3 moving cell window) (Figure 3.2). The slope and aspect are calculated from the first derivatives and the curvatures from the second derivatives of the topographic surface. Several algorithms have been proposed for deriving these derivatives over the years and the most popular are those of Evans (1972), Zevenbergen and Thorne (1987), Shary (1995), and the modified Evans–Young method (Evans, 1972; Young, 1978, Pennock, Zebarth & de Jong, 1987; Shary et al., 2002). The Zevenbergen and Thorne (1987) method is described in some detail here because this algorithm is implemented in ArcGIS, which was used to prepare the sample outputs for Cottonwood Creek scattered throughout this and the next chapter. This description follows that offered by Gallant and Wilson (2000) but the discussion will also highlight some of the differences between this and the other algorithms along with the likely benefits and costs that would follow from using each of the aforementioned algorithms to calculate the slope, aspect and curvature.

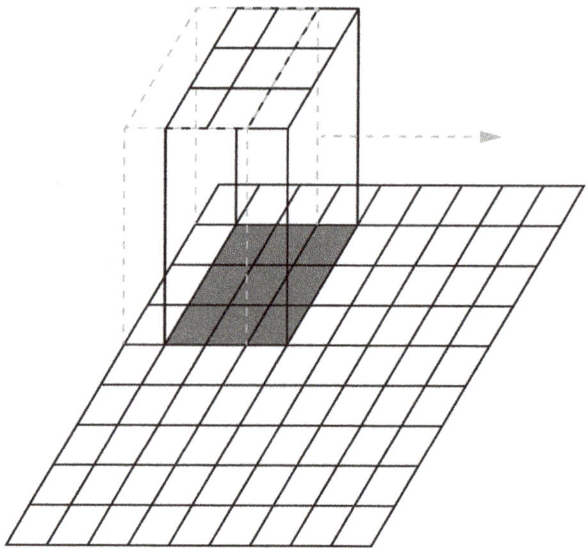

Figure 3.2 A 3×3 moving grid used to calculate selected local land surface parameters. *Source*: Olaya (2009, p. 143). Reproduced with permission of Elsevier.

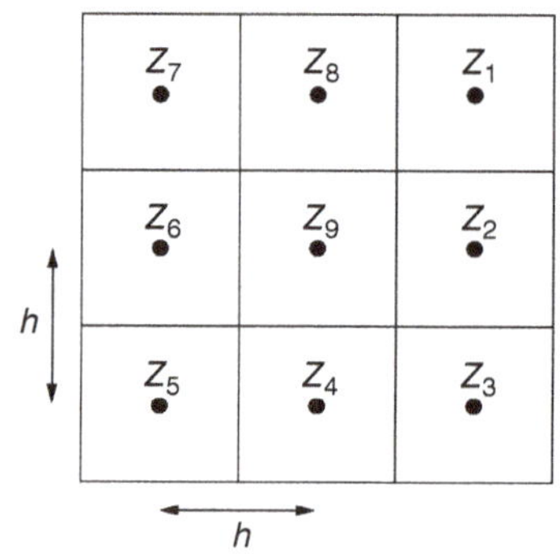


Figure 3.3 Node numbering convention used for calculation of local land surface parameters. *Source*: Gallant & Wilson (2000, p. 52). Reproduced with permission of John Wiley & Sons, Inc.

The first and second derivatives of the topographic surface measure the rate at which elevation changes in response to changes in location (i.e. $x$ and $y$):

$$z_x = \frac{\partial z}{\partial x} \approx \frac{z_2 - z_6}{2h} \tag{3.1}$$

$$z_y = \frac{\partial z}{\partial y} \approx \frac{z_8 - z_4}{2h} \tag{3.2}$$

Figure 3.3 shows the arrangement and numbering of the nine grid points that enter into these two equations and the remainder of the finite-difference equations that follow below, and $h$ is the grid spacing of the DEM. The $y$ axis, which points north, is up in Figure 3.3. There are also special rules for handling the edges where some of the nine points are undefined. Forward and backward finite differences are typically used to avoid reference to non-existent data, as shown below for the left edge, where $z_6$ is not defined, and the $x$ derivative is calculated as:

$$z_x = \frac{z_2 - z_9}{2h} \tag{3.3}$$

Equations 3.4 and 3.5, reproduced from Gallant and Wilson (2000), are the second derivatives that describe the rate of change of the first derivative in the $x$ and $y$ directions, or the curvature in those directions. Equation 3.6 is a mixed second derivative that describes the rate of change of the $x$ derivative in the $y$ direction, or the twisting of the surface. Equations 3.7 and 3.8 are combinations of terms that are used in several places in subsequent equations.

$$z_{xx} = \frac{\partial^2 z}{\partial x_2} \approx \frac{z_2 - 2z_9 + z_6}{h^2} \tag{3.4}$$

$$z_{yy} = \frac{\partial^2 z}{\partial y^2} \approx \frac{z_8 - 2z_9 + z_4}{h^2} \tag{3.5}$$

$$z_{xy} = \frac{\partial^2 z}{\partial x \partial y} \approx \frac{-z_7 + z_1 + z_5 - z_3}{4h^2} \tag{3.6}$$

$$p = z_x^2 + z_y^2 \tag{3.7}$$

$$q = p + 1 \tag{3.8}$$

The slope $S$ measures the rate of change of elevation in the direction of steepest descent and is normally calculated using finite differences:

$$S_{\mathrm{FD}} = \sqrt{p} \tag{3.9}$$

This equation depends only on the elevations in the four cardinal directions. The finite difference equation for slope in ArcGIS and some other software makes use of all eight adjacent cells. Gallant and Wilson (2000) reported there was little difference in the results from these two approaches, but for the fact there is some evidence (e.g. Jones, 1998) that the equation using all eight adjacent cells is slightly more accurate. Slope is the means by which gravity induces the flow of water and other materials and is therefore of great significance in many facets of hydrology, geomorphology and ecology.

Slope can also be calculated using the so-called deterministic eight-node (D8) method which measures the steepest downhill slope to one of the eight neighbors:

$$S_{\mathrm{D8}} = \max_{i=1,8} \frac{z_9 - z_i}{h\phi(i)} \tag{3.10}$$

where $\phi(i) = 1$ for cardinal (north, south, east, west) neighbors ($i = 2, 4, 6$, and 8) and $\phi(i) = \sqrt{2}$ for diagonal neighbors to account for the additional distance to those cells (Figure 3.3) (Gallant & Wilson, 2000).

The finite difference method (Equation 3.9) is usually preferred because of its greater accuracy, but the D8 slope estimate can be useful when the slope of the channel is required, because the finite difference method may be affected by steep slopes adjacent to the channel. Using the D8 slope option guarantees that the slope calculated at a cell corresponds to the slope in the primary flow direction (Gallant & Wilson, 2000). Figure 3.4 shows slope in percent calculated using the finite difference method for the Cottonwood Creek DEM introduced in Chapter 1. The low slopes along valley bottoms and ridgetops as well as the steep slopes on many of the hillsides are clearly visible in this slope map.

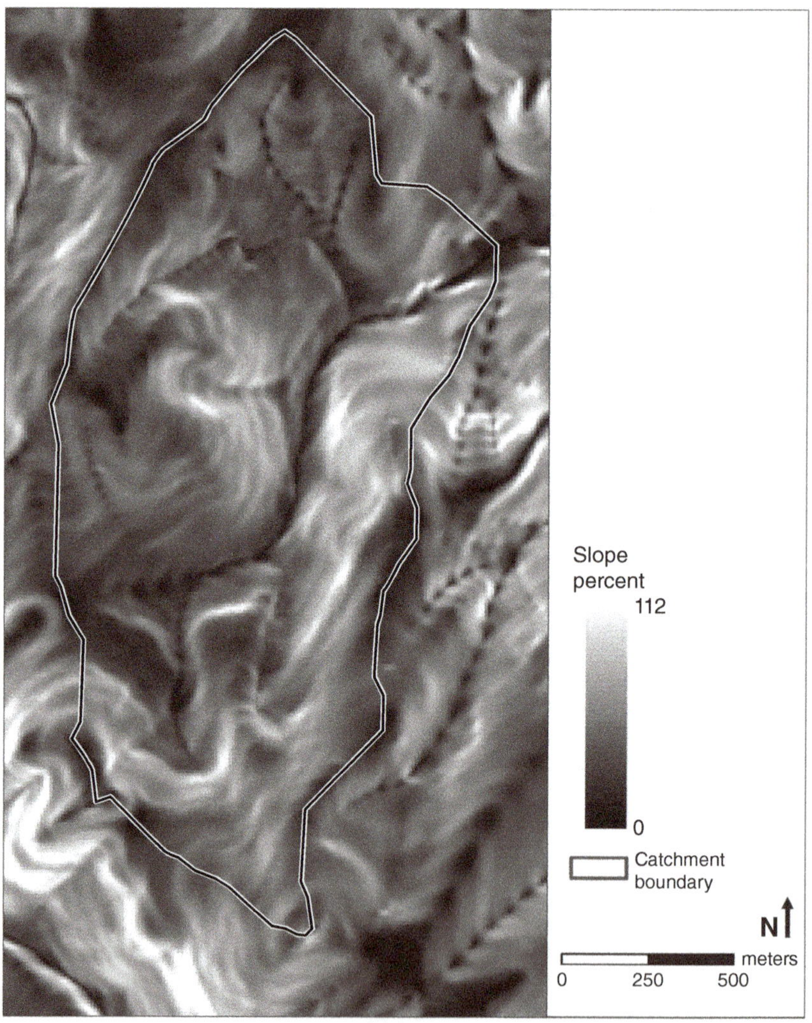


**Figure 3.4** Percent slope grid derived for Cottonwood Creek, MT study site using the finite difference equation, with the catchment boundary overlaid.

The aspect $\varphi_{FD}$ is the orientation of the line of steepest descent and is usually measured in degrees clockwise from north. Aspect in degrees is calculated using finite differences as follows:

$$\varphi_{FD} = 180 - \arctan\left(\frac{z_y}{z_x}\right) + 90\left(\frac{z_x}{|z_x|}\right) \tag{3.11}$$

which is the angle determined by the $x$ and $y$ derivatives via the arc tangent, modified to give degrees clockwise from north (Gallant & Wilson, 2000). Aspect is frequently recorded as a site variable in biological surveys and together with slope can be used to estimate solar radiation since south- and north-facing slopes in the northern and southern hemispheres receive more insolation than those facing north and south, respectively.

Aspect becomes rather meaningless when the slope is small and mathematically undefined when the slope is zero, so cells with slope less than some minimum value may be considered to have undefined aspect (Mitášová & Hofierka, 1993). Aspect is also a circular land surface parameter so that 0 and 360° describe the same slope direction and, as a consequence, some studies have used cos(φ) as slope northness or sin(φ) as slope eastness in place of aspect (Olaya, 2009).

The three maps reproduced in Figures 3.5, 3.6 and 3.7 show the aspect, northness and eastness calculated using the finite differences method in Equation 3.11. The aspect map (Figure 3.5) is dominated by the abrupt changes where aspect switches from 0 to 360° which is an artifact of the linear scale used in this map. Some of these abrupt changes occur along the catchment boundary. The abrupt changes in the final pair of maps, on the other hand, show the changes in orientation across the surface and the hillslope facets that are predominantly north–south (Figure 3.6) and/or east–west facing (Figure 3.7).

The remainder of the land surface parameters in this second group comprise various curvatures (Table 3.1). These land surface parameters are based on second derivatives: the rate of change of a first derivative, such as slope or aspect, usually in a specific direction. Their number has grown tremendously over time but the two most common are the profile (i.e. vertical) and tangential (i.e. horizontal) curvatures, which are used to distinguish locally convex and concave surface features.

Profile curvature ($K_p$) measures the rate of change of slope down a flow line and is important for characterizing changes in flow velocity and sediment transport processes. Plan (or contour) curvature ($K_c$) measures the rate of change of aspect along a contour line and is important for characterizing the topographic convergence and divergence and hence the propensity for water to converge or diverge as it flows across the land surface. This particular parameter takes on extremely large values when slope is small and so the tangential curvature ($K_t$; plan curvature multiplied by the sine of the slope angle) is generally considered more appropriate for studying convergence and divergence nowadays (Mitášová & Hofierka, 1993; Gallant & Wilson, 2000).

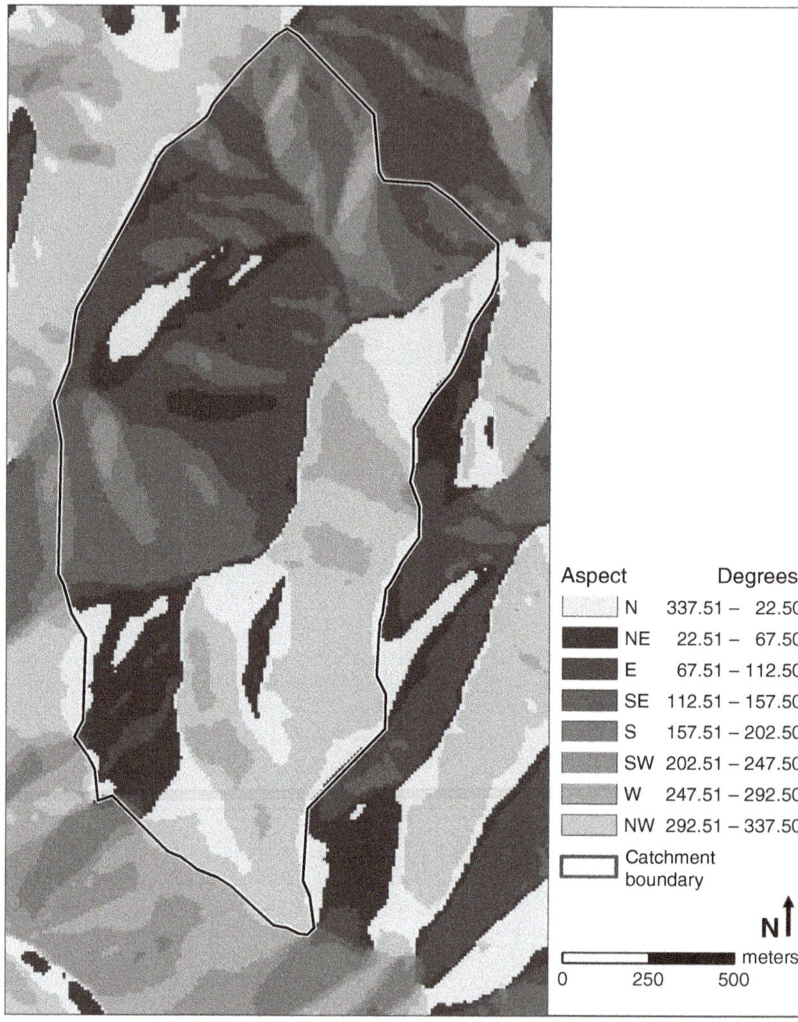


**Figure 3.5** Aspect in degrees from north derived for Cottonwood Creek, MT study site using the finite difference equation, with the catchment boundary overlaid.

For example, Gallant and Wilson (2000, p. 57) described each of these three curvatures as follows:

> Surface curvatures can be thought of as the curvature of a line formed by the intersection of a plane and the topographic surface. The curvature of a line is the reciprocal of the radius of curvature, so a gentle curve has a small curvature and a tight curve has a large curvature value. Plan curvature is the curvature in the horizontal plane of the contour line, while profile curvature is the curvature in the vertical plane of a flow line. Tangential curvature is the curvature in an inclined plane perpendicular to both the direction of flow and the surface.

The units for these three curvatures are radians per meter, the change in orientation from traveling 1 m along the respective line, and they can be calculated using the following equations:

$$K_p = \frac{z_{xx}\, z_x^2 + 2z_{xy}\, z_x\, z_y + z_{yy}\, z_y^2}{pq^{3/2}} \tag{3.12}$$

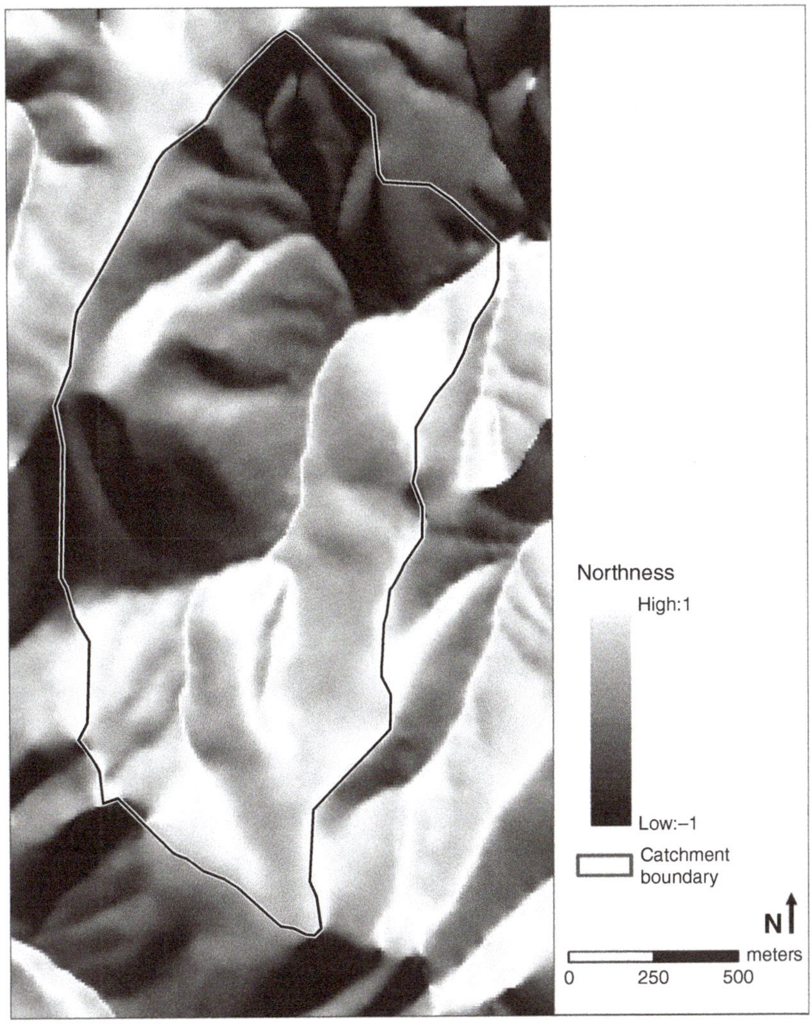


**Figure 3.6** Northness derived for Cottonwood Creek, MT study site, with the catchment boundary overlaid.

$$K_c = \frac{z_{xx}\, z_y^2 - 2z_{xy}\, z_x\, z_y + z_{yy}\, z_x^2}{pq^{3/2}} \tag{3.13}$$

$$K_t = \frac{z_{xx}\, z_y^2 - 2z_{xy}\, z_x\, z_y + z_{yy}\, z_x^2}{pq^{1/2}} \tag{3.14}$$

Using these equations, profile curvature is negative for slope increasing downhill (convex flow profile, typical of upper slopes) and positive for slope decreasing downhill (concave, typical of lower slopes). Plan and tangential curvature, on the other hand, are negative for diverging flow (on peaks and ridges) and positive for converging flow (in valleys).

The total and mean curvature are also used as measures of the surface curvature. The total curvature measures the curvature of the surface itself rather than

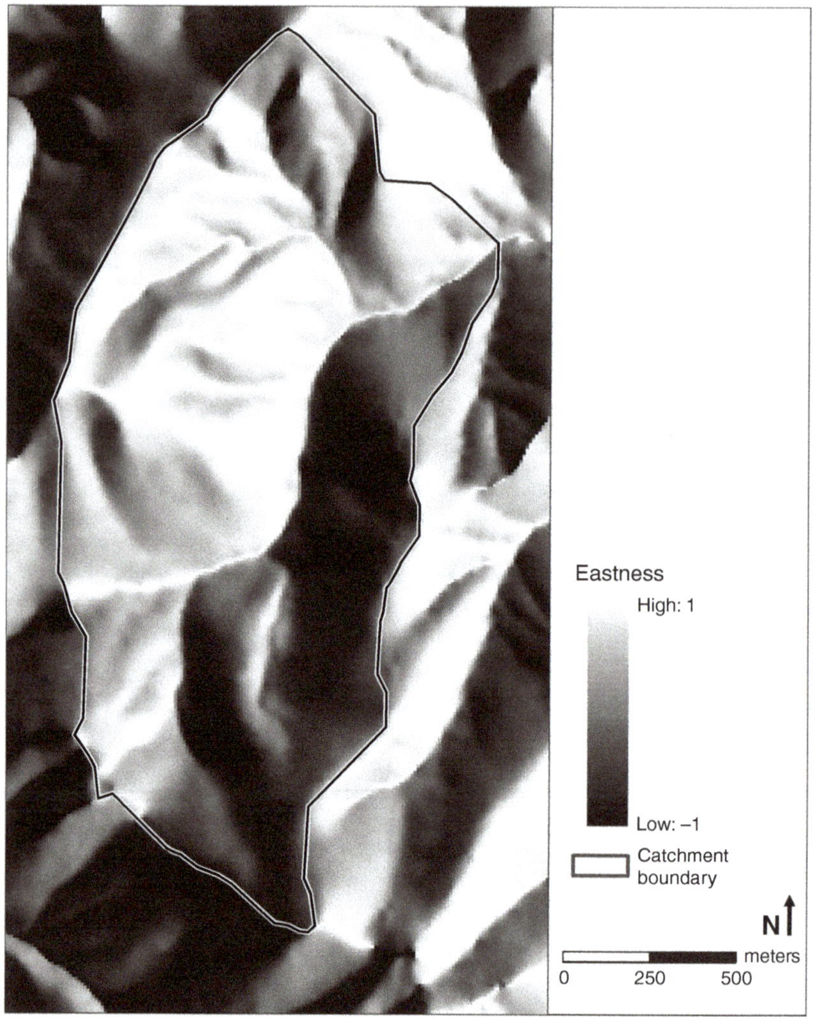


**Figure 3.7** Eastness derived for Cottonwood Creek, MT study site, with the catchment boundary overlaid.

the curvature of a line across the surface in some direction (Gallant & Wilson, 2000). It can be positive or negative, with zero curvature indicating that the surface is either flat or the convexity in one direction is balanced by the concavity in another direction, as would occur at a saddle. Total curvature is calculated as:

$$K = z_{xx}^2 + 2z_{xy}^2 + z_{yy}^2 \tag{3.15}$$

Blaszczynski (1997) used this total curvature to propose a surface curvature index for user-specified groups of grid cells.

The mean curvature is the average of any mutually orthogonal normal sections such as profile and tangential curvature (Olaya, 2009):

$$K_{\mathrm{M}} = \frac{K_{\mathrm{p}} + K_{\mathrm{t}}}{2} \tag{3.16}$$

The land surface parameter in Equation 3.16 describes mean-concave and mean-convex terrain such that positive values of mean curvature are associated with areas that experience mean accumulation and negative values of mean curvature are associated with areas of relative deflection.

Olaya (2009, pp. 152–153) described several additional curvatures proposed by Florinsky (1998) and Shary (1995). Florinsky (1998) proposed using the Rotor, which captures the curvature of flow lines that are perpendicular to contours, as a measure of the twisting of flow lines. Shary (1995), on the other hand, proposed a system of 12 curvatures. The mean curvature along with two other basic curvatures – the unsphericity curvature, which describes how the surface differs from a sphere and the difference curvature, which identifies which of the vertical and horizontal curvatures is formally stronger – were used to derive nine additional curvatures. These curvatures have been used to model flow characteristics and related outcomes (e.g. soil moisture distribution, erosion potential) and/or to delineate geomorphic units (e.g. Dikau, 1989; Shary, Sharaya & Mitusov, 2005).

Figures 3.8 and 3.9 show profile and plan curvature in the Cottonwood Creek catchment using the usual sign convention of positive values for convex curves. Plan curvature shows the locations of ridges and valleys with

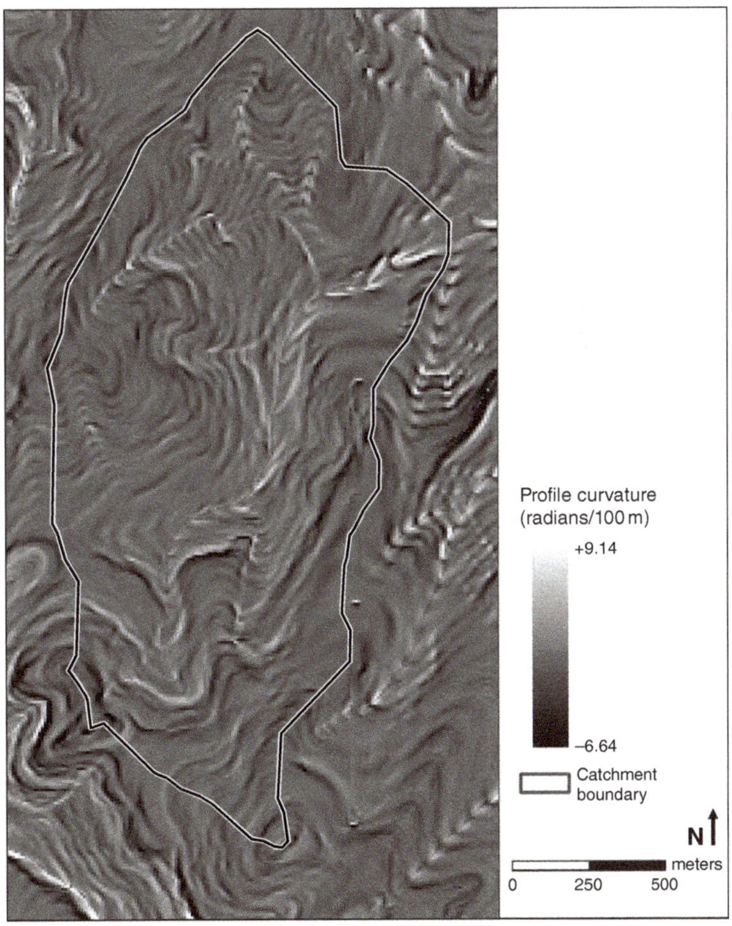


**Figure 3.8** Profile curvature (radians per 100 m, convex curvatures are positive) derived for Cottonwood Creek, MT study site using the finite difference formula, with the catchment boundary overlaid.

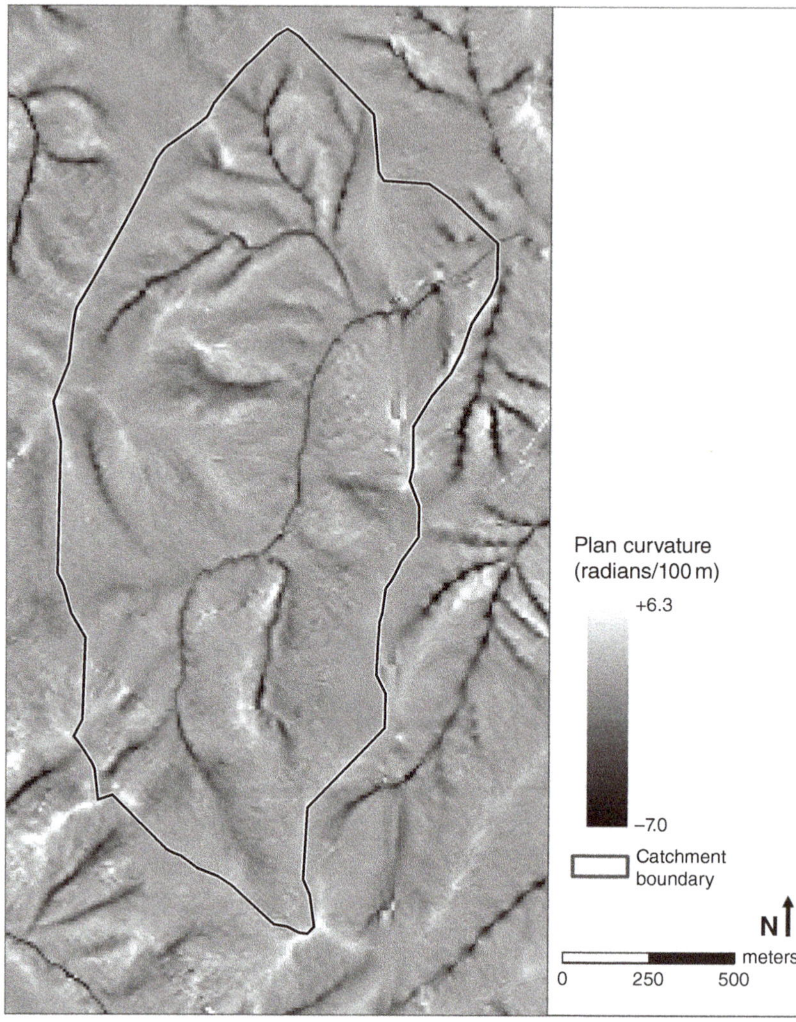


Figure 3.9 Plan curvature (radians per 100 m, convex curvatures are positive) derived for Cottonwood Creek, MT study site using the finite difference formula, with the catchment boundary overlaid.

their large positive and negative values and the more gently curved hillslope regions between them. Profile curvature shows the convex areas around hilltops (negative values) and the concave area in valley bottoms (positive values) and comparison of the patterns on hillslopes in Figure 3.9 with those in Gallant and Wilson (2000, Figure 3.5, p. 59) suggests that the DEM used for the current project was a modest improvement over the 15-m DEM derived from contours that was used for the earlier project.

The interpretation of the meaning of the various curvatures is complicated by one or possibly two potential problems. The first arises because any errors in the source DEM will be accentuated in the calculated curvature values since these land surface parameters are based on second derivatives (Gallant & Wilson, 2000; Schmidt, Evans & Brinkmann, 2003), and the second arises because many of these curvature-based land surface parameters will be correlated with one another and therefore difficult to use in combination for some applications and/or landscapes.

The last question about this group of land surface parameters has to do with the choice of algorithm used to calculate one or more of these parameters. Hengl and Evans (2009) described some of the similarities and differences between the Evans–Young (Evans, 1972; Young, 1978; Pennock et al., 1987), Zevenbergen and Thorne (1987), and Shary (1995) algorithms for deriving first and second derivatives. They explained how the Evans–Young algorithm fits a second-order polynomial to a 3 × 3 neighborhood filter and with only six coefficients, why the resulting polynomial will not necessarily pass through any of the nine original elevations. They also noted that the Shary (1995) polynomial is similar to that of Evans–Young but for the fact it has to pass through the central point. The Zevenbergen and Thorne (1987) algorithm, on the other hand, uses a partial quartic polynomial which passes exactly through all nine data points. And finally, Hengl and Evans (2009) noted how the Evans–Young and Shary algorithms provide a modest smoothing of the input data and, as a consequence, they will be less sensitive to outliers and thus produce more uniform or stable estimates than the Zevenbergen and Thorne (1987) algorithm in some settings.

Shary et al. (2002) have suggested that a parametric isotropic smoothing should be performed prior to calculating the DEM derivatives and proposed a modified version of the Evans–Young algorithm that used a 5 × 5 rather than a 3 × 3 moving cell window (see Figure 3.2). These authors also reported that the smoothing of the various algorithms increased in the following order: (i) Zevenbergen and Thorne; (ii) Evans–Young and Shary; and (iii) modified Evans–Young (Hengl & Evans, 2009).

Skidmore (1989), Guth (1995), Florinsky (1998), and Schmidt et al. (2003) have compared various versions of these algorithms for calculating slope, aspect and curvatures over the years but the results vary and there is no one method that performs best. The optimum choice of algorithm will likely vary with the choice of dataset and/or study area (Hengl & Evans, 2009) and some will therefore welcome the release of the ArcGIS Arc-Geomorphometry add-on (Rigol-Sanchez, Stuart & Pulido-Bosch, 2015) because this provides an opportunity for users to conduct their own comparisons using the Evans–Wood, Shary, and Zevenbergen and Thorne algorithms with their own elevation datasets (see Section 6.3.3 for additional details).

### 3.1.3 Slope Direction and Width

Many approaches have been proposed for calculating slope direction and slope width over the past 30 years because of the difficulty of routing water across square-grids and because these two land surface parameters feature in many of the other flow-based primary and secondary land surface parameters described later in this chapter.

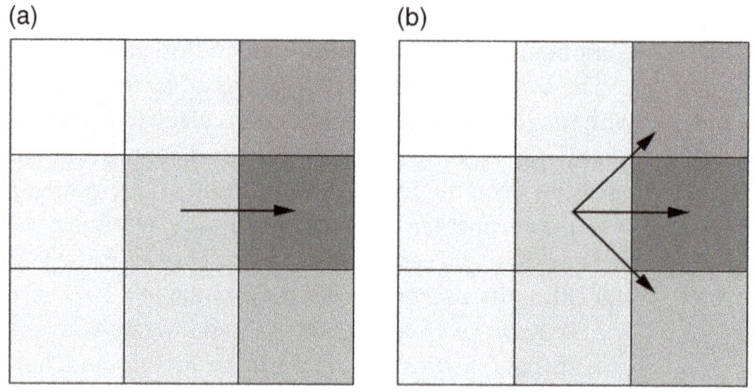


**Figure 3.10** Single- and multiple-flow directions assigned to the central grid cell in a 3 × 3 moving window using the D8 and FMFD flow-direction algorithms. Gray shading represents elevation decreasing with the darkness of the cell. Multiple-flow directions are assigned in (b) and a fraction of the flow of the central cell is distributed to each of the three cells that the arrows point to. *Source*: modified from Gruber & Peckham (2009, pp. 176–177). Reproduced with permission of Elsevier.

The initial development and use of flow-based land surface parameters can be traced to the D8 algorithm (O'Callaghan & Mark, 1984). However, this is now but one or more than 20 possible flow routing algorithms and a distinction is usually drawn between single- and multiple-flow direction algorithms (Figure 3.10). The single-flow direction algorithms, which direct flow to just one downslope or neighboring cell, include the Rho8 (Fairfield & Leymarie, 1991), ADK (Lea, 1992), and D∞ (Tarboton, 1997) algorithms in addition to the D8 algorithm. The multiple-flow direction algorithms, which are capable of directing flow to two or more downslope or neighboring cells, include the FMFD (Freeman, 1991), QMFD1/2 (Quinn et al., 1991, 1995), DEMON (Costa-Cabral & Burges, 1994), Mass Flux (Gruber & Peckham, 2009), and MD∞ (Seibert & McGlynn, 2007) algorithms.

The various approaches that have been used to depict slope (i.e. flow) direction and width are described in turn along with their advantages and disadvantages. The 24 flow-direction algorithms listed in Table 3.2 are split into two groups (i.e. single- and multiple-flow direction algorithms) and introduced in chronological order because problems with earlier algorithms have often been used to motivate the development of the newer flow-direction algorithms and because they will usually generate very different results, influenced in part by the kinds of preprocessing that is performed on the source DEMs (see Section 2.4 for additional details) and in part by the type of landscape (i.e. the landforms and drainage networks present in the study area of interest). The SCA is often used to evaluate the performance of the different flow-direction algorithms using a series of theoretical shapes and/or real-world landscapes, as shown in Sections 3.1.3.1 and 3.1.3.2. However, some new and novel approaches to compare the performance of these algorithms have been proposed in the past few years and examples of

**Table 3.2** List of single- and multiple-flow direction algorithms.

| Flow-routing algorithm | Author(s) | Description |
|---|---|---|
| **Single-flow direction algorithms** | | |
| D8: Deterministic eight-node algorithm | O'Callaghan & Mark (1984) | Permits flow to one of eight neighbors based on the direction of steepest descent on a square-grid surface model |
| Rho8: Randomized eight-node algorithm | Fairfield & Leymarie (1991) | Stochastic version of D8 that directs flow to one of eight neighbors and produces a mean flow direction equal to aspect on a square-grid surface model |
| ADK: Aspect-driven kinematic routing algorithm | Lea (1992) | Specifies the flow direction continuously and assigns flow direction to cardinal cells on a square-grid surface model |
| D∞: Deterministic infinite-node algorithm | Tarboton (1997) | Specifies the flow direction continuously and assigns flow direction to one or at most two downslope cells on a square-grid surface model |
| D8-LAD: Deterministic eight-node least angular deviation algorithm | Orlandini et al. (2003), Orlandini & Moretti (2009) | D8-like method that minimizes the angular path-based deviations between selected and theoretical drainage directions along drainage paths on a square-grid surface model |
| D8-LTD: Deterministic eight-node least transversal deviation algorithm | Orlandini et al. (2003), Orlandini & Moretti (2009) | D8-like method that minimizes the transversal path-based deviations between selected and theoretical drainage directions along drainage paths on a square-grid surface model |
| D∞-LAD: Deterministic infinite least angular deviation algorithm | Orlandini & Moretti (2009) | D∞-like method that minimizes the angular path-based deviations between selected and theoretical drainage directions along drainage paths on a square-grid surface model |
| D∞-LTD: Deterministic infinite least transversal deviation algorithm | Orlandini & Moretti (2009) | D∞-like method that minimizes the transversal path-based deviations between selected and theoretical drainage directions along drainage paths on a square-grid surface model |
| D6: Deterministic six-node algorithm | Wright et al. (2014) | Permits flow to one of six neighbors based on the direction of steepest descent on a hierarchical hexagonal surface model |
| **Multiple-flow direction algorithms** | | |
| FMFD: Freeman multiple flow direction algorithm | Freeman (1991) | Permits flow to multiple downslope neighbors using slope-weighted methods on a square-grid surface model |
| QMFD1: Quinn multiple flow direction algorithm (Variant 1) | Quinn et al. (1991) | Permits flow to multiple downslope neighbors based on the slope and the contour length (i.e. the flow width) on a square-grid surface model |
| DEMON: Digital elevation model network extraction method | Costa-Cabral & Burges (1994) | Represents flow in two directions based on aspect and allows flow width to vary over non-planar topography in order to distribute flow to multiple downslope neighbors on a square-grid surface model |
| MDEMON: Moore digital elevation model network extraction method | Moore et al. (1993e), Gallant & Wilson (1996, 2000) | DEMON-like method that used nodes in place of vertices to define the centroids of the grid cells and distributed flow to multiple downslope neighbors based on aspect using either the original or a depressionless square-grid surface model |
| MMFD1: Moore multiple flow direction algorithm (Variant 1) | Moore et al. (1993e), Gallant & Wilson (1996, 2000) | Permits flow to multiple downslope neighbors based on the slope (similar to FMFD) and D8 flow width (similar to QMFD1) on a square-grid surface model |

(*Continued*)

Table 3.2 (*Continued*)

| Flow-routing algorithm | Author(s) | Description |
|---|---|---|
| MMFD2: Moore multiple flow direction algorithm (Variant 2) | Moore et al. (1993e), Gallant & Wilson (1996, 2000) | Permits flow to multiple downslope neighbors based on the slope and D8 flow width (similar to MMFD1) in upland areas above defined channels and uses either the D8 or Rho8 algorithms below points of channel initiation that were defined with a user-specified maximum cross-grading area threshold on a square-grid surface model |
| QMFD2: Quinn multiple flow direction algorithm (Variant 2) | Quinn et al. (1995) | QMFD1-like method that includes an exponent which varies continuously from 1 (full dispersion) to a large value (single flow direction) as upslope contributing area increases giving a smooth change from multiple- to single-flow directions when moving from summits and ridgelines to valleys and floodplains on a square-grid surface model |
| PMFD: Form-based multiple flow direction algorithm | Pilesjö et al. (1998) | Calculates the topographic form of a 3 × 3 facet and deploys either the D8 or D∞ algorithms to distribute flow to one or two downslope neighbors if the form is concave or flat and the FMFD algorithm to distribute flow to multiple downslope neighbors if the form is convex on a square-grid surface model |
| SDFAA: Spatially distributed flow apportioning algorithm | Kim & Lee (2004) | Calculates relationships between various flow-direction parameters and an expanded channel initiation threshold iteratively and then distributes the flow from upslope to downslope grid cells considering the location of the permanent channel network on a square-grid surface model |
| MFD-md: Multiple flow direction-maximum downslope gradient algorithm | Qin et al. (2007) | Permits flow to multiple downslope neighbors using slope weighted methods that vary as a linear function of maximum downslope gradient on a square-grid surface model |
| MD∞: Triangular multiple flow direction algorithm | Seibert & McGlynn (2007) | Divides each cell into eight triangular facets, calculates local slope directions and gradients around each cell (similar to D∞), and distributes flow in downslope directions using a slope-weighted procedure similar to QMFD1 or QMFD2 on a square-grid surface model |
| MF: Mass Flux algorithm | Gruber & Peckham (2009) | Divides each grid cell into four quarter grid cells, defines a continuous flow direction for each using a grid with twice the dimensions of the original square-grid DEM, and distributes the flow from each quarter grid cell to one or two of its cardinal neighbors on the original square-grid surface model |
| TFN: Triangular facet network algorithm | Zhou et al. (2011) | Divides each cell into four triangular facets and uses these facets to build a triangular facet network and specify flow packages that move flow towards downslope facets based on aspect on a square-grid surface model |
| $D_{trig}$ multiple flow direction algorithm | Shelef & Hilley (2013) | Divides each cell into eight triangular facets and defines the gradient based on the adjacent nodes similar to Tarboton (1997) and then partitions the flow across the edges of the facets and grid cells (rather than through the corners, as was the case with several of the aforementioned flow direction algorithms) |
| TFM: Triangular form-based multiple flow algorithm | Pilsejö & Hasan (2014) | Divides each cell into eight triangular facets similar to Tarboton (1997) and uses a two-step allocation approach to distribute flow to one or more of the eight neighboring cells |

three such approaches are reviewed along with SCA and several other land surface parameters focused on flow accumulation in Section 3.1.4.

It is also worth noting before introducing the individual flow-direction algorithms that several commentators have pointed out that flow lines are not necessarily the same as slope lines because the flow in some landscapes may be affected by significant pressure or stress gradients, making flow dispersion a potentially relevant process, and thereby violating the assumption that overland flow is driven solely by gravity (e.g. Tarboton, 1997; Orlandini et al., 2003, 2012; Gallant & Hutchinson, 2011). This means that the term "flow-direction algorithm" used in the following sections and extensively throughout the remainder of this book may not be appropriate but it is used anyway, partly because of its popularity and partly because of the lack of alternatives. Orlandini et al. (2012) noted the same problem but proceeded in the same way, using the term "flow-direction method" to denote single-flow direction methods (which described non-dispersive methods designed to represent slope and flow lines in gravity-driven overland flows) and multiple-flow direction methods (which describe dispersive methods designed to represent flow lines in flows driven by gravity as well as other forces).

We turn first to the single-flow direction algorithms (Section 3.1.3.1) and then move to multiple-flow direction algorithms (Section 3.1.3.2) and the various methods used to estimate flow width (Section 3.1.3.3) that are typically used in conjunction with one or more of the flow-direction algorithms. Two maps showing the results of using two of the more popular single- and multiple-flow direction algorithms are included in Sections 3.1.3.1 and 3.1.3.2, respectively, to illustrate how the choice of flow-direction algorithm affected the flow directions calculated for the Cottonwood Creek catchment introduced earlier.

#### 3.1.3.1 Single-flow Direction Algorithms

The D8 flow-direction algorithm (O'Callaghan & Mark, 1984) introduced earlier assigns flow to one of eight nearest neighbors based on the path of steepest descent. Because flow can accumulate in any cell from several upslope cells but only flow out into a single cell, this approach can model flow convergence in valleys but not flow divergence in summit and ridge areas. Another problem with this algorithm is that it tends to produce flow in parallel lines along preferred directions that will only concur with the aspect when this parameter is a multiple of 45°. Gallant and Wilson (2000) showed how (i) the valley bottoms and ridge areas were clearly defined in the Cottonwood Creek catchment using a 15-m square-grid DEM derived from contours; (ii) many straight and/or parallel paths were generated, due to the absence of dispersion and the inability of this algorithm to respond to subtle changes in surface orientation; and (iii) the straight-line artifacts and parallel flow paths distorted the spatial pattern of upslope contributing area in numerous parts of the catchment. However, the D8 algorithm is still widely used, notwithstanding the high frequency of one or more of the aforementioned problems, due to its ease of use and because it has proved useful for extracting stream networks, longitudinal profiles, and catchment boundaries (I.D. Moore et al., 1991; Gruber & Peckham, 2009).

The second randomized eight-node (Rho8) single-flow direction algorithm was proposed by Fairfield and Leymarie (1991) and provides a stochastic version of the D8 algorithm. This algorithm seeks to break up the parallel flow paths while maintaining a mean flow direction equal to the aspect by replacing $\varphi(i)$, the distance factor of 1 or $\sqrt{2}$ in Equation 3.10, by $2 - r$ for the diagonal neighbors ($i = 1, 3, 5, 7$ in Figure 3.3), where $r$ is a uniformly distributed random variable between 0 and 1. Gallant and Wilson (2000) implemented this algorithm in Cottonwood Creek using a 15-m square-grid DEM and showed how the Rho8 algorithm did replace the parallel flow lines with a more dendritic structure but at the cost of many more cells with no upslope contributing area (which happens with the D8 algorithm as well). Wilson et al. (2000) also noted that the randomizing of the flow directions produces different flow networks each time the algorithm is executed and given these problems, Gallant and Wilson (2000) suggested that Rho8 is not a useful alternative to the D8 single-flow direction algorithm.

The third single-flow direction algorithm, the aspect-driven kinematic (ADK) single-flow direction algorithm proposed by Lea (1992), specifies flow direction continuously and assigns flow direction to cardinal cells with no dispersion and less grid bias than D8 (Gruber & Peckham, 2009). Flow is routed as though it were a ball rolling on a plane released from the center of each grid cell and the plane was fitted to the elevations of the grid cell corners that were estimated by averaging the elevations of grid cell centers. However, Tarboton (1997) has noted three reasons why this approach might cause problems in certain types of landscapes.

1 The assumption of a plane fitted locally to each grid cell requires approximation because only three points are required to determine a plane.
2 The best fit plane cannot in general pass through the four corner elevations, leading to a discontinuous representation of the surface at grid cell edges.
3 Planes fitted locally to certain elevation combinations will lead to inconsistent or counterintuitive flow directions with this algorithm and with Costa-Cabral and Burges's (1994) DEMON multiple-flow direction algorithm, which is described in Section 3.1.3.3.

This ADK algorithm has proved influential, notwithstanding the aforementioned flaws, and the general approach of characterizing the local topographic form or shape and then using this to guide the calculation of flow directions and widths has been replicated in DEMON and several of the other newer multiple-flow direction algorithms described in Section 3.1.3.2.

The D∞ algorithm proposed by Tarboton (1997) is included here as the fourth single-flow direction algorithm, notwithstanding the fact it behaves like a single-flow direction algorithm in some landscape settings and a

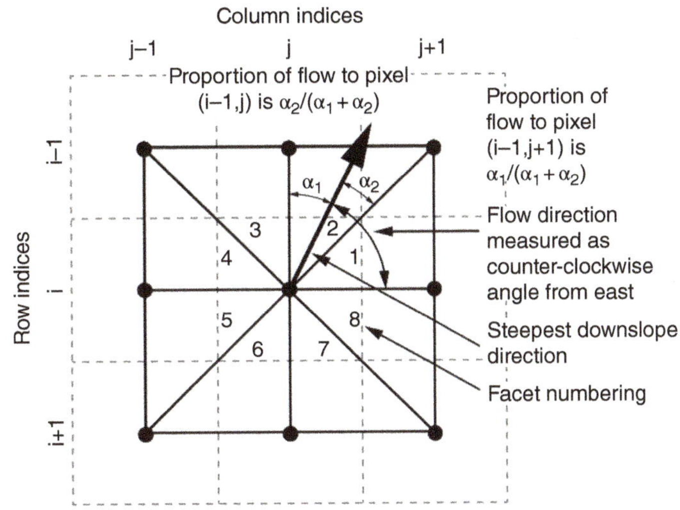


**Figure 3.11** Concept of flow apportioning in D∞. *Source*: following Tarboton (1997) and as shown in Gruber & Peckham (2009). Reproduced with permission of Elsevier.

multiple-flow direction algorithm in others. This particular algorithm assigns a single flow direction but can allocate flow into one or two downslope cells depending on the relationship between this direction and those of the eight surrounding grid cells.

The D∞ algorithm chooses one of an infinite number of possible single flow directions based on triangular facets. This approach represents flow direction as a single angle taken as the steepest downward slope on the eight triangular facets centered on each grid point. Figure 3.11 shows a 3×3 moving cell window with a series of dashed lines and numbers the eight grid cells surrounding the center grid cell 1 through 8. The grid cell centers are represented by solid black circles and the thick lines connecting the grid cell centers form eight triangles over which the drainage direction vector (arrow) is determined. The drainage direction vector is then used to appoint the flow to the one or two grid cells that bound the facet containing the vector. The flow in Figure 3.11 is distributed between grid cells 2 and 3 (see Equation 3.17 where the subscripts 1 and 2 refer to grid cells 2 and 3 in this example) (Gruber & Peckham, 2009, p. 179).

The use of triangular facets in the D∞ algorithm avoids the aforementioned problems with fitting a plane in the methods of Lea (1992) and Costa-Cabral and Burges (1994), and in cases where the direction does not follow one of the cardinal (i.e. 0, 90, 180, and 270°) or diagonal (i.e. 45, 135, 225, and 315°) directions, the upslope contributing area is calculated by apportioning the flow from a grid cell between the two downslope neighbors according to how close the flow angle is to the direct angle to that grid cell (Figure 3.11). The draining proportion $d$ is apportioned to the two grid cells either side of the theoretical drainage direction vector using:

$$d_1 = \frac{4 - \alpha_2}{\pi}, \quad d_2 = \frac{4 - \alpha_1}{\pi} \tag{3.17}$$

where the angles $\alpha_1$ and $\alpha_2$ are measured on a horizontal planar surface between the drainage direction vector and the vectors to the two pixels on either side of it ($\alpha_1 + \alpha_2 = 45°$) (Gruber & Peckham, 2009, p. 178).

When introducing the D∞ algorithm, Tarboton (1997) noted five issues which had influenced the design and performance of the D∞ algorithm and several of the single- and multiple-flow direction algorithms described in this and Section 3.1.3.2:

1. the need to avoid or minimize dispersion;
2. the need to avoid grid bias, due to the orientation of the numerical grid;
3. the precision with which flow directions are resolved;
4. a simple and efficient grid-based matrix storage structure; and
5. the ability to cope with "difficult" data, such as in saddle areas, pits, and flat areas.

Tarboton (1997) also conducted a series of quantitative and qualitative tests using outward and inward draining circular cones, a planar surface, and DEMs for the Walnut Gulch Experimental Watershed in southeastern Arizona and the Tennessee Valley in Marin County, California to compare the performance of D∞ and several of the other flow-direction algorithms. The results showed how D∞ did better than D8 (O'Callaghan & Mark, 1984) on the second and third of the aforementioned criteria, better than the FMFD (Freeman, 1991) and QMFD1 (Quinn et al., 1991) multiple-flow direction algorithms on the first and fourth criteria, and better than the ADK (Lea, 1992) and DEMON (Costa-Cabral & Burges, 1994) methods on the fifth criterion.

A few years later, Orlandini and colleagues proposed two variants for each of the D8 and D∞ algorithms to delineate non-dispersive and dispersive flow directions on square-grid DEMs (Orlandini et al., 2003; Orlandini & Moretti, 2009). The first two of these hybrid methods extended the descriptive capabilities of the classical D8 method by first summing and then minimizing the path-based deviations between selected theoretical drainage directions along drainage paths. Two methods, designated D8-LAD (eight drainage directions, least angular deviation) and D8-LTD (eight drainage directions, least transversal deviation), were developed and tested on four synthetic drainage systems of known geometry and on the Liro catchment in the central Italian Alps and were used to demonstrate that the proposed methods provided significant improvement over the D8 method for the determination of flow directions and upslope contributing areas. The results also showed that the D8-LTD method was the more versatile of the two methods, and did better than D8-LAD in delineating the long drainage paths that reach valleys and in other settings where artificial dispersion is not desirable.

The final pair of methods among this quartet are variants of the D∞ (Tarboton, 1997) algorithm. Two methods, designated D∞-LAD and

D∞-LTD (Orlandini & Moretti, 2009), were developed using (i) Tarboton's (1997) method for calculating steepest flow directions; (ii) the same methods for first summing and then minimizing the path-based deviations between selected threshold drainage directions along drainage paths that were used with the D8 algorithm and described by Orlandini et al. (2003); and (iii) either single- and multiple-flow directions based on the calculation of the local plan (or contour) curvature given by Zevenbergen and Thorne (1987). Orlandini and Moretti (2009) used elevation data for the Col Rodella area in the eastern Italian Alps in a series of tests to determine the plan curvature threshold $K_{ct}$ which would support the use of both single (i.e. non-dispersive) and multiple (i.e. dispersive) flow directions across a terrain surface (since D∞ supports flow to one or two neighboring cells depending on the path of steepest descent and the orientation of the underlying square grid). The results showed the following.

1. The multiple-flow direction (i.e. dispersive) methods [LAD with $\lambda = 0$ or LTD with $\lambda = 1$, $K_{ct} \geq \max(K_c)$] provided a better representation of the spatial pattern of drainage areas than those provided by single-flow direction (i.e. non-dispersive) methods [LAD with $\lambda = 0$ or LTD with $\lambda = 1$, $K_{ct} < \min(K_c)$], which displayed numerous source grid cells scattered across the terrain surface in addition to those that were expected near ridges, saddles, and peaks.
2. The path-based non-dispersive methods [LTD with $\lambda = 1$, $K_{ct} < \min(K_c)$] were equally or even more accurate than the dispersive methods in the description of surface flow paths along both convergent drainage basins and divergent slopes.

Further comparisons of the performance of both single- and multiple-flow algorithms are provided in Section 3.1.3.2.

The ninth and final single-flow direction algorithm, the deterministic six-node (D6) algorithm was proposed by Wright et al. (2014) and implemented on top of hierarchical hexagonal surface models (HHSMs). This relatively new algorithm mimics the D8 algorithm but relies on both hexagonal sampling and the adaptive cell sizes supported by HHSMs. The first factor is straightforward because the flow is simply directed toward the steepest downslope neighbor of the six nearest neighbors in a hexagonal sampled space as opposed to either the four or eight nearest neighbors in a flat rectangular array like a square-grid DEM. However, the use of the D6 flow-direction algorithm in multilevel analysis and modeling is complicated by the variations in the angle of rotation between hexagonal imaging processing (HIP) system levels. The directions of the neighbors at a given level of detail (LoD) may not be represented in the neighborhoods of coarser and finer LoDs. Wright et al. (2014) adjusted the direction to the nearest valid direction in the coarser LoD, while noting that adjusting the flow direction in this way was just as likely to improve the result or worsen it without additional sources of information and assuming no bias is

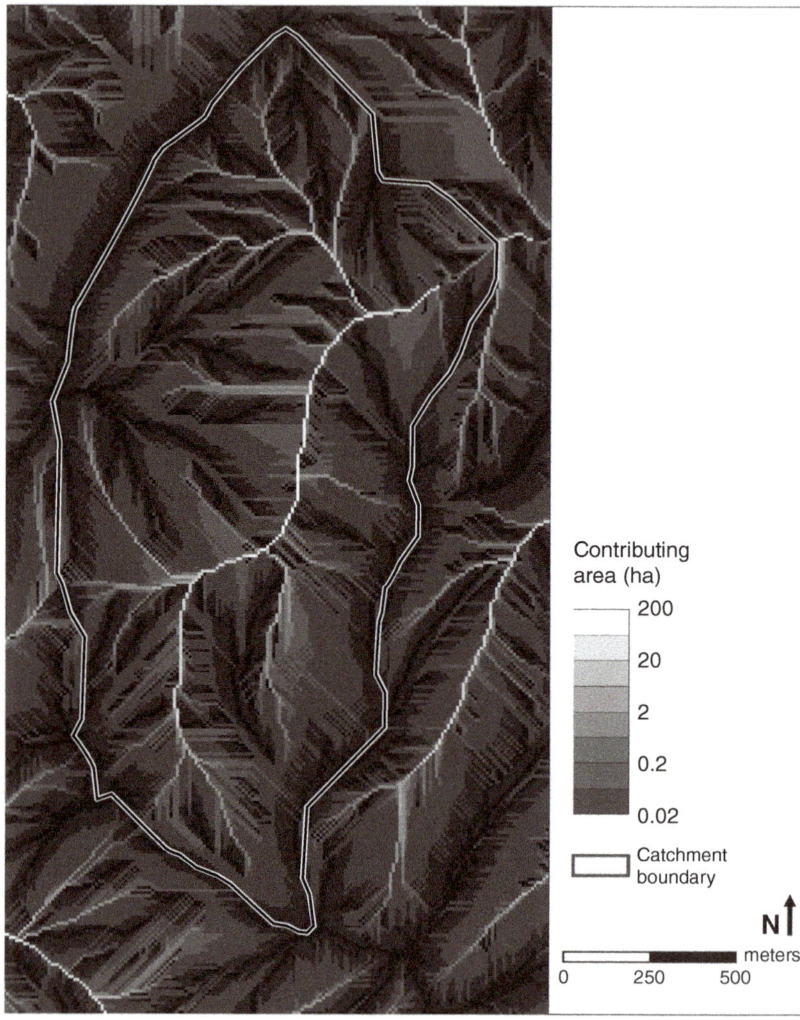


Figure 3.12 Upslope contributing area (ha) derived for Cottonwood Creek, MT study site using the D8 single-flow direction algorithm, with the catchment boundary overlaid.

present, and concluded that hybrid- and/or multiple-flow direction algorithms, such as D∞ (Tarboton, 1997) and MD∞ (Seibert & McGlynn, 2007), would be required to generate adaptive hexagonal flow-direction arrays and support realistic applications, such as the provision of hydrologically adaptive representations of urban environments for hydrologic modeling.

Figures 3.12 and 3.13 show the uplope contributing area calculated for the Cottonwood Creek catchment using the D8 (O'Callaghan & Mark, 1984) and D∞ (Tarboton, 1997) flow-direction algorithms, respectively. The valley bottoms are clearly defined as a line of white high-contributing area cells, and the ridges both within the catchment and at the divide are visible as dark cells with low-contributing area in the D8 upslope contributing area map reproduced in Figure 3.12. However, many straight-line artifacts are visible, due to the absence of dispersion and the inability of this flow-direction algorithm to respond to subtle changes in

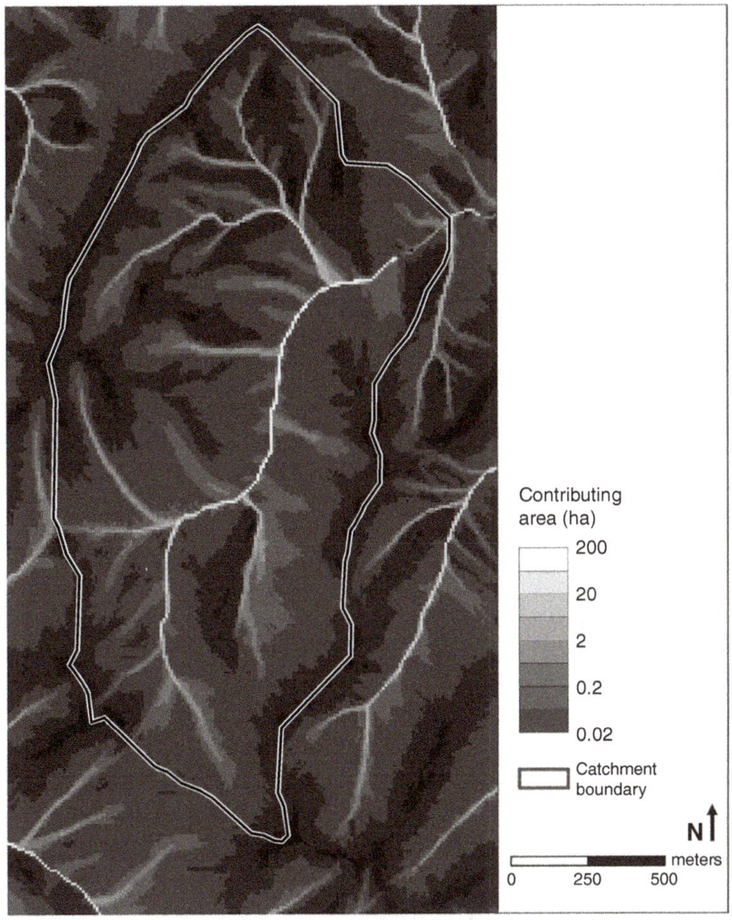


**Figure 3.13** Upslope contributing area (ha) derived for Cottonwood Creek, MT study site using the D∞ single-flow direction algorithm, with the catchment boundary overlaid.

surface orientation. Gallant and Wilson (2000, pp. 61–63) provided a similar map and have described the problems accompanying the use of the D8 flow-direction algorithm and their causes in greater detail. The second upslope contributing area map for the Cottonwood Creek catchment reproduced in Figure 3.13 shows that the D∞ flow-direction algorithm did considerably better than the D8 algorithm. The valley bottoms and ridges, both within the catchment and at the divide, are clearly visible in this map and some but not all of the straight-line artifacts have been eliminated as well. There are still problems on the upper parts of slopes and these will be discussed when the MD∞ (Seibert & McGlynn, 2007) multiple-flow direction algorithm is introduced later in this chapter.

That said, we turn our attention next to the 15 multiple-flow direction algorithms described in Section 3.1.3.2, which demonstrate, among other things, the growth in the popularity of this group of algorithms over time. The results of several studies that have compared the performance of two or more single- and multiple-flow direction algorithms are discussed as well.

#### 3.1.3.2 Multiple-flow Routing Algorithms

The fundamental goal with multiple-flow direction algorithms is to find a method or sequence of methods that move water into one or more downslope cells. This approach might be justified by actual divergence (i.e. the need to treat flows across convergent and divergent land surfaces) and/or the attempt to overcome the limits of having only eight adjacent cells (i.e. methods to overcome limitations generated when we represent a continuous flow field with a regular grid that only has eight possible directions in multiples of 45°) (Gruber & Peckham, 2009).

The first of these multiple-flow direction algorithms, FMFD, was proposed by Freeman (1991) and distributed flow to two or more downslope neighbors based on the slope:

$$F_i = \frac{(\tan\beta_i)^p}{\sum_{i=1}^{n}(\tan\beta_i)^p} \quad (\tan\beta > 0; n \leq 8) \tag{3.18}$$

where $F_i$ is the proportional flow to be distributed to the $i$th neighboring cell, and $\tan\beta_i$ denotes the slope between the center cell and the $i$th neighboring cell. This approach did not consider the flow width like those of Quinn et al. (1991, 1995), and based on tests on a conical surface Freeman suggested that $p = 1.1$ produced the most accurate results. However, Holmgren (1994) reported that much higher values of $p$ in the range 6–8 might be more appropriate and Pilesjö and Zhou (1997) later tested this algorithm on a spherical surface and found that $p = 1.0$ would be more appropriate for that case. What is clear is that an exponent of 1 provides for full dispersion and that the balance shifts towards a single-flow direction for very large values of $p$.

The second multiple-flow direction algorithm (QFMD1) was proposed by Quinn et al. (1991) and distributed flow to two or more downslope neighbors based on the slope and the contour length (i.e. the flow width):

$$F_i = \frac{L_i \tan\beta_i}{\sum_{i=1}^{n} L_i \tan\beta_i} \quad (\tan\beta > 0; n \leq 8) \tag{3.19}$$

where $F_i$ was the proportional flow to be distributed to the $i$th neighboring cell, $L_i$ was the flow width, and $\tan\beta_i$ denoted the slope between the center cell and the $i$th neighboring cell. The algorithm is very similar to FMFD (given its implied exponent of 1) but for the inclusion of flow width which was calculated using one or other of two approaches for cardinal (i.e. edge adjacent) and diagonal downslope cells, respectively as described in Section 3.1.3.3.

The third multiple-flow direction algorithm was proposed by Costa-Cabral and Burges (1994) and offered a completely different approach for modeling flow direction and flow width. The flow in the digital elevation model network extraction (DEMON) stream-tube method is generated at each source pixel and routed down a stream tube until the edge of the DEM or a pit is encountered. The stream tubes are constructed from the points of intersection of a line drawn in the gradient (i.e. aspect) direction and a grid cell edge, but they are not constrained to coincide with the edges of cells and can expand and contract as they traverse divergent and convergent regions of the DEM surface. The amount of flow in DEMON is expressed as a fraction of the area of the source grid cell entering each grid cell downstream of the source grid cell and this is added to the flow accumulation value of that grid cell as part of the approach to construct the final flow accumulation grid (see Sections 3.1.3 and 3.1.4 for additional details). This method is conceptually similar to the stream-tube approach used with contour-based DEMs by Moore and Grayson (1991).

Around the same time, Moore proposed a variant of the original DEMON stream-tube method in TAPES-G that incorporated three minor modifications (Moore et al., 1993e). Moore's implementation of DEMON is listed as the fourth multiple-flow direction algorithm in Table 3.2 and the modifications took the following forms: (i) this version could be applied to either the original or a derived depressionless DEM; (ii) the nodes of the DEM were used to define the centroids of the grid cells rather than the vertices; and (iii) the flow direction of the stream tube for each grid cell was defined by the aspect (as calculated using Equation 3.11). Gallant and Wilson (2000) showed how this approach produced smooth variation of contributing area in upland areas, similar to FMFD, and well-defined channels in the lowlands that are more similar to D8, when it was used with a 15-m square-grid DEM to calculate flow direction, width and accumulation for the Cottonwood Creek, MT catchment. These authors also noted how this TAPES-G version of DEMON took a similar amount of processing time to FMFD, gave results of similar or better quality and produced few artifacts, but for a slight preference for flow in the cardinal north, south, east, and west directions in some instances.

Moore also implemented two variants of the multiple-flow direction algorithms proposed by Freeman (1991) and Quinn et al. (1991) in TAPES-G at about the same time (Moore et al., 1993e). The first variant, and fifth multiple-flow direction algorithm listed in Table 3.2, incorporated both flow width, similar to Quinn et al. (1991), and the exponent $p = 1.1$, similar to Freeman (1991). This particular variant incorporated the D8 flow width and, like most multiple-flow direction algorithms, gave more realistic distributions of flow accumulation in upslope areas than the single-flow direction algorithms while also eliminating the parallel flow paths that have often accompanied the deployment of D8. However, this algorithm (like the FMFD and QMFD1 flow-direction algorithms that were modified) also caused considerable dispersion of flows in valleys, which will often be undesirable since streamlines are frequently well defined in valleys.

To overcome these kinds of problems, the multiple-flow direction algorithm was disabled in the second variant (i.e. the sixth multiple-flow direction algorithm listed in Table 3.2) and replaced by D8 (or Rho8) whenever the upslope contributing area exceeded some user-specified threshold, which was labeled the "maximum cross-grading area" (Gallant & Wilson, 1996, 2000). Gallant and Wilson (2000) suggested setting a maximum cross-grading area of approximately 10 ha or 100,000 m$^2$ as a reasonable starting place and noted how the optimum value depends on the valley density as represented in the DEM. Finer textured landscapes with high valley density would require a smaller value of maximum cross-grading area to reflect the smaller hillslopes, whereas coarser landscapes with larger hillslopes would require a higher value (cf. Güntner, Seibert & Uhlenbrook, 2004). However, the transition from multiple- to single-flow direction algorithms sometimes caused sharp variations in the distribution of upslope contributing area if the maximum cross-grading area was set to too small a value, and this particular method could not accommodate those cases where the dispersion of the flow in valleys might be desirable (i.e. on floodplains or in riparian areas) without setting the maximum cross-grading area to a very large value in these instances. The fundamental problem with this and similar kinds of global strategies is, of course, that they might work well in some parts of a landscape of interest but not others.

The seventh multiple-flow direction algorithm (labeled QMFD2 in Table 3.2) was proposed by Quinn et al. (1995) who modified their earlier approach summarized in Equation 3.19 by adding an exponent $p$ as follows:

$$F_i = \frac{L_i\left(\tan\beta_i\right)^p}{\sum_{i=1}^{n} L_i\left(\tan\beta_i\right)^p} \quad \left(\tan\beta > 0; n \leq 8\right) \tag{3.20}$$

The exponent in this equation varied continuously from 1 (full dispersion) to a large value (single flow direction) as upslope contributing area increased, giving a smooth change from multiple to single flow directions when moving from summits and ridgelines to valleys and floodplains (Gallant & Wilson, 2000). An "optimal" channel initiation threshold (CIT) was also proposed for computing the flow-partition exponent using the following equation:

$$p = \left(\frac{A}{thresh + 1}\right)^h \tag{3.21}$$

where $A$ is the pre-computed flow accumulation in the cell of interest, *thresh* is the accumulation threshold for the region of interest, the same as the CIT in Moore et al. (1993e), and $h$ is a positive constant. The larger the value of $h$, the faster this method approaches the D8 single-flow direction result.

We will shift our focus for a moment here and switch our attention to the work of Zhou and Liu (2002) which compared the performance of five of the single- and multiple-flow direction algorithms introduced thus far – D8 (O'Callaghan & Mark, 1984), Rho8 (Fairfield & Leymarie, 1991), FMFD (Freeman, 1991), DEMON (Costa-Cabral & Burges, 1994), and D∞ (Tarboton, 1997) – using four mathematical surfaces and the five criteria noted by Tarboton (1997) (see Section 3.1.3.1 for additional details). The four mathematical surfaces proposed by Zhou and Liu (2002) were based on an

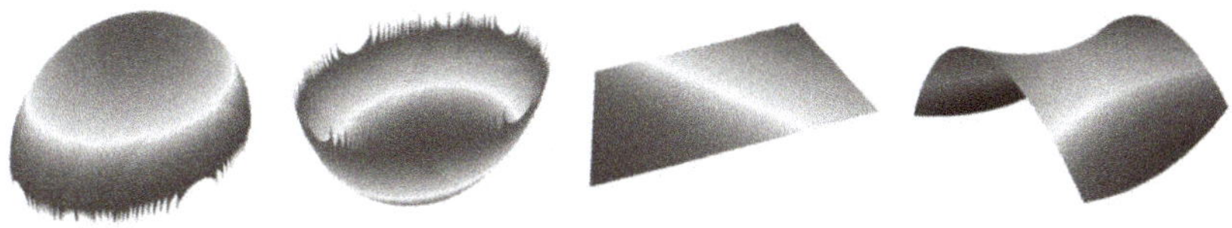


**Figure 3.14** The four mathematical surfaces commonly used for data-independent assessment of different flow-direction algorithms. *Source*: Pilesjö & Hasan (2014, p. 114). Reproduced with permission of John Wiley & Sons, Inc.

ellipsoid (representing convex slopes), an inverse ellipsoid (representing concave slopes), a plane and a saddle, and these four surface models were used to compare the errors generated from these algorithms when they were used to extract hydrologic parameters from gridded DEMs (Figure 3.14). This approach is introduced here because it has been used frequently during the past decade to assess the efficacy of various flow-direction algorithms and the theoretical true value of the SCA can be calculated at any point on these surfaces using mathematical inference and compared with the values estimated with one or more of the flow-direction algorithms.

The results of Zhou and Liu (2002) showed how all five of the grid-based methods that they examined produced some level of error that could be related to the morphology of the surface as well as the difficulties that are encountered trying to route overland flow across a terrain surface presented by square grids. From the perspective of accuracy, Zhou and Liu (2002, p. 840) concluded that:

1 DEMON, D∞, and FMFD performed well with generally low RMSE;
2 DEMON achieved the best overall performance across all four tests notwithstanding some errors caused by the gridded data structure;
3 D∞ and FMFD also delivered good results, particularly on the saddle surface, but showed weaknesses on convex and planar surfaces; and
4 D8 and Rho8 produced unacceptable errors in all four tests.

Zhou and Liu (2002) also noted how the poor performance of the two single-flow direction algorithms might be a cause for concern if they are used in real-world applications where the algorithm accuracy is of concern, while noting that for many of the hydrologic applications using real-world DEMs, the outcome might also be influenced by the errors inherent in the DEM itself, with the result that the errors produced by the flow-direction algorithm may be masked by these DEM errors.

That said, the eighth form-based multiple-flow direction (PMFD) algorithm proposed by Pilesjö et al. (1998) calculates the topographic form of a $3 \times 3$ facet (i.e. moving cell window) and uses this result to dictate the flow-direction algorithm which is applied. The D8 algorithm is used if the form is concave or flat and the calculated flow direction matches the direction to one of the eight neighbors; otherwise the flow distribution for concave and flat forms is split between two adjacent cells using a similar approach to D∞, or the FMFD algorithm is applied if the form is convex. Pilesjö et al. (1998) demonstrated how the PMFD algorithm performed better than the D8 algorithm (as implemented in ArcGIS) in estimating flow accumulation but worse in delineating the drainage network for a small watershed near Bredbo in New South Wales, Australia. Qin et al. (2007) later criticized this method, noting that:

1 it was very sensitive to the quality of the DEM because small errors in the DEM could shift the form from convex to concave and vice versa; and
2 the flow on convex forms was still modeled with a fixed flow-partition exponent (see the description of the MFD-md algorithm that follows for a more detailed discussion of this limitation).

The ninth spatially distributed flow-apportioning algorithm (SDFAA) proposed by Kim and Lee (2004) calculated the relationships between various flow-direction parameters and an expanded channel initiation threshold (ECIT) iteratively to distribute the contributing area from upslope to downslope grid cells considering the locations of the permanent channel network. A spatially varying flow apportioning factor, *H(I,J)*, was introduced, where *I* and *J* are spatial coordinates and the values of *H(I,J)* are determined with ECIT which starts from the CIT proposed by Quinn et al. (1995) and decreases until an optimum solution is obtained. A genetic algorithm (Goldberg, 1989) is integrated into SDFAA and applied to estimate the small number of additional parameters included in this new algorithm using the river cell locations to evolve candidate solutions of parameter values. The power function describing the relationship between *H(I,J)* and ECIT is expressed as:

$$H(I,J) = c_1 \times A_{\text{ECIT}}^{n+1} + c_2 \tag{3.22}$$

where the coefficients $c_1$ and $c_2$ can be determined from the boundary conditions and the expanded channel initiation threshold ($A_{\text{ECIT}}$), and $n$ can be determined with the aforementioned optimization process. The SDFAA method needs an additional input, the channel locations, compared with the aforementioned algorithms but these kinds of datasets are readily available in many parts of the world nowadays, for example the NHDPlus dataset in the USA (Moore & Dewald, 2016).

Kim and Lee (2014) applied the SDFAA method to the Sulmachun watershed in the headwaters of the Imjin River in South Korea and showed that this spatially distributed flow-apportioning scheme offered several advantages over the D8 (O'Callaghan & Mark, 1984), FMFD [Freeman (1991), with the flow partitioning factor, *p* in Equations 3.18 and 3.20, set to 5 as per Holmgren's (1994) recommendation], QMFD1 (Quinn et al., 1991) and QMFD2 (Quinn et al., 1995) flow-direction algorithms, including (i) the relaxation of over-dispersion problems near channel cells; (ii) the improved connectivity of river cells; and (iii) the inclusion of a more robust method for parameter determination compared to the preceding multiple-flow direction algorithms.

The tenth multiple-flow direction algorithm based on the local maximum downslope gradient (MFD-md for short) was proposed by Qin et al. (2007) and builds on the methods previously proposed by Freeman (1991) and Quinn et al. (1991, 1995). The selection of a single value for $p$ in Equations 3.18 and 3.20 determined the flow-partition scheme in each of these algorithms and meant that both of these approaches modeled the terrain conditions, no matter how flat or steep or convergent or divergent they were across different parts of the landscape, in the same way. Qin et al. (2007) therefore proposed an adaptive method for determining this flow-partition exponent $p$ based on local maximum downslope gradient:

$$F_i = \frac{\left(\tan\beta_i\right)^{f(e)} \times L_i}{\sum_{j=1}^{8}\left(\tan\beta_i\right)^{f(e)} \times L_i} \tag{3.23}$$

where $f(e)$ is the function for determining the flow-partition exponent using the maximum downslope gradient, and the definition of the other terms is the same as in Equations 3.18 and 3.20. Qin et al. (2007) chose a linear function for the maximum downslope gradient as follows:

$$f(e) = \begin{cases} p_1 & \left(e \le e_{\min}\right) \\ \dfrac{e - e_{\min}}{e_{\max} - e_{\min}} \times \left(p_u - p_1\right) + p_1 & \left(e_{\min} < e < e_{\max}\right) \\ p_u & \left(e \ge e_{\max}\right) \end{cases} \tag{3.24}$$

where $f(e)$ is the flow-partition function, $e$ is the tangent value of the maximum downslope gradient (tan β and β being the maximum downslope gradient), $p_u$ and $p_1$ are the upper and lower bounds of $f(e)$ and are used as the $p$ values representing completely divergent and convergent flows, respectively, and $e_{min}$ and $e_{max}$ are the $e$ values which are associated with $p_u$ and $p_1$, respectively. The values of 1.1 and 10 were chosen for $e_{min}$ and $e_{max}$, respectively based on the $p$ values proposed by Freeman (1991), Holmgren (1994), and Quinn et al. (1995).

The MFD-md algorithm ensures that the flow-partition exponent can be modeled with large and small values on steep and gentle terrain surfaces, respectively. Qin et al. (2007) used the RMSE of the SCA along with the four mathematical surfaces proposed by Zhou and Liu (2002) and their theoretical "true" values of SCA (see Figure 3.14) to show how this new algorithm produced smaller errors than the D8 (O'Callaghan & Mark, 1984) and QMFD2 (Quinn et al., 1995) algorithms. They also showed how this new method produced flow accumulation patterns that were better adapted to local terrain conditions, based on visual judgment, for a small watershed in northeastern China.

The eleventh triangular multiple-flow direction (MD∞) algorithm proposed by Seibert and McGlynn (2007) represents an evolution of two existing algorithms in that it allows multidirectional flow in any downslope

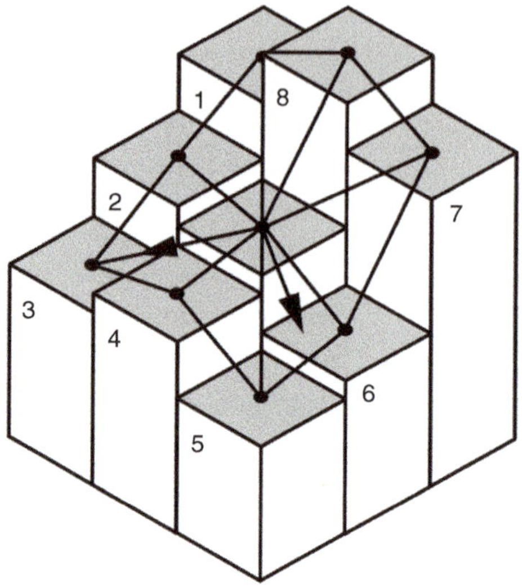


**Figure 3.15** Concept of flow apportioning in MD∞ based on the construction of triangular facets around one cell. *Source*: Seibert & McGlynn (2007, p. 2). Reproduced with permission of John Wiley & Sons, Inc.

direction, thereby combining the benefits of D∞ (Tarboton, 1997) and QMFD1 (Quinn et al., 1991). The part of the accumulated area for each grid cell that is distributed to its neighboring cells is computed by first determining the flow directions which receive area with this particular method. Triangular facets are next used to calculate local slope directions and gradients around each cell similar to Tarboton (1997). Eight planar triangular facets are constructed using adjacent neighboring cells ($P_1$ and $P_2$, $P_2$ and $P_3$, etc.) and the midpoint ($M$) of the center grid cell and the direction of the steepest gradient is calculated for each of these local planes. The direction of the steepest slope for the triangular facets may point between two neighboring cells (i.e. cells 5 and 6 in Figure 3.15) or point to a direction outside the 45° angle range of the particular triangular facet (e.g. the triangular facet defined by the midpoint *M* and cells 4 and 5 in Figure 3.15). The direction is set to the steeper of the directions toward the two neighboring cells, but this direction only receives area if the same direction has been determined for two adjacent triangular facets (i.e. cells 5 and 6 in this instance). This is the case for the direction pointing toward cell 3 but not for the directions pointing towards 5 and 6 in Figure 3.15. Once all the downslope directions have been calculated for a cell, its accumulated area is distributed to these directions using a slope-weighted procedure similar to that proposed by Quinn et al. (1991) in QMFD1. A flow-partition parameter can be used to weight steeper directions more heavily to manage flow dispersion (similar to Quinn et al., 1995) and if the downslope direction falls between two neighboring cells, the area is further distributed to these two cells based on the relative differences in direction (similar to Tarboton, 1997).

Seibert and McGlynn (2007) demonstrated the efficacy of MD∞ by comparing its performance with that of the D8 (O'Callaghan & Mark, 1984), QMFD1 (Quinn et al., 1991), QMFD2 (Quinn et al., 1995), and D∞ (Tarboton, 1997) flow-direction algorithms using synthetic elevation data

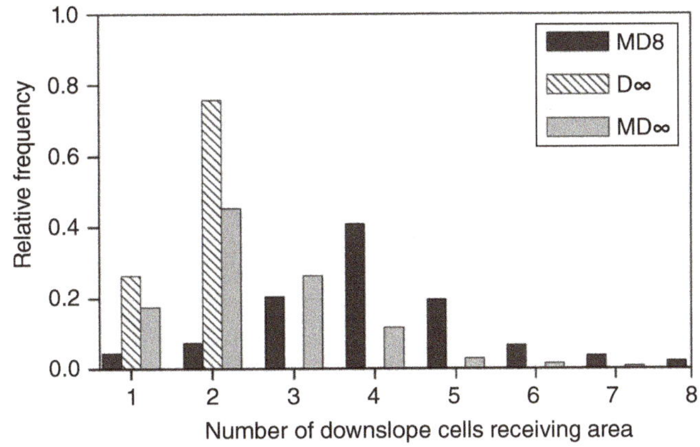


**Figure 3.16** Distribution of the number of cells that receive accumulated area (i.e. flow) from one cell in a sample DEM for an area in central Sweden. *Source*: Seibert & McGlynn (2007, p. 5). Reproduced with permission of John Wiley & Sons, Inc.

and low- and high-relief real-world DEMs from central Sweden and southwestern Montana, respectively. Four synthetic datasets representing convergent, planar and divergent hillslopes as well as saddles showed how:

1. D8 directs flow to a single downslope cell in each case (as would be expected);
2. both QMFD1 and QMFD2 direct flow to all, but varying numbers of the downslope cells, in each case (as would be expected); and
3. the D∞ and MD∞ algorithms provide exactly the same results on planar and convergent hillslopes but different results in cases where there is more than one locally steepest downslope direction from a cell such as occurs on divergent hillslopes or along ridges and near summits.

The histogram reproduced in Figure 3.16 shows the numbers of cells that received accumulated area (i.e. flow) from one cell for the low-relief area in central Sweden. These results show how the flow routing using D∞ was limited to one or two of the neighboring cells, in contrast to the MD∞ algorithm which distributed flow to more than two cells for about 40% of all cells and the QMFD1/2 algorithms which distributed flow among one to eight cells and to more than two cells for about 90% of all cells (Figure 3.16). The MD∞ algorithm identified more than one downslope direction in about 30% of the cells and three or more downslope directions in 8% of the cells for this watershed. These results were more or less repeated in the Montana watershed, and the availability of 2-, 10- and 20-m DEMs in this watershed showed how the QFMD1/2 and MD∞ multiple-flow direction algorithms tend to predict flow into greater numbers of downslope cells with increasing cell size.

The final pair of maps provided by Seibert and McGlynn (2007, Figure 10, p. 8) demonstrated the impact of incorporating a single flow direction in the D∞ approach because D∞ will allow an area to flow in just one direction on watershed divides (i.e. ridges and shoulder slopes) in contrast to MD∞ which supports flows in these areas in multiple flow directions. The convergent areas on their difference maps are colored red (the accumulated area predicted by D∞ was greater than that predicted by MD∞) or blue (the accumulated area predicted by MD∞ exceeded that predicted by D∞) depending on which downslope direction was chosen by D∞ in upslope areas, illustrating how the differences between flow-direction algorithms in specific locations in a watershed can be propagated downslope and therefore throughout the landscape.

The twelfth multiple-flow direction method is the proprietary Mass Flux method, which is provided in RiverTools™ (see Section 6.4.6 for additional details). This method endeavors to characterize the local topographic form (i.e. shape) by dividing each grid cell into four quarter grid cells and defining a continuous flow-direction angle for each, using a grid that has twice the dimensions of the DEM (Gruber & Peckham, 2009). The elevations of the whole grid cell and two of its cardinal neighbors uniquely determine a plane and the corresponding slope and aspect (Figure 3.17). This approach removes the ambiguity of plane fitting and the associated problems and the ambiguity of flow direction for grid cells that mark peaks or ridges, since it allows flow from these grid cells to be routed in different directions (i.e. similar to Seibert and McGlynn, 2007 with the MD∞ flow-direction algorithm). Flow from each quarter pixel is permitted to one or two of its cardinal neighbors, and the flows into these neighbors are determined by treating each grid cell as a control volume (see Gruber & Peckham, 2009, pp. 179–180 for additional details).

The thirteenth multiple-flow direction algorithm proposed by Zhou et al. (2011) sought to capitalize on the advantages of the TIN-based approach by creating a TFN and using this to specify the surface flow paths over a square-grid DEM. This particular method first partitions the grid cells by creating triangular facets between the cell centers based on their surface form, and then combines the facets for all the grid cells to form a TFN to eliminate

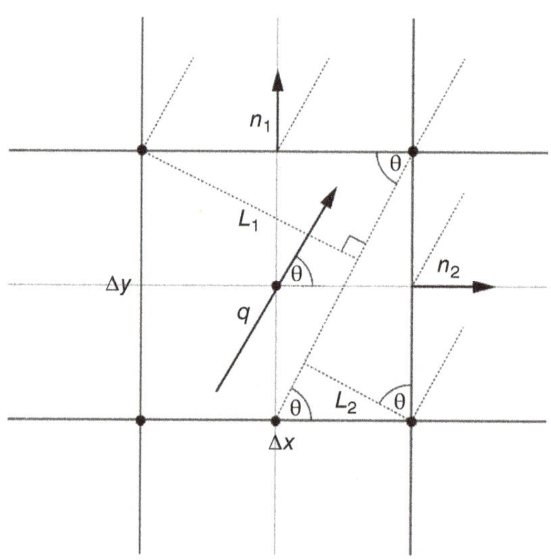


**Figure 3.17** Flow apportioning between two cardinal neighbors in the Mass Flux method. $L_1$ and $L_2$ denote the projected flow widths into the upper and right neighbor and together equal the projected flow width ω. $n_1$ and $n_2$ are vectors normal to the cell boundaries, $q$ is the flow vector and θ is the flow direction. Source: Gruber & Peckham (2009, p. 180). Reproduced with permission of Elsevier.

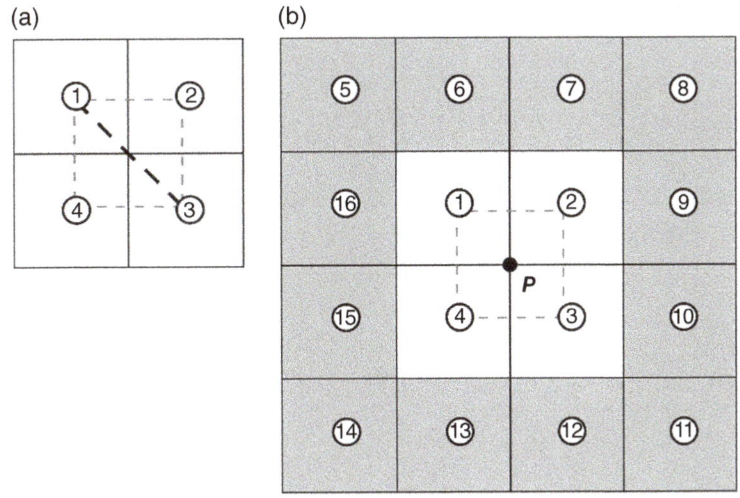


**Figure 3.18** (a) Two triangular facets are formed in a 2 × 2 cell moving window using the spot heights at the center of each grid cell; (b) a 4 × 4 cell moving window is used to estimate elevation at *P* by fitting a bivariate cubic spline surface. *Source*: Zhou et al. (2011, p. 3). Reproduced with permission of Elsevier.

gaps between adjacent cells and/or facets. The initial triangular facets are formed using a 2 × 2 cell moving window and using two diagonal lines (the points labeled 1 and 3 and 2 and 4, respectively, in Figure 3.18a) to divide the surface into two pairs of triangular facets (Figure 3.18a). The elevation at the intersection of the four grid cells (*P* in Figure 3.18b) is then estimated by fitting a bivariate cubic spline surface using a 4 × 4 cell moving window (Figure 3.18b) and the diagonal line that is closest to the surface at the central point (i.e. *P* in Figure 3.18b) is taken as the common edge of the triangular facets. The individual facets that comprise the final TFN have constant slopes and aspects and the flow direction of each facet is determined using its aspect. The flow generated on a facet is next assigned to its center of gravity to form a flow package which moves toward a downslope facet based on its flow direction. The flow package may alter its direction when it crosses the edge of the facet to follow the aspect of the downstream facet (see Figure 3.19 for additional details) and will continue from its source to the final destination (i.e. a local sink or the catchment outlet). These flow packages clearly share some similarities with the stream tubes calculated by DEMON (Costa-Cabral & Burges, 1994; Gallant & Wilson, 1996) and the stream-tube approach used by Moore and Grayson (1991) with contour-based DEMs.

The fourteenth $D_{trig}$ multiple-flow direction algorithm proposed by Shelef and Hilley (2013) also divides the land surface into a series of triangular facets, whose extent is defined within their associated grid cell (Figure 3.20). However, the gradient is defined based on the adjacent grid nodes, similar to the D∞ approach of Tarboton (1997). This particular algorithm uses the eight triangular facets reproduced in Figure 3.20 to describe the internal topography of the grid cell and to partition the flow between these facets and those associated with neighboring grid cells. The novelty of this method is that $D_{trig}$ partitions flow across the edges of the facets and grid cells rather than through the corners (as was the case in several of the algorithms described earlier).

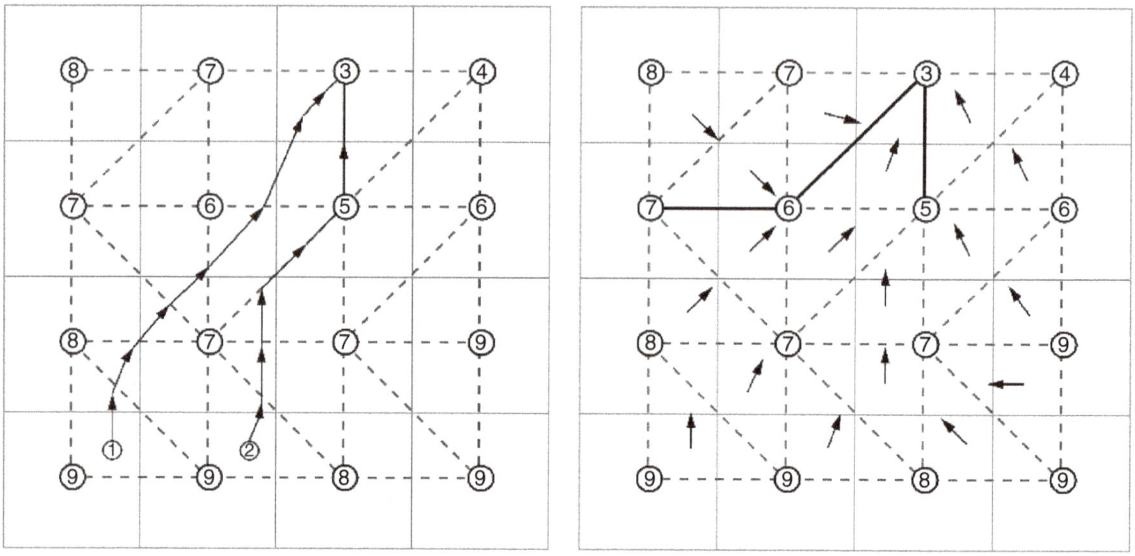


Figure 3.19 Flow line over a TFN: the numbers at the nodes of triangles represent elevation, the light lines show the original grid cells, and the flow lines represented by the arrow chains are formed by tracking the movement of flow (i.e. the flow directions). *Source*: Zhou et al. (2011, p. 4). Reproduced with permission of Elsevier.

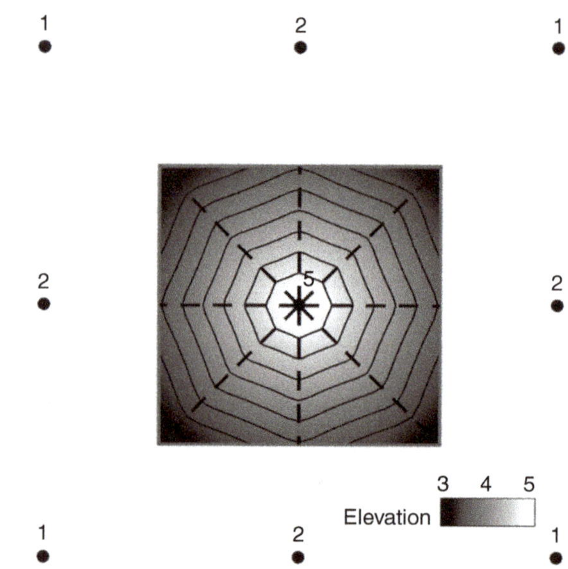


Figure 3.20 The decomposition of grid cells into a set of eight triangular facets defined by the nine-cell kernel nodes (black circles) in $D_{trig}$. The node's elevations are listed next to each node and facet boundaries are denoted by dashed lines. The surface extent is limited to the central cell so that the only node within this domain is the element-centered node. The contours and gray scale illustrate the elevation variability within the element and the rounding of the contours adjacent to facet boundaries is an artifact of the contouring algorithm. *Source*: Shelef & Hilley (2013, p. 2107). Reproduced with permission of John Wiley & Sons, Inc.

The routing of flow in $D_{trig}$ is accomplished in two steps: (i) the spurious pits in the DEM are filled first and the land surface is then decomposed into a set of triangular facets for which gradient and mean elevation are calculated; and (ii) the facets are ordered by their mean elevation and the drainage area is progressively tabulated and partitioned from upslope facets to those at lower elevations in either the same grid cell or adjacent grid cells. The flow is partitioned from each facet into surrounding facets by intersecting a line parallel to the facet's gradient vector $(\vec{g})$ with the node of the facet that allows this line to traverse the facet's area (Figure 3.21). If $\vec{g}$ points toward this node, the intersecting line delineates the facet's drainage divide and area will be partitioned between the neighboring facets according to the subareas

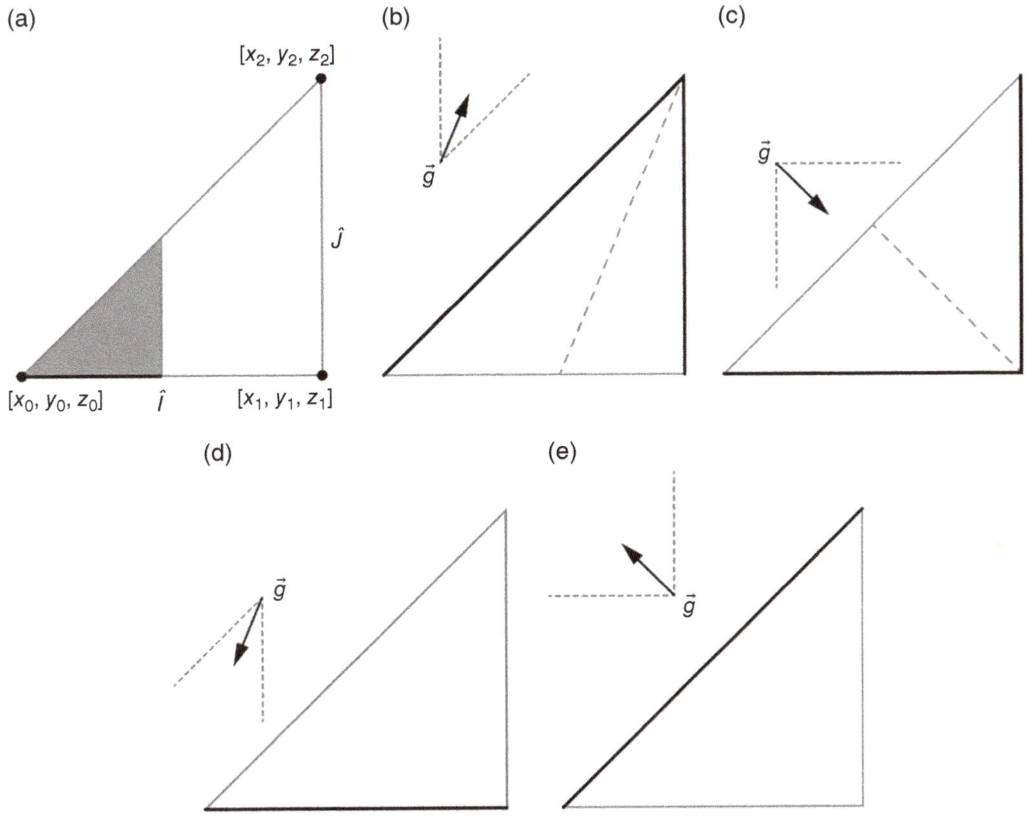


**Figure 3.21** Examples of flow partitioning from a triangular facet. (a) A triangular facet, the local coordinates, and the $\hat{\imath}, \hat{\jmath}$ directions. (b) The case where the line oriented in the direction of $\vec{g}$ intersects node $[x_2, y_2, z_2]$ and is plunging toward this node. The dashed lines that bound $\vec{g}$ denote the range of orientations where $\vec{g}$ intersects this node and divides the area into two triangles. In this case, the facet's drainage area is partitioned proportionally to the area of each of the triangles bounded by the facet's drainage divide (i.e. the dashed intersecting line) and the facet's bounding legs. The area is partitioned into the two facets sharing the bold colored facet legs. (c) Same as (b) except that $\vec{g}$ is dipping toward node $[x_1, y_1, z_1]$. (d) Same as (b) except $\vec{g}$ is plunging away from node $[x_2, y_2, z_2]$. (e) Same as (d) except that $\vec{g}$ is plunging away from node $[x_1, y_1, z_1]$. *Source*: Shelef & Hilley (2013, p. 2108). Reproduced with permission of John Wiley & Sons, Inc.

defined by the drainage divide (Figure 3.21b,c). However, when $\vec{g}$ plunges away from the node, the whole area is partitioned through the leg of the triangle opposite this node into the adjoining facet (Figure 3.21d,e). This approach means that flow is directed exclusively through the orthogonal boundaries of the grid cells and therefore no part of the flow is distributed directly to diagonal nodes in a DEM (Shelef & Hilley, 2013).

Shelef and Hilley (2013) compared the performance of this new algorithm with the D4, D8 (O'Callaghan & Mark, 1984), FMFD (Freeman, 1991), and D∞ (Tarboton, 1997) flow-direction algorithms in three different experiments. The D4 algorithm is a five-cell kernel method that directs flow to one of the four cardinal directions based on steepest descent. This method favors convergent flows along paths aligned with the cardinal flow directions, whereas the D8 flow-direction algorithm that succeeded it favors convergent flows aligned with the four cardinal and four diagonal flow directions.

Shelef and Hilley (2013) used outward-facing cones, inward-facing cones, and tilted planes that were modified to capture the accuracy of drainage area calculations produced by individual flows along with flows in natural landscapes for the first test. A series of plots was used to compare the results of the five flow-direction algorithms using three flow paths with different orientations. Bar plots of the RMSE values showed that the $D_{trig}$, FMFD, and D∞ multiple-flow direction algorithms performed better than the D4 and D8 single-flow direction methods in eight of the nine experiments. Two maps were next used to show how flows starting from one or five grid cells were routed over the Eel River basin in northern California. The results showed how the flows predicted with the D4 and D8 single-flow routing algorithms traversed a single-thread set of downslope grid cells whereas the three multiple-flow direction algorithms distributed flow between adjacent grid cells. The single-flow direction algorithms predicted parallel flows in some parts of the landscape and FMFD sometimes directed flows across catchment boundaries into adjacent watersheds.

The second experiment compared the impact of different flow direction and slope calculation rules on the calculated extent of advective (e.g. convergent channelization) versus diffusive (e.g. dispersive hillslope creep) processes within the Eel River DEM. To do this, Shelef and Hilley (2013) calculated the Péclet number (Pe):

$$\mathrm{Pe} = \frac{K(3A)^{m+1/2}}{D} \tag{3.25}$$

where $K$ is the bedrock erodibility coefficient, $A$ is the drainage area, $m$ is the area scaling component for the detachment limited erosion rule, and $D$ is the soil diffusion coefficient. The experiments used a prescribed $K/D$ ratio ($3.3 \times 10^{-4}\ \mathrm{m}^{-2}$) and $m$ value (0.5) where $A$ was calculated with different flow-direction algorithms. A series of maps was used to plot the points in the landscape at which Pe >1 and the results showed the expected patterns. Hence, the reduced flow convergence produced by D4 restricted Pe >1 to large valleys, D8 did better and allowed advective flows to dominate in convergent areas that extended further upslope, and D∞ and $D_{trig}$ did better than FMFD because of less dispersion.

However, the previously described maps also showed how the multiple-flow direction algorithms produced isolated areas with Pe >1 because the flow can disperse and reconverge downslope depending on the local topography. The predictions of the extent of saturation overland flow (SOF) with the D8 single- and FMFD multiple-flow direction algorithms followed the predictions of flow convergence generated with these two algorithms and this ensured that the extent of landsliding in the landscape showed a similar sensitivity to the flow direction and slope rules, because the local slope and extent of subsurface saturation are thought to exert an important control on the distribution of landslides (e.g. Dietrich et al., 1992, 1993; Montgomery & Dietrich, 1994).

The third and final experiment examined the effect of flow direction and slope calculation rules on the outcomes of geomorphic transport law (GTL) landscape evolution models. The steady-state topography predicted by these models showed significant differences in the geometry of the modeled landscapes and suggests that the outcomes of these kinds of models will be influenced by the choice of flow direction and slope calculation rules.

Shelef and Hilley (2013) concluded that the difficulty of identifying the "best" flow-direction algorithm for a specific application and study area remains, because the three aforementioned sets of experiments could not identify which of the five flow-direction algorithms and associated slope calculation rules best mirrored what happens in real-world landscapes.

That said, we turn our attention to the fifteenth and final multiple-flow direction algorithm. This particular algorithm, proposed by Pilesjö and Hasan (2014), divides each cell into eight triangular facets (similar to Tarboton, 1997) and uses the triangular form-based multiple flow algorithm (TFM) to distribute flow to one or more of the eight neighboring cells. The TFM method treats the local topographic form or shape in more detail than many of the previous methods and can accommodate all terrain types (i.e. convex, concave and planar slopes as well as their combinations). Pilesjö and Hasan (2014) described this TFM method as a further development of the TFN method proposed by Zhou et al. (2011) a few years earlier.

The division of the grid cells into triangular facets with constant slopes and aspects (Figure 3.22) in the TFM method means that the flow direction from each facet can be uniquely determined and used to track and distribute surface flow to one or more of the other facets covering the center grid cell and one or more of the eight cells surrounding the center grid cell. Pilesjö and Hasan (2014, pp. 112–113) described how different aspect values lead to one or other of three different actions – *stay*, *move* or *split* – as follows.

1 If the aspect of facet 1 in Figure 3.22 is between 0 and 45°, then the amount of water to be transported from this facet to a neighboring cell will *stay* as it is (i.e. one-eighth of a cell).

2 Alternatively, if the aspect is between 90 and 180° or between 225 and 270°, then all the water on facet 1 will be *moved* to one adjacent facet within the center cell (i.e. facets 2–8 in Figure 3.22).
3 Or, if the aspect is between 45 and 90° or between 270 and 360°, then the water on facet 1 is *split* and partly routed to an adjacent facet within the center cell (i.e. facets 2–8 in Figure 3.22) as well as to a neighboring cell.
4 And finally, if the aspect is between 180 and 225°, then the water on facet 1 will be *split* between two adjacent facets (i.e. facets 2–8 in Figure 3.22).

The flow is redistributed across the surface in two stages using the TFM algorithm because the various transfers among the eight triangular facets are resolved first and the flow is then routed from these facets to one or more of the neighboring cells. Each of the eight triangular facets in the center cell has two neighboring cells (e.g. cells *C1* and *C2* for facet number 1 in Figure 3.22) and flow is distributed using one of the following cases (Pilesjö & Hasan, 2014, p. 113).

1 If one neighboring cell is lower in elevation than the current center cell facet and the other is higher or equal in elevation, then all the water accumulated in this facet will be distributed to the neighboring cell.
2 If both neighboring cells are lower in elevation than the current center cell facet, then the accumulated water in the facet will be proportionally distributed to both lower cells based on the slope:

$$f_i = \frac{(\tan \beta_i)^x}{\sum_{j=1}^{2} (\tan \beta_i)^x} \qquad \text{for all } \beta > 0 \tag{3.26}$$

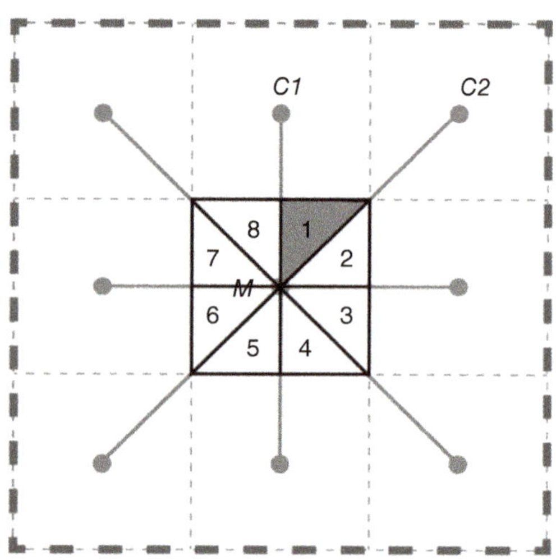


Figure 3.22 The center cell in a 3×3 grid cell window divided into eight triangular facets (1–8) with each facet formed from three points; one is the center of the central grid cell (*M*) and the other two are the centers of two adjacent grid cells (e.g. *C1* and *C2*). *Source*: Pilesjö & Hasan (2014, p. 110). Reproduced with permission of John Wiley & Sons, Inc.

where $i, j$ are the flow directions (1,…, 2) to lower cells (i.e. cell facets), $\tan \beta_i$ indicates the slope gradient between the center cell (facet) and the cell (facet) in direction $i$, and $x$ is a variable exponent.

3 If both neighboring cells are higher than or equal in elevation to the central cell facet and there are other lower elevation cells, then the accumulated water in the facet will be proportionally distributed to all lower cells.

The various calculations incorporated in this particular method start from cells with no incoming water (i.e. summits, ridgelines, and cells at the edge of the DEM) and the flow directions and flow accumulation are estimated moving downslope.

Pilesjö and Hasan (2014) also used the same four mathematical surfaces as Zhou and Liu (2002) (Figure 3.14) to compare the performance of the D8 (O'Callaghan & Mark, 1984), D8-LTD (Orlandini et al., 2003), D∞ (Tarboton, 1997), MD∞ (Seibert & McGlynn, 2007), FMFD (Freeman, 1991), QMFD1 (Quinn et al., 1991), DEMON (Costa-Cabral & Burges, 1994), and TFN (Zhou et al., 2011) algorithms with TFM and a 1-m DEM interpolated from LiDAR for a peatland area near Abisko in northern Sweden to compare the SCA values calculated with the D8 and TFM algorithms. Table 3.3 summarizes the ranking of the RMSEs (one of three metrics used to compare the performance of these algorithms) across all four mathematical surfaces and the mean rankings in the final column. The results show how (i) the TFM and TFN algorithms performed better than the others; (ii) the D∞ and four multiple-flow direction algorithms (MD∞, FMFD, QMFD1, and DEMON) fell in the middle; and (iii) the two single-flow direction algorithms (D8 and D8-LTD) performed particularly poorly.

**Table 3.3** Rankings of RMSEs for the TFM and eight other flow-direction algorithms on the four mathematical surfaces illustrated in Figure 3.14 (with 1 assigned to the flow-direction algorithm with the lowest RMSE and 9 to the flow-direction algorithm with the largest RMSE).

| | Ellipsoid | Inverse ellipsoid | Plane | Saddle | Mean |
|---|---|---|---|---|---|
| D8 (O'Callaghan & Mark, 1984) | 8 | 9 | 9 | 8 | 8.50 |
| D8-LTD (Orlandini et al., 2003) | 9 | 8 | 6 | 9 | 8.00 |
| D∞ (Tarboton, 1997) | 7 | 7 | 7= | 6 | 6.75 |
| MD∞ (Seibert & McGlynn, 2007) | 6 | 6 | 7= | 7= | 6.50 |
| QMFD1 (Quinn et al., 1991) | 3 | 5 | 5 | 4 | 4.25 |
| DEMON (Costa-Cabral & Burges, 1994) | 5 | 3 | 2 | 7 | 4.25 |
| FMFD (Freeman, 1991) | 2 | 4 | 4 | 3 | 3.25 |
| TFN (Zhou et al., 2011) | 4 | 2 | 1 | 1 | 2.00 |
| TFM (Pilesjö & Hasan, 2014) | 1 | 1 | 3 | 2 | 1.75 |

*Source*: Pilesjo & Hasan (2014, Table 1, p. 117).

The visual comparison of the SCAs calculated with the D8 and TFM algorithms (Pilesjö & Hasan, 2014, Figure 7, p. 122) clearly shows how the TFM algorithm produced a more plausible simulation of the flow distribution on this real-world terrain surface.

Figures 3.23 and 3.24 show the upslope contributing area calculated for the Cottonwood Creek catchment using the MD∞ (Seibert & McGlynn, 2007) and TFM (Pilesjö & Hasan, 2014) multiple-flow direction algorithms, respectively. The valley bottoms are clearly defined as a line of white high-contributing area cells, and the ridges both within the catchment and at the divide are visible as dark cells with low-contributing area in both of these maps (similar to the D8 and D∞ single-flow direction upslope contributing area maps reproduced in Figures 3.12 and 3.13). However, most of the straight-line artifacts and problems observed with arbitrary flow directions

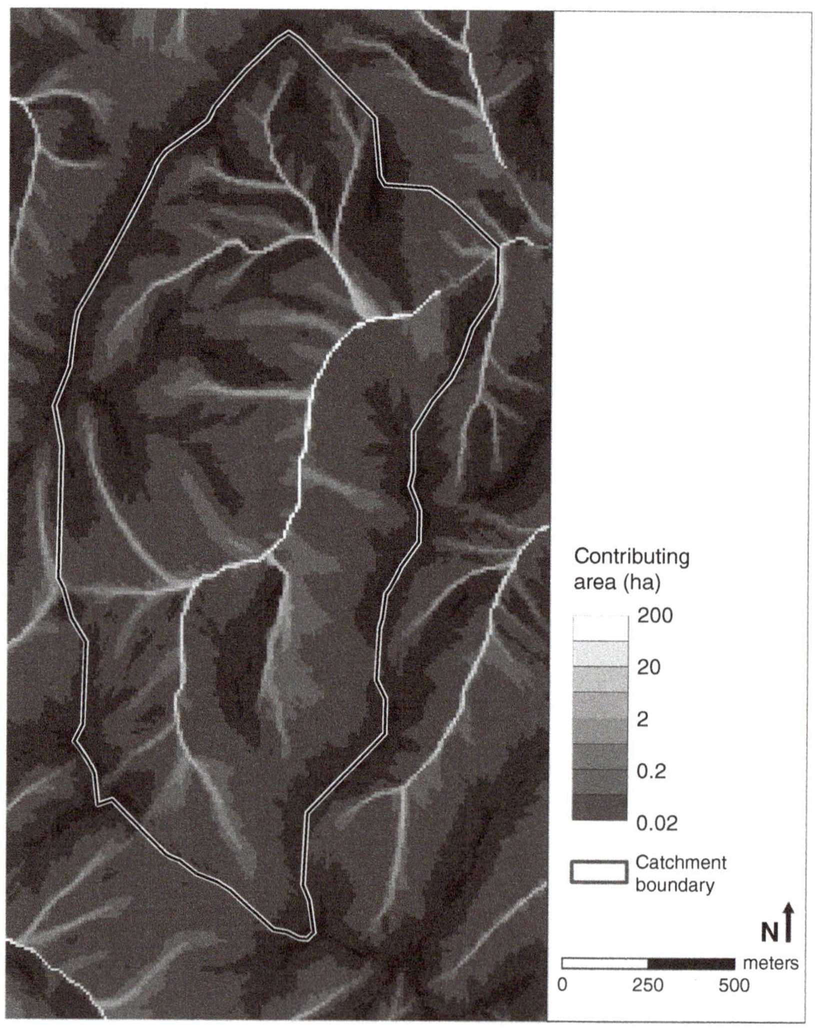


**Figure 3.23** Upslope contributing area (ha) derived for the Cottonwood Creek, MT study site using the MD∞ multiple-flow direction algorithm, with the catchment boundary overlaid.

near ridgelines in Figures 3.12 and 3.13 have been eliminated in these new maps, verifying the claims made by the authors of this pair of multiple-flow direction algorithms. The maps reproduced in Figures 3.23 and 3.24 also generated less dispersion in valleys than the upslope contributing area map for the Cottonwood Creek catchment generated by Gallant and Wilson (2000, Figure 3.10, p. 66) with the MMFD2 (Moore et al., 1993e; Gallant & Wilson, 1996, 2000) multiple-flow direction algorithm and similar results to the upslope contributing area map for the Cottonwood Creek catchment generated by Gallant and Wilson (2000, Figure 3.11, p. 67) with the DEMON (Costa-Cabral & Burges, 1994) multiple-flow algorithm. The progression evident in the four upslope contributing area maps reproduced in Figures 3.12, 3.13, 3.23, and 3.24 shows not only how different flow-direction algorithms are likely to generate substantially different results but also how some of the more recent algorithms can eliminate the most serious and

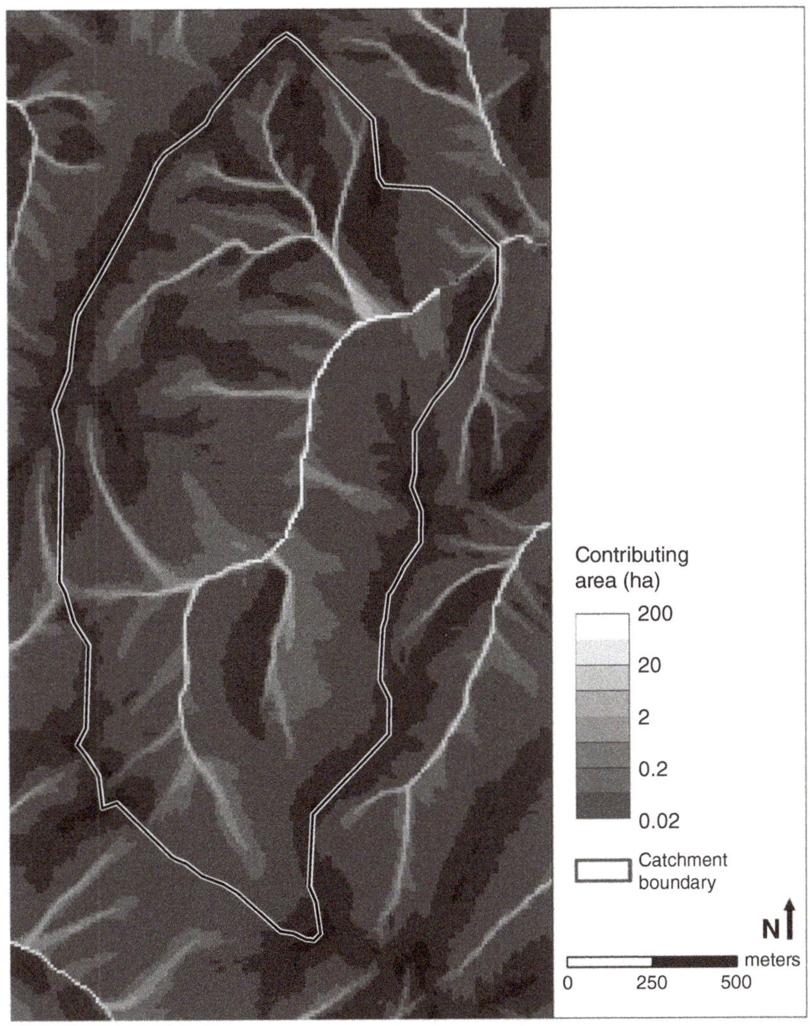


Figure 3.24 Upslope contributing area (ha) derived for the Cottonwood Creek, MT study site using the TFM multiple-flow direction algorithm, with the catchment boundary overlaid.

persistent artifacts that accompanied some of the early flow-direction algorithms.

That said, several of the regional primary topographic attributes that rely on these flow-direction algorithms and measure flow accumulation – such as the upslope contributing area and SCA – will be described in more detail in Section 3.1.4, but first we quickly review the various ways in which the flow width has been calculated in conjunction with one or more of the aforementioned single- and multiple-flow direction methods because this local primary land surface parameter is also needed to calculate the SCA.

#### 3.1.3.3 Flow Width

The flow width or effective contour length orthogonal to the outflow ($w$) is another important concept in hydrology and for the flow-based land surface parameters listed in Table 3.1 (Gallant & Wilson, 2000; Gruber & Peckham, 2009). The flow width is usually reported in terms of cell width and is calculated in different ways for different flow-direction algorithms.

For the D8 and Rho8 algorithms, the flow width is equal to the cell width ($h$) for flow in cardinal directions and the cell diagonal for flow in diagonal directions:

$$w = \begin{cases} h & i = 2, 4, 6, 8 \\ \sqrt{2}h & i = 1, 3, 5, 7 \end{cases} \tag{3.27}$$

This approach seems reasonably simple and straightforward; however, Chirico et al. (2005) suggested using the cell width for flow in both the cardinal and diagonal directions. These authors compared the performance of D8 (O'Callaghan & Mark, 1984) and D∞ (Tarboton, 1997) using the standard grid cell width ($h$) for flows in both the cardinal and diagonal directions with the two values summarized in Equation 3.27, and found that the versions of D8 and D∞ incorporating the standard grid cell width performed better based on the magnitude and pattern of the absolute bias, mean absolute error, and local relative error than the versions which used varying flow widths depending on the direction of flow when calculating SCA. Notwithstanding these results, most implementations of D8 thus far have used the two flow widths summarized in Equation 3.27.

The flow width for the QMFD1 (Quinn et al., 1991) and QMFD2 (Quinn et al., 1995) flow-direction algorithms is also difficult to model because the flow is dispersed in different fractions to two or more downslope cells. Quinn et al. (1991, 1995) used a geometric construction to determine that flow widths in the cardinal directions should be $0.5h$ and those in the diagonal directions should be $0.354h$. The total flow width for a cell is the sum of the flow widths in the direction of all downslope cells, and ranges from $0.354h$ (i.e. flow to one diagonal cell) to $3.416h$ (i.e. flow to four cardinal cells and four diagonal cells as would occur if the center cell was a peak) (Gallant & Wilson, 2000).

The two variants of the multiple-flow direction algorithms proposed by Freeman (1991) and Quinn et al. (1991) that were implemented by Moore in TAPES-G used the D8 flow width in the early years, but this was later changed to account for the amount of dispersion of flow. The effective number of flow directions was first calculated from the weights as follows:

$$n_{\text{flow}} = \frac{1}{\sum_{i=1}^{8} F_i^2} \tag{3.28}$$

where $n_{\text{flow}}$ varied from a minimum of 1 when flow was in a single direction to a maximum of 8 when flow was dispersed equally in all eight downslope directions. The flow width was then estimated as:

$$w = \begin{cases} h & n_{\text{flow}} < 3 \\ h\left[1 + 3\dfrac{n_{\text{flow}} - 3}{5}\right] & n_{\text{flow}} \geq 3 \end{cases} \tag{3.29}$$

This approach took no account of whether the flow was in a cardinal or diagonal direction and calculated a flow width equal to $h$ for flow on convergent and planar slopes when $n_{\text{flow}}$ is small and gradually increased the flow width to a maximum of $4h$ when $n_{\text{flow}}$ grew larger on divergent slope areas (Gallant & Wilson, 2000).

And finally, the flow width can be estimated for those single- and multiple-flow direction algorithms that calculate flow direction using a single, continuous flow angle, such as the ADK, D∞, DEMON, Mass Flux, and TFN algorithms as:

$$\omega = |\sin\varphi| . \Delta x + |\cos\varphi| . \Delta y \tag{3.30}$$

where $\varphi$ is the aspect angle, $|x|$ denotes the absolute value of $x$, and $\Delta x$ and $\Delta y$ are the grid cell sizes along the two coordinate axes which will equal one another for most applications using planar coordinates (Gallant & Wilson, 2000; Gruber & Peckham, 2009).

#### 3.1.3.4 Summary

It is clear from the previous discussion throughout Section 3.1.3 that the number of flow-direction algorithms has grown steadily during the past 30 years and that the choice of flow direction and/or flow width algorithms may have important consequences for both the flow-based primary land surface parameters that focus on flow accumulation and operate over regional scales (Section 3.1.4), and several of the flow-based secondary land surface parameters that are described in Section 3.2. The flow accumulation parameters are described next because (i) the upslope contributing area and/or SCA has been frequently used to compare and contrast the performance of the various flow direction and flow width algorithms that have been proposed to date; and (ii) this pair of land surface parameters was already introduced when describing several of these algorithms (see Figures 3.12, 3.13, 3.23, and 3.24 for examples of the upslope contributing areas for the Cottonwood Creek catchment calculated with four different single- and multiple-flow direction algorithms).

### 3.1.4 Flow Accumulation

We saw in Sections 3.1.3.1 and 3.1.3.2 that the upslope contributing area and SCA have often been used to evaluate the performance of two or more flow-direction algorithms. The SCA is one of several regional land surface parameters that focus on flow accumulation and the opportunity is taken here to describe this and several similar attributes before introducing several novel approaches for comparing the results of using different flow-direction algorithms in Section 3.1.4.3 so that one might know which of these algorithms to use going forward.

#### 3.1.4.1 Upslope Contributing Area and Specific Catchment Area

The upslope contributing area ($A$) is the area above a certain length of contour that contributes flow across the contour, as shown in Figure 3.25. The upslope contributing area is often referred to as drainage or catchment area. The specific catchment area ($A_s$) is the ratio of the contributing area to the contour length, $A/l$. This pair of land surface parameters is important because the upslope and specific contributing areas will be directly proportional to the discharge and specific discharge in the idealized case of a constant, spatially uniform rainfall rate (Gruber & Peckham, 2009).

When calculating these parameters from gridded DEMs, the contributing area is determined by the number of cells contributing flow to a cell and the contour length is assumed to be the same as the flow width. Both of these quantities are dependent on the estimation of flow direction(s) from a given node (i.e. node 9 in Figure 3.3), as described in Section 3.1.3.

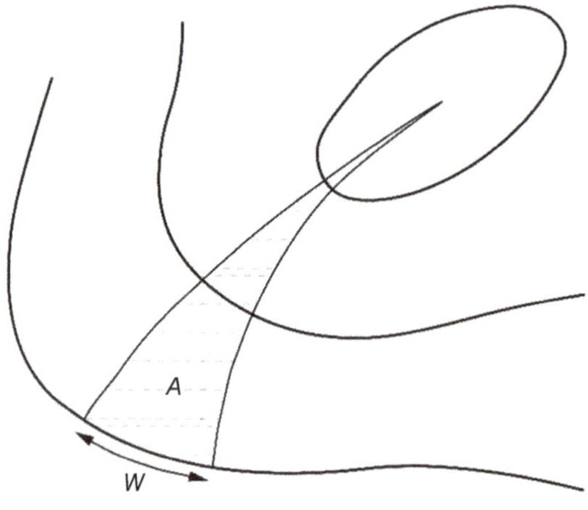


**Figure 3.25** An idealized stream tube originating at a hilltop and terminating at a contour on a hillslope. The average specific catchment area $a$ along the contour segment is the ratio of contributing area $A$ to flow width $w$. *Source*: Gallant & Hutchinson (2011, p. 2). Reproduced with permission of John Wiley & Sons, Inc.

#### 3.1.4.2 Rate of Change of Specific Catchment Area

The rate of change of SCA along the flow path ($dA_s/ds$) is computed as the specific catchment area of the flow leaving the cell less the average of the specific catchment area entering the cell, divided by the flow length across the cell (Gallant & Wilson, 2000). This parameter has been used in several soil erosion models where erosion is a function of slope, profile curvature and $dA_s/ds$ (Moore, 1996).

#### 3.1.4.3 Influence of Choice of Flow Direction Algorithm on Flow Accumulation Estimates

The single- and multiple-flow directions described in Section 3.1.3 have more often than not been evaluated using SCA. The methods used to compare and contrast these algorithms can be divided into two groups – those using real-world terrain and/or those using mathematical surfaces – and the goal here is to review the strengths and weaknesses of these approaches and to describe a couple of recent studies that have followed a third approach, comparing the spatial patterns of flow predicted with one or more of these flow-direction algorithms with flow observations on real-world terrain surfaces. Three relatively new studies, one from each group, are highlighted as a part of this review.

The first group, which uses real-world terrain surfaces to assess the performance of two or more of these flow-direction algorithms and the land surface parameters that are derived from them, is difficult to execute well because most real-world terrain surfaces contain a complicated mix of terrain forms that might span multiple scales and because the errors caused by the choice of flow-direction algorithm may be difficult to separate from and/or sometimes completely obscured by the errors in the DEMs themselves.

However, the spread and growing popularity of geographic information systems with their powerful spatial query, analysis and mapping tools have provided new opportunities for conducting these kinds of evaluations. Seibert and McGlynn (2007), for example, offered a series of simple but powerful metrics to compare the performance of four flow-direction algorithms in two real-world terrain settings, and used these results to suggest that the MD∞ (Seibert & McGlynn, 2007) algorithm is more appropriate than the D8 (O'Callaghan & Mark, 1984), D∞ (Tarboton, 1997), and QMFD1 (Quinn et al., 1991) algorithms when the performance of these algorithms was compared across a range of landscapes, DEM resolutions, and applications.

Wilson et al. (2007) took this approach a step further by calculating flow direction and SCA with five flow-direction algorithms and comparing their performance across six landscape classes that were delineated with a fuzzy classifier and eight topo-climatic attributes from a 10-m DEM in the Santa Monica Mountains of southern California. The matched paired $t$-test was used to compare the performance of pairs of SCA grids calculated with the D8 (O'Callaghan & Mark, 1984), Rho8 (Fairfield & Leymarie, 1991), D∞ (Tarboton, 1997), DEMON (Costa-Cabral & Burges, 1994), and MMFD2 (Moore et al., 1993e; Gallant & Wilson, 1996, 2000) algorithms across six user-defined fuzzy landscape classes representing coastal plains and gentle slopes, moderately steep lower valley slopes, stream channels, north-facing slopes, south-facing slopes, and hilltops and ridgetops. The results showed the following.

1. The "source" cells predicted with the D∞, DEMON, and the MMFD2 algorithms were confined to hilltops and ridgelines, whereas those predicted with the two single-flow direction algorithms (D8, Rho8), produced many more low flow cells that were scattered across a much larger area and a greater range of landscape settings. The former, of course, is the preferred result.
2. The D8 algorithm predicted the largest numbers of stream channel cells but the other algorithms followed closely behind – the MMFD2 results were expected to mirror those for D8 because this algorithm switches to D8 when a user-specified maximum cross-grading threshold is exceeded (see Section 3.1.3.2 for additional details).
3. The results of the *t*-test pairwise comparisons were mixed but they did show how the various flow-direction algorithms can be expected to generate different patterns of flow (based on upslope contributing areas) on different parts of a hillslope or watershed.
4. The choice of flow-direction algorithm has potentially important consequences for the calculation of upslope contributing area, sediment transport capacity, topographic wetness, and several other primary and secondary land surface parameters.

Taken as a whole, the methods used and results generated in this particular study are noteworthy because they show how the variation in the performance of different flow-direction algorithms across one or more different landscapes might be evaluated in a reproducible manner.

The second group has used SCA with a variety of mathematical surfaces to compare and contrast the performance of several flow-direction algorithms. In one of the studies highlighted earlier, Pilesjö and Hasan (2014) used the four mathematical surfaces reproduced in Figure 3.14 to compare the performance of the D8 (O'Callaghan & Mark, 1984), D8-LTD (Orlandini et al., 2003; Orlandini & Moretti, 2009), D∞ (Tarboton, 1997), MD∞ (Seibert & McGlynn, 2007), FMFD (Freeman, 1991), QMFD1 (Quinn et al., 1991), DEMON (Costa-Cabral & Burges, 1994), TFN (Zhou et al., 2011), and TFM (Pilesjö & Hasan, 2014) algorithms. The criticism of the basic approach used in this and similar studies is that none of the surfaces used for these kinds of assessments actually occurs in real-world landscapes and so the results may not be indicative of how one or more of these algorithms might perform in a specific landscape setting.

Gallant and Hutchinson (2011) recently proposed a differential equation for SCA that may offer a possible way to avoid this type of criticism. They delineated one or more stream tubes (i.e. a tract of land defined by two adjacent slope lines that delivers water to a channel; Onstad & Brakensiek,

1968) and derived a simple, non-linear differential equation that described the rate of change of SCA along the flow path. They then showed how this differential equation can be solved numerically along a flow path to calculate SCA at any point on a DEM without requiring the usual estimates of upslope contributing area ($A$) and flow width ($\omega$). Gallant and Hutchinson (2011) acknowledged that this method is more computationally intensive than the 20 or so conventional grid-based methods for calculating SCA (assuming that each of the single- and multiple-flow direction algorithm described in Sections 3.1.3.1 and 3.1.3.2 leads to a different estimate of upslope contributing area and/or flow width and therefore SCA), and that its main application is as a reference against which the performance of these other approximate methods can be evaluated.

This is the first method that provides a benchmark against which to evaluate the performance of the less computationally demanding methods in complex terrain with both convergent and divergent areas because it (i) goes beyond the usual approach in which comparisons between methods on real-world landscapes are coupled with an intuitive expectation of what the results should look like; and (ii) it avoids simple surfaces for which analytical solutions are known. A preliminary evaluation of four commonly used methods supported previous work which showed that (i) the more sophisticated algorithms [MMFD2 (Moore et al., 1993e; Gallant & Wilson, 1996, 2000), DEMON (Costa-Cabral & Burges, 1994), and D∞ (Tarboton, 1997) in this instance] performed better than D8 on hillslopes; (ii) D∞ is the most accurate of these four methods for estimating SCA overall; and (iii) all four methods overestimated SCA on ridges and hilltops because of the underestimation of flow width.

The third group comprises a small number of studies that have compared the flow paths predicted with these flow-direction algorithms to observations of overland flow on real-world terrain surfaces. In what may have been the first study of this kind, Endreny and Wood (2003) calculated a series of topographic attributes for two small fields in New Jersey and compared the spatial congruence of overland flow paths observed in the field with those predicted with five flow-direction algorithms. The observed flow paths were sketched in the field and then mapped on the DEM grid cells for comparison with the calculated flow paths. The lowest congruence ratings were produced with D8 (O'Callaghan & Mark, 1984) (which constrains flow to a single neighbor) and FMFD (Freeman, 1991) (which directs flow into all neighbors that are lower in elevation). The ADK (Lea, 1992) and D∞ (Tarboton, 1997) algorithms, which direct flow to one or at most two downslope neighboring cells, produced the highest congruence ratings. However, modified versions of D8 (i.e. using a buffer to direct flow to neighboring cells that overlapped the buffered area) and FMFD (limiting flow to three or five downslope cells) produced congruence rankings that rivaled those of the more sophisticated ADK and D∞ algorithms.

A few years later, Moretti and Orlandini (2008) showed how drainage basins can be automatically delineated from contour elevation data using skeleton construction techniques and later they used a detailed contour and flow line network to compare the performance of several of the grid-based flow routing methods (Orlandini & Moretti, 2009; see Section 3.1.3.1 for additional details).

Fortunately, the rapid development and application of new remote sensing systems will provide new opportunities for detailed assessments like those of Endreny and Wood (2003) and Orlandini and Moretti (2009) in the years ahead. For example, Orlandini et al. (2012) recently evaluated six of the single- and multiple-flow direction algorithms described in Sections 3.1.3.1 and 3.1.3.2 against novel field observations of overland flow dispersion. The field data were gathered when thin flows of cold water were released at selected points on three warmer slopes and individual overland flow patterns were observed using a terrestrial laser scanner and a thermal imaging camera. The land surface was determined by using laser returns from the dry land surface (with 26, 28, and 21 points per square centimeter, respectively) and the overland flow patterns were determined by using laser returns and infrared emissions from the wetted parts of the three natural slopes chosen for the experiments.

Three pairs of flow-direction methods – the D8 (O'Callaghan & Mark, 1984) and D8-LTD (Orlandini et al., 2003; Orlandini & Moretti, 2009) single-flow direction, non-dispersive methods; the D∞ (Tarboton, 1997) and D∞-LTD (Orlandini et al., 2003; Orlandini & Moretti, 2009) single-flow direction, moderately dispersive methods (this pair of hybrid flow-direction algorithms calculate a single flow direction but retain the ability to split the flow between two adjacent grid cells in certain situations); and the FMFD (Freeman, 1991) and QMFD1 (Quinn et al., 1991) multiple-flow direction, fully dispersive methods – were applied to a series of specially built, grid-based DEMs with resolutions ($h$) of 1, 2, 5, 10, 20, 50, 100, and 200 cm to evaluate the ability of these algorithms to reproduce the overland flow patterns observed with the aforementioned technologies.

Orlandini et al. (2012) also verified that the observed overland flow patterns were not significantly affected by infiltration (the infiltration rates were at least two orders of magnitude smaller than the surface flow velocities for the three study slopes) and introduced two new terms to facilitate the evaluations.

1. The term "sensitivity" was used to denote the method's ability to predict flow in areas where it was observed (and thus to not allow false negatives);.
2. The term "specificity" was used to denote the method's ability to not predict flow in areas where it was not observed (and thus not to generate false positives).

Orlandini et al. (2012) also described several assumptions and workarounds that were required to generate their findings, and a summary of these follows.

The findings are as novel as they are instructive. The first was that the overland flow dispersion was observed to play an important role immediately downslope of the point sources but that it attenuated rapidly as flow was propagated downslope on all three study slopes. The second was that none of the six flow-direction methods was able to reproduce the observed overflow patterns in a consistently sensitive and specific manner and their performance was found to vary significantly with the size $h$ of the grid cells involved. For example, all six methods were found to be poorly sensitive in very fine-resolution grids (i.e. $h \leq 2$ cm) and to be poorly specific in coarse-resolution grids (i.e. $h \geq 2$ m). Satisfactory results were reported in grids with resolutions approximating the average flow width (~50 cm), with the best performances displayed by the QMFD1 method in the finest grids ($5 \leq h \leq 20$ cm) and by the MD∞, D∞, and D∞-LTD methods in the coarsest grids (20 cm < $h \leq 1$ m). However, Orlandini et al. (2012) also showed that the single-flow direction, non-dispersive methods (D8 and D8-LTD) and the hybrid or single-flow direction, moderately dispersive methods (D∞-LTD and D∞) varied dramatically with grid orientation and sometimes provided inaccurate solutions. These kinds of problems were noted with some of these same algorithms in Section 3.1.3.1 and in this particular study, their magnitude diminished with coarser resolutions that matched or exceeded the average flow width (i.e. $h \geq 50$ cm).

The results from this last study confirm in many ways what is already known, most notably that the final choice of flow-direction method for a specific landscape and/or application will be a compromise. The single-flow direction algorithms cannot represent divergent flow but for the same reason have no problem with over-dispersal (i.e. the dispersal of the available flow over too many cells or too large an area) whereas the multiple-flow direction algorithms can represent divergent flow but usually also suffer from some over-dispersal. The ways in which individual methods treat ambiguous flow directions (as for example occurs along ridgelines or saddles and across flat plains or valley bottoms) and the reconciliation of DEM-delineated flow lines and drainage lines acquired from some other source are also likely to influence the results across large parts of the terrain surface.

This said, we turn next to the final two groups of local primary land surface parameters which ignore the hydrologic connectivity of the surface topography and focus instead on the relationship of a grid cell or a point to the surrounding landscape. The first group focuses on the elevation residuals first described by Gallant and Wilson (2000, pp. 73–77) and the second group focuses on the statistical parameters described by Olaya (2009, pp. 157–163).

### 3.1.5 Elevation Residuals

This group of land surface parameters includes seven parameters for characterizing the relationship of a point to the surrounding landscape (see Table 3.1). A key issue with these land surface parameters is the choice of a suitable analysis window. The shape and/or size might be varied since these parameters can be calculated in angular, circular or square neighborhoods of different sizes. Gallant and Wilson (2000) suggested the window size should match the length scale of the process under study, and that this will be the hillslope length in many instances. In the equations that follow, $C$ denotes the set of cells included in the local neighborhood around the point under consideration, and $N_c$ denotes the number of cells in this local neighborhood.

The mean elevation in the local neighborhood is useful when a smoothed DEM is required and as a reference against which to compare the elevation at the central point of the local neighborhood:

$$\bar{z} = \frac{1}{N_c} \sum_{i \in C} z_i \tag{3.31}$$

The difference between the elevation at the center of the local neighborhood and the mean elevation in this window is a measure of the relative topographic position of the central point:

$$\text{Diff} = z_0 - \bar{z} \tag{3.32}$$

The range of this parameter will depend on the range in elevation within the local neighborhood window chosen for the analysis.

The standard deviation (SD) of elevation is a measure of the variability of elevation within the local neighborhood. It is useful as a measure of local relief at the scale specified by the size and shape of the local neighborhood:

$$\text{SD} = \sqrt{\frac{1}{N_c - 1} \sum_{i \in c} \left(z_i - \bar{z}\right)^2} \tag{3.33}$$

The standard deviation of elevation is a useful measure of the local relief (assuming a window size similar to the hillslope length is chosen) or the roughness of the landscape at the scale specified by the radius of the window (so long as a much larger window size is chosen and the elevation differences with respect to the regression plane are accounted for in mountainous terrain and/or continental-scale applications).

The elevation range measures the full range of elevations within the local neighborhood:

$$\text{Range} = \max_{i \in C} z_i - \max_{i \in C} z_i \tag{3.34}$$

This parameter is similar to the standard deviation of elevation in that it measures local relief, but it has the disadvantage of sometimes changing abruptly from one location to the next as individual high and low points enter and leave the analysis window (Gallant & Wilson, 2000).

The deviation from the mean is the difference from the mean divided by the standard deviation:

$$\text{Dev} = \frac{z_0 - \bar{z}}{\text{SD}} \tag{3.35}$$

This parameter, whose values range mostly between −1 and +1, measures the relative topographic position as a fraction of local relief and so is normalized to the local surface roughness. Lindsay, Cockburn and Russell (2015) used this land surface parameter to illustrate how an integral image approach can be used to perform multiscale topographic position analysis (see Section 6.3 for additional details).

The percentile is a ranking of the point at the center of the local neighborhood relative to all the points in the analysis window. It is measured by simply counting the number of points lower than the central point:

$$\mathrm{PCTL} = \frac{100}{N_c} \mathrm{count}_{i \in C} \left( z_i - z_0 \right) \tag{3.36}$$

The elevation percentile ranges from 0 to 100, with a value of 0 indicating that the center point is the lowest in the analysis window, 100 indicating that it is the highest, and 50 indicating that half of the points in the analysis window are lower. The elevation percentile is similar to the deviation from mean elevation, but its straightforward and well-defined range makes its application simple and robust, and it is sometimes used as an index of local topographic position in studies that have endeavored to explain spatial patterns of vegetation cover (e.g. Mackey et al., 2000).

The seventh and final land surface parameter in this group is the local elevation percentage which measures the elevation as a percentage of the elevation range:

$$\mathrm{PCTG} = 100 \frac{z_i - z_{\min}}{z_{\max} - z_{\min}} \tag{3.37}$$

This parameter, like the elevation percentile, ranges from 0 to 100. However, it is not as sensitive to the shape of the surface because it does not account for the distribution of values within the analysis window and it suffers from the same artifacts as elevation range as individual high and low points enter and leave the analysis window.

Figures 3.26, 3.27 and 3.28 show the difference from mean elevation (Equation 3.32), elevation percentile (Equation 3.36), and the standard deviation of elevation (Equation 3.33) for the Cottonwood Creek, MT study site.

Figure 3.26, difference from mean elevation, shows the general structure of the summits, ridges, and valleys. The brightness or darkness indicates the intensity of highness or lowness and shows several low hilltops around the catchment boundary that do not protrude much above the surrounding landscape and the deeply incised valley of the main channel extending in a northeasterly direction towards the catchment outlet.

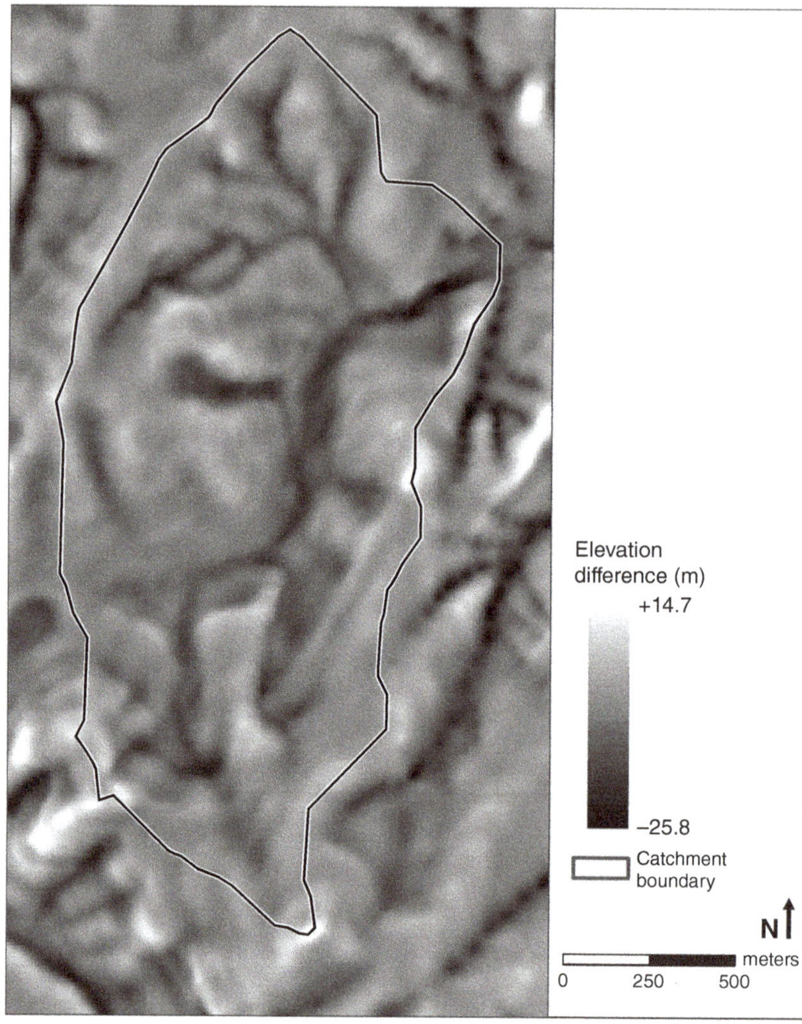


Figure 3.26 Difference from mean elevation for the Cottonwood Creek, MT study site using a 15 × 15 cell moving window, with the catchment boundary overlaid.

Figure 3.27, elevation percentile, shows the structure of the catchment (similar to Figure 3.26) but picks out different features because it is based on rankings rather than elevation differences (i.e. relief). The map clearly delineates the ridges and valleys and shows several summits and ridges that divide the catchment into a series of sub-catchments and direct flow in a variety of directions based on landscape position and surface orientation or aspect.

Figure 3.28, standard deviation, shows the pattern of local relief across the catchment. The valley bottoms and to a lesser extent the catchment boundary, are areas of lower relief while the steep hillslopes to the east of the main channel and to the north and south of the large hill that divides the area to the west of the main channel in the center of the catchment into north and south-facing slopes are areas of high relief.

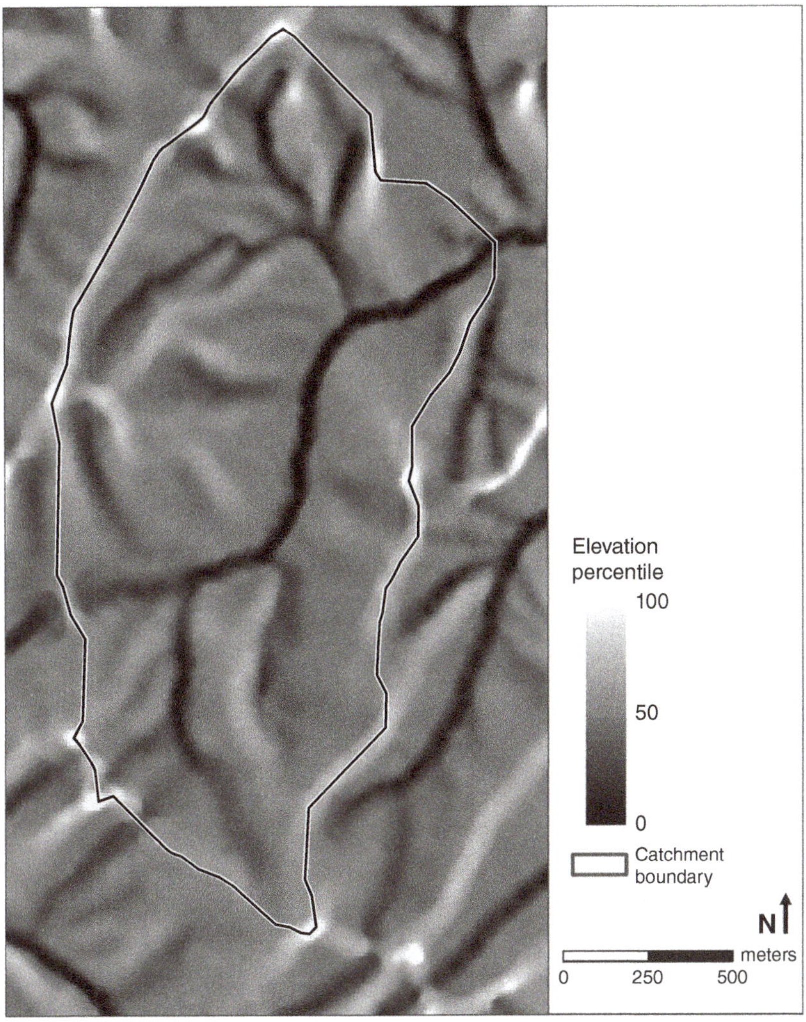


**Figure 3.27** Elevation percentile for the Cottonwood Creek, MT study site using a 15 × 15 cell moving window, with the catchment boundary overlaid.

## *3.1.6 Statistical Parameters*

Olaya (2009, pp. 157–163) described two broad groups of statistical land surface parameters. The first group relied on the calculation of averages, standard deviations as well as skewness and kurtosis coefficients for a series of square, circular, and angular neighborhoods of various sizes, and shared some similarities with the list of elevation residuals proposed by Gallant and Wilson (2000, pp. 73–76) and described in Section 3.1.5. However, the second group comprised land surface parameters that were created specifically for the analysis of DEMs. Several of these parameters provide methods for measuring terrain roughness, which seeks to describe how undulating or complex the terrain is.

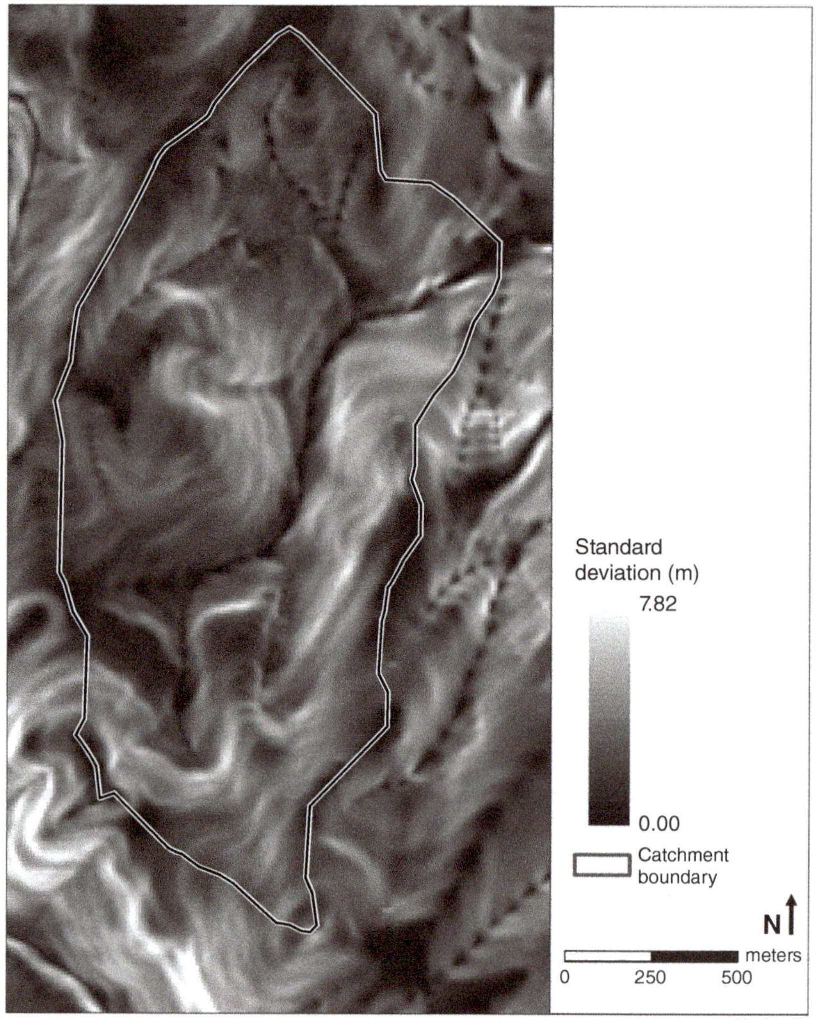


**Figure 3.28** Standard deviation of elevation for the Cottonwood Creek, MT study site using a 15 × 15 cell moving window, with the catchment boundary overlaid.

The practice of using the standard deviation or coefficient of variation of elevation in some form of circular window as a measure of the local relief or roughness of the landscape at the scale specified by the radius of the analysis window was noted in Section 3.1.5. High values of standard deviation indicate that the terrain in this window is irregular while low values reflect a smooth land surface (Olaya, 2009). However, this is not a very precise method (it varies with the size of the analysis window and the grid resolution) and care needs to be taken to account for any trend(s) that may be present in mountainous terrain and/or continental-scale applications (Gallant & Wilson, 2000).

Olaya (2009) also listed several additional measures that might be used in place of standard deviation for measuring the terrain roughness. For example, Melton (1965) proposed a ruggedness index:

$$\text{RUGN} = \frac{\text{Range}}{\sqrt{a}} \tag{3.38}$$

where "range" is the difference between the highest and lowest elevation and $a$ is the area covered by the local region being analyzed. Hobson (1972), on the hand, proposed a vector approach to define a surface roughness factor (SRF):

$$\text{SRF} = \frac{\sqrt{\left(\sum_{i=1}^{n} X_i\right)^2 + \left(\sum_{i=1}^{n} Y_i\right)^2 + \left(\sum_{i=1}^{n} Z_i\right)^2}}{n} \tag{3.39}$$

where $n$ is the sample size (i.e. the number of cells in the analysis window) and $X_i$, $Y_i$, and $Z_i$ are the components of the unit vector normal to the land surface at each of the cells in the analysis window, which can be calculated from slope and aspect using the following expressions:

$$X_i = \sin(s) \cdot \cos(a) \tag{3.40}$$

$$Y_i = \sin(s) \cdot \sin(a) \tag{3.41}$$

$$Z_i = \cos(s) \tag{3.42}$$

Riley, DeGloria and Elliot (1999) later proposed a topopgraphic ruggedness index to express the amount of elevation variability between adjacent cells in a DEM. This method calculates the difference in elevation between the center cell and the eight cells immediately surrounding it, squares these values to make them all positive and averages the squares, and then takes the square root of this average to represent the average elevation change between each individual grid point and its eight neighbors.

One possible concern with these last three indices is that they give information about the DEM itself and it is likely that the accuracy of the DEM in representing the actual land surface will vary with its complexity (i.e. roughness). Olaya (2009) also described another (different) surface roughness index (SRI) which can be derived using geostatistical analysis from LiDAR point clouds and other dense elevation datasets as the ratio between the fitted nugget variation and the local variance:

$$\text{SRI} = \frac{C_0}{\sigma_{\text{NB}}^2} \% \tag{3.43}$$

where $C_0$ is the locally fitted nugget parameter and $\sigma_{\text{NB}}^2$ is the local variance in the predetermined neighborhood. The nugget parameter needs to be estimated using a robust variogram fitting method and a value of SRI of 0% would indicate that the terrain surface is perfectly smooth.

Olaya (2009) also described in considerable detail how variogram analysis can be repeated in various directions to derive (i) an anisotropy index (ANI) that depicts how isotropic the land surface is locally; (ii) an anisotropic coefficient of variation, a simplified version of the ANI, which can be derived from the difference in the first derivative in four directions; and (iii) how the fractal dimension can be used as an indicator of terrain complexity.

The final example offered by Olaya (2009, pp. 162–163) showed how a DEM could be converted into a discrete layer and the shape and structure of each class (which defines one or more polygons) can be analyzed and the results used to derive new land surface parameters. Hengl et al. (2003), for example, used the shape complexity index (SCI), which was first used to describe polygons, on DEMs sliced into elevation classes to measure how compact (i.e. oval) a feature is. The SCI is derived using the perimeter-to-boundary ratio:

$$\mathrm{SCI} = \frac{P}{2\pi r}, \qquad r = \sqrt{\frac{A}{\pi}} \tag{3.44}$$

where $P$ is the perimeter of the polygon, $A$ is the area of the polygon, and $r$ represents the radius of a circle with the same area. This particular land surface parameter shows how the general and specific forms of terrain modeling noted by Evans (1972) and Pike et al. (2009), among others, are connected to one another given the direct connection between SCI and some of the landform objects described in Chapter 4 and the knowledge that pits and peaks are typically more oval and valleys and ridges are more linear in terms of their form (Figure 3.29).

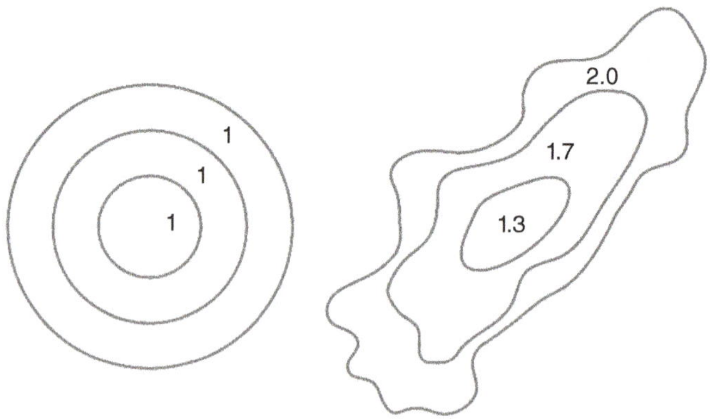


**Figure 3.29** A comparison of the shape complexity index values for a perfectly oval shape (left) and for different levels of complexity (right). *Source:* Olaya (2009, p. 163). Reproduced with permission of Elsevier.

### 3.1.7 Upslope Parameters

A number of upslope land surface parameters have also been proposed in addition to those focused on flow accumulation and introduced in Section 3.1.4.

The first, and probably most important, is the maximum upslope flow-path length, which records the length of the longest flow path from a peak or ridge marking the catchment boundary to the cell of interest (Table 3.1). There is also the minimum and mean upslope flow-path lengths, with the first recoding the length of the shortest flow path from a peak or ridge marking the catchment boundary and the second recording the mean length of all the flow paths from a peak or ridge on the catchment boundary to the cell of interest (Table 3.1). The values calculated for these land surface parameters will likely vary in magnitude and/or accuracy depending on the flow-direction algorithm that is chosen and its ability to model the flow of water across the land surface. The popular D8 (O'Callaghan & Mark, 1984) flow-direction algorithm, for example, will use the cell width $h$ for flow in a cardinal direction ($i=2, 4, 6$, or 8 in Figure 3.3) and $\sqrt{2}\,h$ for flow in a diagonal direction ($i=1, 3, 5$, or 7 in Figure 3.3) to accumulate flow distances across cells but still fail because the flow paths calculated with this method are restricted to the four cardinal and four diagonal directions and the result will likely be flow directions with a zigzag appearance (Gallant & Wilson, 2000).

There are several other local parameters (number of upslope neighbors, number of neighboring cells contributing flow) and regional upslope parameters (the elevation of a cell of interest relative to the elevation of the nearest peak and/or ridgeline, upslope averages of slope and curvature) that will avoid most if not all of these kinds of problems (Table 3.1). The regional parameters listed above may help to describe processes connected with the accumulation and dispersal of materials across the surface, such as soil development, for which the average characteristics of the landscape in a site's upslope contributing area may be more significant than the terrain attributes at the site itself (Wilson & Gallant, 2000). McKenzie et al. (2000), for example, highlighted several studies in which the mean slope or curvature of the site proved to be a useful index of processes linked to the sorting and transport of material to the site (i.e. the grid cell of interest), and Bell et al. (2000) used the elevation above each grid cell in a soil-terrain model for estimating spatial patterns of soil organic carbon in the Cedar Creek Natural History Area in northern Minnesota.

Finally, there are several hypsometric curves or area–altitude functions that have a long history (Strahler, 1952; Pike & Wilson, 1971; Howard, 1990) and are particularly useful for the morphometric characterization of watersheds and upslope catchment areas (Luo, 2000; Olaya & Conrad, 2009). The width (Kirby, 1976; Gupta, Waymire & Wang, 1980; Troutman

& Karlinger, 1984) and closely related area–distance functions, which measure the fraction of a catchment that is at some user-specified distance from the outlet, and the cumulative area function (Rigon et al., 1993; Peckham, 1995), which measures the fraction of a watershed that has a contributing area greater than some user-specified value for example, could be adapted and used to measure the equivalent values for one or more grid cells of interest. The RiverTools™ software suite that is described in Chapter 6 can extract these and several related functions from a DEM (Peckham, 2009).

### 3.1.8 Downslope Parameters

In some other applications, it is the shape of the landscape downslope of each grid cell, rather than upslope, that is of particular interest (Gallant & Wilson, 2000). Table 3.1 lists three local and nine regional land surface parameters which more or less mirror those used to describe the landscape in upslope areas. The specific dispersal area, which is the area downslope of a unit length of contour to a channel and/or downslope depression, represents the available area into which water can drain. Speight (1980), for example, described how the soil-water content was influenced by specific dispersal area as well as SCA in some landscapes. These same land surface parameters, along with one or more of the three downslope path lengths listed in Table 3.1, could also be used where proximity to a stream is considered a useful attribute. These land surface parameters might be used to evaluate the efficacy of grassed waterways, buffer strips and other best management practices which seek to mitigate non-point source pollution from sediment and absorbed pollutants, for example.

### 3.1.9 Visibility and Visual Exposure

The ninth and final group of regional primary land surface parameters focus on visibility. Visibility is relatively straightforward and, in its most basic form, calculates what other points can be seen from a single point or the reverse, from what other points can a single point be seen by drawing the line of sight from the point of interest to all other points and checking whether or not the relief (i.e. landforms) that occur between them block visibility. The result is nearly always a simple Boolean raster showing the cells that are visible and those that are not visible. Fisher (1991, 1992, 1993, 1995, 1996) wrote a series of landmark papers on viewshed analysis that describe the basic approach, types of applications that can be supported, and some of the pitfalls that must be avoided. There are various ways to extend this viewshed analysis, for example we can add the height of the observer or the height of the observer plus some structure (i.e. a building or tower) before performing the viewshed analysis or we could conduct this analysis for a series of points representing a road or wildlife corridor to assess from what other points these features can be seen, or the reverse, what other points can be seen from these features. The result in all of the aforementioned cases is a viewshed which may or may not be a single continuous polygon (like a catchment or watershed for example) because it will be made up of many disjointed locations (i.e. grid cells) as happen to be in the view (Olaya, 2009).

There are two other visibility-related land surface parameters that have been proposed as well. The first is a visual exposure or visibility index that represents the number of grid cells that can be seen from each cell or, alternatively, the number of different viewsheds to which a cell belongs (Table 3.1). This index can be derived by simply calculating the visibility for the whole DEM (Olaya, 2009).

The second visibility-related land surface parameter is the openness index, which represents the visual dominance or enclosure of locations on a landscape based on their local exposure (Yokoyama, Shirasawa & Pike, 2002). Openness is an angular measure of the relation between surface relief and horizontal distance which incorporates the terrain line of sight, or viewshed, and is calculated from multiple zenith and nadir angles along eight azimuths. This openness index offers two viewer perspectives: (i) positive values, expressing openness above the surface, which are high for convex forms; and (ii) negative values describe this parameter below the surface and which are high for concave forms (Yokoyama et al., 2002). The reader can find an openness map for the Baranja Hill case study that was used throughout the geomorphometry book edited by Hengl and Reuter (2009) to illustrate key concepts and applications along with the script used to calculate openness in LandSerf in Wood (2009b).

There are several other kinds of exposures which are related to climate and use two or more of the primary land surface parameters described earlier in this chapter. These are taken up in the next section, which describes 15 of the most popular secondary land surface parameters.

## 3.2 Secondary Land Surface Parameters

The secondary land surface parameters are derived from DEMs using two or more of the foregoing primary land surface parameters and additional inputs in some instances. The origins of many of the secondary land surface parameters described here can be traced to a series of pioneering papers authored by Ian D. Moore and his colleagues 25–30 years ago (see Moore & Burch, 1986a–c and Moore et al., 1988a,b, 1991, 1993a,d,e for additional details) and they have found widespread use in a large number and variety of hydrologic, geomorphologic, and ecological applications during the past few decades.

The growing popularity of the secondary land surface parameters is in some ways easy to understand. Most of the environmental research of the past century has been conducted at the global and nano- or micro-scales identified in Figure 1.1. The meso- and topo-scales have received much less attention, and yet these scales are important because many of the solutions to environmental problems, such as accelerated soil erosion and non-point source pollution, will require changes in management strategies at these landscape scales (Wilson et al., 2000).

The influence of geologic substrate on soil chemistry (e.g. Likens et al., 1977) and impact of prevailing weather systems and elevation-dependent lapse rates on long-term average monthly climate (e.g. Hutchinson, 1995) exemplify the kinds of controls operating at the meso-scale. The influence of surface morphology on catchment hydrology and impact of slope, aspect, and horizon shading on insolation represent the most important controls operating at topo-scales. Numerous studies have shown how the shape of the land surface can affect the lateral migration and accumulation of water, sediment, and other materials (e.g. I.D. Moore et al., 1991). These variables, in turn, may influence soil development (e.g. Kreznor et al., 1989) and exert a strong influence on the spatial and temporal distribution of light, heat, water, and mineral nutrients required by photosynthesizing plants (e.g. Franklin, 1995; Mackey, 1996).

Table 3.4 lists 15 of the most frequently used secondary land surface parameters. These parameters are assigned to two groups in this table – those quantifying water flow and soil redistribution and those quantifying the energy and heat regimes operating at the land surface – as a way to organize the discussion which follows and references are given for parameter descriptions where appropriate. Some of these secondary land surface parameters can be calculated with a variety of software platforms but others can only be computed using one or perhaps two specialist packages (see Chapter 6 for additional details). And last, but not least, nearly all of these secondary land surface parameters aim to capture regional interactions (as illustrated in Figure 3.1) in one way or another.

### *3.2.1 Water Flow and Soil Redistribution*

The movement of water is primarily driven by gravity and to some degree modified by the properties of the material it flows through or over (Gruber & Peckham, 2009). The effect of gravity can be approximated well and easily with a DEM but the surface and subsurface properties and conditions are cumbersome to describe and treat. There are steadily improving regional and national datasets describing the spatial variability of selected soil and land surface characteristics, for example the National Land Cover Database (NLCD; Homer et al., 2015) and the USDA-NRCS Soil Survey Geographic Database (SSURGO; US Department of Agriculture, Natural Resources Conservation Service, 2016), but these have a much coarser resolution than modern DEMs and seldom include the various properties needed for specific applications.

**Table 3.4** List of secondary land surface parameters and their significance.

| Land surface parameter | Type | Description |
|---|---|---|
| **Group 1: Water flow and soil redistribution** | | |
| Topographic wetness index | Regional | The index assumes steady-state conditions and describes the spatial distribution and extent of zones of saturation (i.e. variable source areas) for runoff generation as a function of upslope contributing area, soil transmissivity, and slope (Beven & Kirkby, 1979) |
| Topographic wetness index | Regional | This variant of the topographic wetness index also assumes steady-state conditions and describes the spatial distribution and extent of zones of saturation (i.e. variable source areas) for runoff generation as a function of upslope contributing area and slope (O'Loughlin, 1986; I.D. Moore et al., 1991) |
| Network index | Regional | This index incorporates a simple network modification by which saturated areas only contribute to runoff when one or other of the first two steady-state topographic wetness indices listed above indicates continuous saturation along the length of a flow path to the point where the path becomes a stream (Lane et al., 2004) |
| Quasi-dynamic topographic wetness index | Regional | This index substitutes effective upslope contributing area for the upslope contributing area term used in the steady-state topographic wetness index and thereby overcomes the limitations of the steady-state assumption that was used in the first two versions of this particular index (Barling, 1992; Barling et al., 1994) |
| Slope-adjusted topographic wetness index | Regional | This index assumes steady-state conditions and describes the spatial distribution and extent of zones of saturation for runoff generation as a function of upslope contributing area and slope over some user-specified downslope distance (Hjerdt et al., 2004) |
| Modified topographic wetness index | Regional | This index assumes steady-state conditions and describes the spatial distribution and extent of zones of saturation for runoff generation as a function of upslope contributing area, slope, soil-saturated hydraulic conductivity, and the soil erodibility factor (Yi et al., 2017) |
| Integrated moisture index | Regional | This index provides a relative measure of the soil moisture as a function of hillshade, flow accumulation, curvature, and total water-holding capacity (Iverson et al., 1997) |
| Multi-resolution valley bottom flatness index | Regional | This index uses slope and elevation percentile calculated across multiple resolutions to delineate valley bottoms based on their topographic signature as flat low-lying areas (Gallant & Dowling, 2003) |
| Stream power index | Regional | Measure of erosive power of water based on the assumption that the discharge is proportional to the specific catchment area (I.D. Moore et al., 1991) |
| Sediment transport indices | Regional | Measure of the spatial patterns of erosion and deposition derived from unit stream power theory (I.D. Moore et al., 1991) |
| Mass transport and deposition index | Regional | Measure of the sediment routed through individual grid cells as a function of the flow received from its upslope neighbors plus its own contribution (i.e. source term) minus deposition in the same grid cell (Gruber, 2007) |
| **Group 2: Energy and heat regimes** | | |
| Site exposure index | Regional | Measure of relative exposure ranging from coolest to warmest locations as a function of slope and aspect (Balice et al., 2000) |
| Heat load index | Regional | Measure of relative heat load ranging from coolest to warmest locations as a function of slope, aspect, and latitude (McCune & Keon, 1992) |
| Topographic radiation indices | Regional | These indices (in their most complete forms) estimate net radiation for specific grid cells as a function of surface albedo, surface emissivity, direct, diffuse, and reflected short-wave irradiance, and the incoming and outgoing long-wave irradiance (Moore et al., 1993d; Wilson & Gallant, 2000c) |
| Topographic temperature indices | Regional | These indices estimate temperatures for specific grid points as a function of the mean monthly minimum and maximum air temperatures, surface temperature, minimum, maximum and average temperature lapse rates, land cover, and elevation for a reference station (Moore et al., 1993d; Wilson & Gallant, 2000c) |

The typical approach therefore relies on a series of pedotransfer functions or soil parameter estimation equations (e.g. Rawls, 1983; Abdulla & Lettenmier, 1997; Seybold, Grossman, & Reinsch, 2005; Saxton & Rawls, 2006; Hollis, Hannam & Bellamy, 2012) which will introduce some additional error and/or uncertainty to the terrain modeling workflow (Band, 1993; Zhu & Mackay 2001; Quinn, Zhu & Burt, 2005). Various strategies have been adopted to improve the performance of these pedotransfer functions during the past two decades, including some that combine soil data (i.e. texture, soil organic carbon) with environmental data such as percent slope, annual average rainfall, and vegetation indices (e.g. Leij et al., 2004; Romano & Chirico, 2004; Sharma, Mohanty & Zhu, 2006; Jana & Mohanty, 2011; Akpa et al., 2016). These strategies have provided some improvement, especially when used to predict soil properties over small areas with relatively uniform soil properties and topography.

Given this state of affairs, we can assume that the DEM-based parameters will do better where the relative importance of gravity is greatest (i.e. in headwater areas and on steep slopes). This said, the 11 secondary land surface parameters focused on water flow and soil redistribution listed in Table 3.4 are described separately in the following sections, starting with the topographic wetness index in its multiple forms.

#### 3.2.1.1 Topographic Wetness Indices

The topographic wetness index (TWI) is the most popular of the secondary land surface parameters and unlike the catchment (basin) boundaries and some of the other parameters described in this and the next chapter, this land surface parameter can be calculated with both single- and multiple-flow direction algorithms. Some of the implications of choosing one of the flow-direction algorithms described in Section 3.1.3 are discussed along with the more popular variants that have been proposed for characterizing topographic wetness next.

The first TWI was proposed by Beven and Kirkby (1979) to describe the effects of topography on the location and size of saturated source areas of runoff in the semi-distributed TOPMODEL hydrologic model as follows:

$$W_T = \ln\left(\frac{A_s}{T\tan\beta}\right) \tag{3.45}$$

where $A_s$ is the specific catchment area (m$^2$ m$^{-1}$), $T$ is the soil transmissivity when the soil profile is saturated, and $\beta$ is the slope in degrees. TOPMODEL partitions the rainfall between three components – recharge in the unsaturated zone, overland flow ($Q_o$), and flow in the saturated zone ($Q_b$) – and handles these three components separately. The rainfall is directed to storage in the unsaturated zone if the soil is not saturated. However, there will be no recharge and the rainfall will enter the channel network as overland flow, with an appropriate delay function, when the soil is saturated. And finally, the flow in the saturated zone can be estimated making two important assumptions: (i) that the runoff in the saturated zone is spatially uniform; and (ii) the hydraulic gradient within the saturated zone can be approximated by the local topgraphic slope, tan β. The division and transfer of water among these three states is handled by a lumped accounting system in which the local propensity to saturation is controlled by the TWI summarized in Equation 3.45. The soil transmissivity function in the standard TOPMODEL and the aforementioned TWI is an exponential function of the storage deficit that has a shape controlled by a "soil parameter" $m$, which is constant within each flow strip and the transmissivity, $T_0$, at saturation (i.e. with zero deficit when the soil is saturated to the surface) (Lane et al., 2004).

Several variants of this original index have been proposed and used over the years. For example, the following form of the original TWI has also been used extensively to describe the effects of topography on the location and size of saturated source areas of runoff:

$$W = \ln\left(\frac{A_s}{\tan\beta}\right) \tag{3.46}$$

where $A_s$ and β are the same as in Equation 3.45 (O'Loughlin, 1986; I.D. Moore et al., 1991). This particular variant of the original TWI equation contains one less term because soil transmissivity is assumed to be constant and equal to unity throughout the catchment or area of interest. Wood, Sivapalan and Beven (1990) have shown that the variation of the topographic component is often far greater than the local variability of the soil transmissivity component and that Equation 3.46 can be used in place of Equation 3.45 in many landscapes.

The results from applying one or other of these first two TWI equations will be similar, since both will predict zones of saturation where the SCA is large (which typically occurs in converging areas of the landscape), the slope is small (which typically occurs at the base of concave slopes where slope gradient is reduced), and for Equation 3.45 the soil transmissivity is low (which often occurs in areas with impervious geology and/or shallow soils). They describe the spatial distribution of the saturation deficit or soil moisture because the lumped predictions of storage deficit generated with these two equations can be mapped back onto a distributed map of locations where the saturation deficit is locally zero or negative and overland flow is likely to occur (Western et al., 1999; Lane et al., 2004).

These first two TWIs, which are sometimes referred to as either the compound topographic index (CTI) or the topographic index (TI), were quickly adopted and used in both TOPMODEL and a variety of other kinds of environmental assessments during the early years. The popularity of TOPMODEL (Beven & Kirkby, 1979) grew quickly because it provided an opportunity to characterize the spatial distribution and extent of zones of saturation and variable source areas for runoff generation. These two steady-state topographic indices were also quickly adopted and used outside of the TOPMODEL framework to characterize saturation zones since these areas are frequently encountered along drainage flow paths and in water concentration zones (e.g. Burt & Butcher, 1985; Jones, 1986; O'Loughlin, 1986; Moore et al., 1988a). Some of the other early applications went further and broke new ground: Sivapalan et al. (1987), for example, used the TWI to characterize hydrologic similarity; Phillips (1990) used it to delineate wetlands in a coastal watershed; Moore et al. (1993a) used slope and topographic wetness to characterize the spatial variability of soil properties for a toposequence in Colorado; and Montgomery and Dietrich (1994) used the semi-distributed TOPOG (O'Loughlin, 1986) hydrologic model to predict the degree of soil saturation in response to a steady-state rainfall for topographic elements defined by the intersection of contour and stream-tube boundaries. This last measure of relative saturation was then used to analyze the stability of each topographic element for the case of cohesionless soils of spatially uniform thickness and saturated hydraulic conductivity in three California, Oregon and Wasington study areas.

However, these types of static indices must be used carefully to predict the distribution of dynamic phenomena like soil water content because surface saturation is a threshold process that is subject to the presence of hysteric effects, and the implicit assumptions that are made in the derivation of the two TWIs noted earlier. Most notably, the gradient of the piezometric head, which dictates the direction of subsurface flow, is assumed to be parallel to the land surface and steady-state conditions are assumed to apply (Moore et al., 1993e). Some authors were therefore quick to describe the potential pitfalls of using these kinds of indices in inappropriate ways: Jones (1986, 1987), for example, documented some of the advantages and limitations of using TWIs to describe spatial patterns of soil water content and drainage; Moore et al. (1993c), Wolock and Price (1994), and Zhang and Montgomery (1994) examined the effects of grid resolution on TWI and/or TOPMODEL watershed model predictions; and Wolock and McCabe (1995) examined the effects of the choice of flow routing method on the calculation of the topographic parameters in TOPMODEL. Quinn et al. (1995) noted some of these same problems as a precursor to describing how the steady-state TWI can be calculated and used effectively with either the D8 (O'Callaghan & Mark, 1984) single- or the QMFD2 (Quinn et al., 1995) multiple-flow direction algorithms as part of the TOPMODEL hydrologic modeling framework. More recently, Schröter et al. (2015) have used terrain data and sparse time-domain reflectometry (TDR) data with a fuzzy c-means clustering (FCM) approach to estimate the soil moisture pattern. The results showed how this FCM sampling and estimation approach was able to reproduce a substantial part of the soil moisture pattern under relatively wet conditions but that the transition between wet and dry states was marked by a complete reorganization of the soil moisture pattern. All of these studies point to the challenges of using static indices to describe dynamic phenomena like soil moisture.

That said, the rapid growth in the numbers of slope and flow-direction algorithms, along with the shift from DEMs with resolutions ranging from

30 to 100 m (which represented the state of the art when TOPMODEL was first proposed) to the 1–3 m LiDAR-based DEMs that cover large areas of the world today, bring new challenges as well. The likely impact of the choice of grid resolution as well as different slope and flow-direction algorithms are described next and the discussion then shifts to a series of proposals for expanding or modifying the various inputs used to calculate the TWIs themselves.

The increasing availablility of precise, fine-resolution elevation datasets may complicate comparisons over time because the studies conducted 20–30 years ago used much coarser DEMs and/or different algorithms for managing pits, flat areas and hollows, among other phenomena. The reality, however, is that many things have changed and clarifying the role of individual methods and datasets may be difficult. Many studies have tried to estimate the errors in DEMs and how these may be propagated in specific land surface parameters, including SCA, slope, and TWI (e.g. Wechsler & Kroll, 2006; Temme et al., 2009). The final decisions (interpretations) should be guided by fitness-for-use and, in this regard, the findings of Martinez et al. (2010) may be the most helpful with respect to topographic wetness, since these authors concluded that a 5-m or better DEM was required for a study site in a small agricultural catchment in the Upper Hunter Valley region of New South Wales, Australia to accurately capture catchment geomorphology and hydrology and to model the spatial distribution of soil moisture. The good news is that 1–3 m LiDAR DEMs with submeter vertical accuracy are now standard in many parts of the world (see Section 2.5 for a description of the 3DEP project currently underway in the USA for one such example). In addition, the following commentary around the network index proposed by Lane et al. (2004) offers a concise summary of what has changed and why we may need to go about some tasks using these new precise, fine-resoluton elevation datasets differently moving forward.

Numerous studies have examined the impacts of choosing one or more single- and multiple-flow direction algorithms on TWI values (e.g. Güntner et al., 2004; Pan et al., 2004; Erskine et al., 2006; Sørensen et al., 2006; Kopecký & Čížková, 2010; Lewis & Holden, 2012; Shelef & Hilley, 2013). Sørensen et al. (2006), for example, used various combinations of methods for calculating slope, SCA, and the CIT, and examined the TWI predictions using spatially distributed field observations of plant species richness, soil pH, groundwater level, and soil moisture to evaluate the different methods. They did not find a flow-direction algorithm that performed best for all measured variables; rather the best methods were site-specific and varied with the chosen variable (i.e. outcome). Kopecký and Čížková (2010) calculated 11 variations of TWI with different flow-direction algorithms for 521 georeferenced plots with vegetation data. Ten of the eleven algorithms [D8 (O'Callaghan & Mark, 1984), Rho8 (Fairfield & Leymarie, 1991), ADK (Lea, 1992), DEMON (Costa-Cabral & Burges, 1994), D∞ (Tarboton, 1997), Mass Flux (Gruber & Peckham,

2009), QMFD1 (Quinn et al., 1991), FMFD (Freeman, 1991), QMFD2 (Quinn et al., 1995), and MD∞ (Siebert & McGlynn, 2007)] were described in Section 3.1.3 and the eleventh was the seldom used Braunschweiger relief model (BR; Bauer, Rohdenburg & Bork, 1985) which allows flow to a maximum of three neighboring cells. The results showed that (i) the choice of flow-direction algorithm had a considerable effect on the performance of the TWI; (ii) the multiple-flow direction algorithms generally peformed better than the single-flow direction algorithms; and (iii) the QMFD2 and FMFD algorithms were recommended for vegetation ecology applications. Shelef and Hilley (2013), however, showed how many of the multiple-flow direction algorithms produce isolated areas with high or low values because flow can disperse and reconverge downslope depending on the local topography which may lead to spotty patterns of surface saturation and saturation overland flow. Lane et al. (2004) had earlier proposed a network index to use in conjunction with TOPMODEL to eliminate or alleviate these kinds of problems in TOPMODEL and possibly other applications, as noted next.

The simple network index modification to the standard TOPMODEL proposed by Lane et al. (2004) ensures that saturated areas will only contribute to overland flow when the TWI indicates continuous saturation throughout the length of the flow path to the point where the path becomes a stream. This new approach was combined with an enhanced method for dealing with the problems of pits and hollows which various authors (e.g. Quinn et al., 1995) have shown become more acute with higher resolution topographic data as well. There are two causes for concern here. The first is the realization that the assumption that all areas with a zero or negative saturation are connected to the channel network is not necessarily correct (Lane et al., 2004). This will lead to certain areas contributing runoff too quickly and it may also change the rate at which the catchment becomes saturated and therefore predict overland flow when water might otherwise re-infiltrate in zones of lower TWI before the channel is reached (see Lane et al., 2004, pp. 194–195 for additional details). The second complication is that there is a greater probability of error in slope as grid spacing is reduced even if there is no change in the quality of the source elevation data (Lane, Westaway & Hicks, 2003). This is particularly likely in low slope areas and will increase the difficulty of separating areas of natural depression storage from spurious or artificial depressions that result from random errors in the topography defined by the estimated elevation precision (Lane et al., 2004).

The guiding principle for this new network index approach is straightforward: a saturated area predicted with either Equation 3.45 or 3.46 can only connect to the drainage network when all cells in the model between the saturated area and the channel network are saturated as well (Lane et al., 2004). The implementation of the network index approach begins with a simple assumption, that each grid cell in the DEM has a single flow path that can be defined using the path of steepest descent. The authors next used a three-step conditioning process for handling pits (i.e. single cells that are local elevation minima), flats (i.e. adjacent cells with equal elevation) and hollows (i.e. contiguous cells that have no external drainage) in that order and then traced the subwatershed network upslope from each channel cell, recording the lowest value of the TWI for each cell (this was possible because the steepest direction flow algorithm meant that each cell has a unique downslope direction which is visited only once). This approach ensures that each grid cell has a TWI value, which can be used to identify when the saturation deficit is zero or negative and overland flow can occur, and a network index value, which defines when a cell with a zero or negative saturation deficit is connected to the drainage network. The runoff in this TOPMODEL network index-based approach only becomes overland flow if all cells along the flow path to the channel network can also generate overland flow.

Lane et al. (2004) used this new network index-based approach with the standard TOPMODEL (Beven & Kirkby, 1979) and a 2-m LiDAR DEM to show how many saturated areas were never connected with the drainage network during an extreme flood event in a 13.8 $km^2$ upland North Yorkshire, UK catchment. The lesson for the reader here focuses on the interpretation of the TWIs rather than the method of calculation because the latter is unchanged [since Lane et al. (2004) could have used either Equation 3.45 or 3.46 in this instance] and the additional value of calculating the network index accrues because it helps to predict how rainfall is likely to be parsed between soil moisture storage and overland flow in various parts of the landscape. Lane, Reaney and Heathwaite (2009) have subsequently demonstrated this by showing that the static network index can be used to generalize a significant proportion of the time-averaged spatial variability in connectivity, measured in terms of both the propensity to and the duration of connection, since the level of catchment wetness required to maintain connectivity should be greater for flow paths that have a lower value of the topographically controlled local wetness.

This network index is listed as the third entry in Table 3.4 for the sake of completeness since Lane et al. (2004) did not propose a new variant of the TWI and instead simply changed the way in which one or other of the first two steady-state TWIs listed in Table 3.4 can be deployed in TOPMODEL and possibly other applications.

Turning next to the slope term in Equations 3.45 and 3.46, a smaller number of studies have examined the impacts of calculating and using various slope terms to estimate the TWI (e.g. Sørensen et al., 2006; Qin et al., 2011; Buchanan et al., 2014). Buchanan et al. (2014) calculated TWIs using more than 400 unique approaches that considered different DEM resolutions, the vertical precision of the DEM, different flow-direction and slope algorithms, smoothing versus low-pass filtering and the inclusion of relevant soil properties, and compared the resulting topographic wetness maps with observed patterns of soil moisture across numeorus agricultural fields in the state of New York. This particular study considered the following options.

1. Two forms of the TWI: the first form followed Equation 3.46 and the second followed Equation 3.45 but for the reliance on a different method to estimate soil transmissivity. This particular study computed the soil transmissivity ($m^2\,d^{-1}$) as the product of the average saturated hydraulic conductivity ($m\,d^{-1}$) and the depth to the restrictive layer (m) (Lyon et al., 2004; Schneiderman et al., 2007). Soil properties were derived from the SSURGO database using the Soil Data Viewer application (US Department of Agriculture, Natural Resources Conservation Service, 2009).
2. Three DEM resolutions: the NED 10-m DEM and 3- and 10-m DEMs derived from LiDAR points clouds.
3. Six slope distributions: the (a) maximum triangle slope (MTS; Tarboton, 1997); (b) least squares fitted plane (LSFP; Horn, 1981); (c) second-degree polynomial (SDP; Zevenbergen & Thorne, 1987); and (d) downslope index (DSI: Hjerdt et al., 2004) with $d = 2$ m; (e) DSI with $d = 5$ m; and (f) DSI with $d = 10$ m.
4. Six single- and multiple-flow direction algorithms: (a) D8 (O'Callaghan & Mark, 1984); (b) Rho8 (Fairfield & Leymarie, 1991); (c) FMFD (Freeman, 1991); (d) D∞ (Tarboton, 1997); (e) BR (Bauer et al., 1985); and (f) MD∞ (Siebert & McGlynn, 2007).
5. Filtered versus unfiltered maps of TWI recognizing that high-resolution DEMs tend to result in high local variations in TWI values, which may translate to unrealistically irregular predictions of soil mositure and water table depths (Hjerdt et al., 2004; Lanni, McDonnell & Rigon, 2011). A 3 × 3 pixel, low-pass mean filter was therefore applied to the TWI values in this particular study.

The results showed how the TWIs were moderately well correlated with observed soil moisture patterns (in the best case the relationship between TWI and soil moisture had average $R^2$ and Spearman correlation coefficients of 0.61 and 0.78, respectively) and in all cases the fine-scale 3-m LiDAR-derived DEMs worked better than the 10-m USGS and LiDAR-based DEMs and, in general, the approaches incorporating the local slopes and soil properties improved the correlations.

Turning next to more substantial modifications of Equations 3.45 and 3.46, three sets of authors have proposed new variables for calculating TWI values that rely on improved representations for variable contributing areas and downslope effects.

In the first of these approaches, Barling (1992) proposed the following quasi-dynamic TWI to overcome the limitations of the steady-state assumption:

$$W_{QD} = \ln\left(\frac{A_e}{\tan\beta}\right) \tag{3.47}$$

where $A_e$ is the effective SCA which is estimated using the steady-state SCA incorporated in Equations 3.45 and 3.46 plus three additional variables: the saturated hydraulic conductivity, effective porosity, and the date since the last precipitation event (Table 3.4). Barling et al. (1994) then calculated steady-state and quasi-dynamic indices for a semi-arid catchment near Wagga Wagga in New South Wales, Australia and showed how the quasi-dynamic TWI correctly predicted how the topographic hollows and not the drainage channels themselves determined the hydrologic response of the catchment. This particular variant has not received much attention but for Borga, Dalla Fontana and Cazorzi (2002), who used a quasi-dynamic TWI to examine topographic and climatic controls on rainfall-triggered shallow landslides and Nguyen and Wilson (2010), who calculated the quasi-dynamic TWI with several flow-direction algorithms (D8 was used in the original work) and showed how the results varied depending on the flow-direction algorithm that was adopted.

In the second example, Hjerdt et al. (2004) proposed a slope-adjusted TWI to better represent the downslope control imposed by local groundwater gradients:

$$W_s = \ln\left(\frac{A_s}{\tan\beta_d}\right) \tag{3.48}$$

where $W_s$ is the slope-adjusted TWI, $A_s$ is the SCA, and $\tan\beta_d$ is the downslope gradient up to the distance $d$ (Table 3.4). This particular index permits the slope of the land surface and the groundwater table to vary (Figure 3.30) to

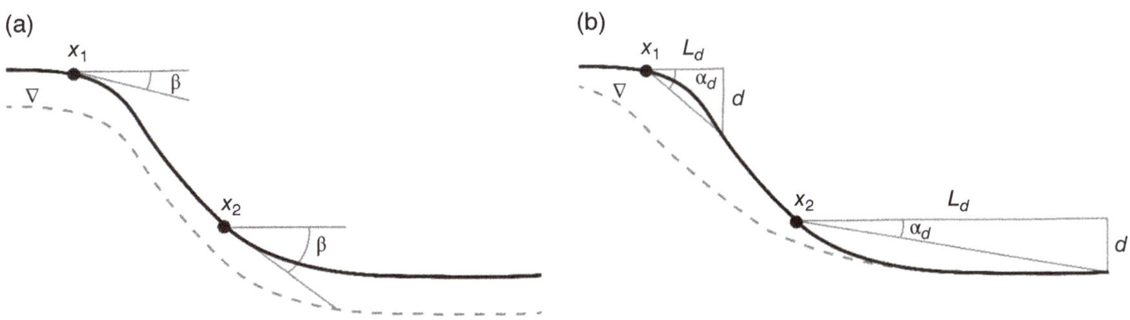


**Figure 3.30** (a) The local gradient in the original topographic wetness index and (b) with the new slope term proposed by Hjerdt et al. (2004). The dotted lines represent the gradient of the groundwater table that is constant in the original topographic wetness index (a) and variable in the slope-adjusted topographic wetness index (b). *Source*: modified from Hjerdt et al. (2004, p. 2). Reproduced with permission of John Wiley & Sons, Inc.

handle the case where the local form of tan β cannot adequately represent the local drainage impedance, which controls the hydraulic gradient when the slope diminishes in a downslope direction. Hjerdt et al. (2004) modified the original TWI to include a distance which represents the drainage power degradation due to the decline in the amount of potential energy. The distance to this user-specified point can be calculated as either a straightline distance or the distance along the flow path following the path of steepest descent. This new slope estimate will give similar results to the local slope for small values of $d$, but the results will likely differ for larger values of $d$ (i.e. long downslope distances). Sørensen et al. (2006) started with a square grid measuring 20 m on a side and conducted a series of experiments to evaluate different methods for calculating the TWI that, among other things, compared the performance of various slope estimates using the local slope and values of 2, 5, 10, 15, and 20 m calculated with both distance definitions for $\tan \beta_d$ in Equation 3.48.

For the third of these approaches and the sixth variant of the TWI listed in Table 3.4, Yi, Zhang and Yan (2017) incorporated the soil saturated hydraulic conductivity ($K_s$) and the soil erodibility factor ($K$) into the original TWI to construct a modified TWI ($W_3$):

$$W_3 = \ln\left(\frac{A_s}{\tan\beta \cdot K_s \cdot K}\right) \tag{3.49}$$

The performance of the TOPMODEL (Beven & Kirkby, 1979) and TOPX (Yong et al., 2009) topography-based models with the modified index as input was evaluated in three catchments with varying climate conditions and the results showed how the performance of both of these models was improved when the modified TWI was used in place of the original index and that the performance improvements increased along with the spatial heterogeneity of $K_s$ and $K$.

However, two recent studies show how difficult it will be to find variants of the existing TWIs that would substantially improve one or other of the first two steady-state TWIs listed in Table 3.4. The first study by Grabs et al. (2009) calculated three model-based TWIs based on simulations of distributed catchment models and two estimates of the static steady-state TWI summarized in Equation 3.45. The model-based indices used two spatially distributed versions of the HBV model (Bergeström, 1976, 1992, 1995) using a square grid measuring 50 m on a side and the static indices were calculated using the QMFD1 (Quinn et al., 1991) and MD∞ (Seibert & McGlynn, 2007) multiple-flow direction algorithms. The model-based approaches required additional field data but avoided the assumption that the groundwater gradient mimics the surface gradient, which is incorporated in the steady-state TWI. The modified distance-based slope approach of Hjerdt et al. (2004) that was described earlier sought to relax this same assumption but with less need for additional input data. The results from Grabs et al. (2009) showed that the model-derived wetness indices predicted the spatial distribution of wetlands significantly better than the steady-state TWIs and the authors advocated using this approach to derive TWI maps in catchments with topographically driven groundwater regimes.

The second of these studies by Wilson, Western and Grayson (2005) used ground-based measurements of soil moisture in the root zone across six agricultural fields in Australia and New Zealand to examine the predictive power of seven land surface parameters – elevation, SCA, slope, TWI, potential solar radiation, lowness, and the multi-resolution valley bottom flatness index (MRVBF; described in more detail in Section 3.2.1.3) – with respect to soil moisture. Three cases were specified for the simulations as follows: (1) site-specific terrain information; (2) site-specific terrain information plus additional site-specific pattern information (i.e. soil moisture measurements); and (3) site-specific terrain information, pattern information, and the preserving statistical characteristics (i.e. the error term which accounts for the measurement error and the unresolved variability in the observed maps, to provide patterns that preserved the overall variance). The results showed that the maps created using the three best performing land surface parameters in Case 1 were spatially unrealistic due to the weak correlation between topography and root zone moisture in these catchments. The TWI appeared in models for four fields and was positively correlated with soil moisture in all four of these fields. The maps created using Case 2 were significantly better because they accounted for the actions of temporally persistent non-topographic controls. The residuals constituted the most significant attribute in all six fields, and TWI was included in models used for two of the six fields. The maps for Case 3 were the closest to reality because the inclusion of an error term simulated the observed small-scale fluctuations realistically. The authors concluded that this method was an advance on the foregoing approaches because it explicitly accounted for the changes in controls on soil moisture patterns with changes in mean soil moisture content, and that this approach could be generalized once a theory or generalized empirical basis is provided for forecasting the spatial distribution of soil moisture in different environments. It will take some time to accomplish this outcome, so the use of the variants of the TWI listed in Table 3.4 will likely continue in the immediate future.

The realization that the first two variants have continued to find widespread use, notwithstanding the previously noted shortcomings and various efforts to revise or propose new TWIs, supports the aforementioned prognosis. The TWI has been used for identifying the locations of gully initiation (e.g. Momm et al., 2013) and urban flood risk (e.g. Pourali et al., 2016), mapping soil organic matter content (e.g. Pei et al., 2010), species distribution modeling (e.g. Van Niel et al., 2004; Evans & Cushman, 2009), prediction of vegetation types (e.g. Taverna, Urban & MacDonald, 2005; Dobrowski et al., 2008), modeling of alpine treeline (e.g. Bader & Ruijten, 2008), and assessment of factors controlling species performance (e.g. MacDonald & Urban, 2004; Bunn, Waggoner & Graumlich, 2005), richness (e.g. Moody &

Meentemeyer, 2001; Zinko et al., 2005) and composition (e.g. Svenning et al., 2004; Kopecký & Vojta, 2009) during the past decade.

In addition, this index has featured prominently in three noteworthy works featured in other parts of this book: Leempoel et al. (2013) used TWIs as one of several variables to help delineate habitat in the western Swiss Alps (see Section 1.2 for additional details), Zhang et al. (2016) used TOPMODEL and the TWI to simulate global wetlands (see Section 5.2.5 for additional details), and Marthews et al. (2015) seized the opportunity provided by the release of several new global elevation datasets to produce a new high-resolution, spatially consistent data layer for steady-state TWI for all ice-free pixels worldwide. Finally, the TWI concept has been integrated into numerous hydrologic models, for example TOPMODEL (Beven & Kirkby, 1979), VSLF (Schneiderman et al., 2007) and SWAT-VSA (Easton et al., 2008), and in pollution risk indices (e.g. Agnew et al., 2006; Marjerison et al., 2011; Reaney et al., 2011; Buchanan et al., 2013).

Figure 3.31 shows the steady-state TWI calculated for the Cottonwood Creek catchment using Equation 3.46. The values vary from 2.85 to 16.35 and the mean computed for the catchment (6.20) shows that the distribution was positively skewed (as is the case for most catchments). Wilson and Gallant (2000c) calculated both the steady-state and quasi-dynamic TWIs for this catchment. The two maps showed varying patterns because the user-specified soil depth, drainable porosity, saturated hydraulic conductivity, and drainage time values of 1.3 m, 0.4 $cm^3 cm^{-3}$, 200 $mm h^{-1}$, and 20 days, respectively, used with the quasi-dynamic TWI meant that this index predicted the largest wetness values in topographic hollows at higher elevations (i.e. in local areas with convergent flows) and immediately above gently sloping areas near channels (i.e. on footslopes). The steady-state TWI values and patterns displayed in Figure 3.31 more or less match those predicted by Wilson and Gallant (2000c, Figure 4.10, p. 130) notwithstanding the use of a different DEM source and spacing.

#### 3.2.1.2 Integrated Moisture Index

Iverson et al. (1996, 1997) proposed an integrated moisture index (IMI) to capture the topographic and soil features of the landscape in a single index that can be statistically related to a variety of ecological processes operating in a landscape. This index uses four inputs to provide a relative rating of moisture as follows:

$$\mathrm{IMI} = \left(HS \cdot 0.40\right) + \left(FA \cdot 0.30\right) + \left(C \cdot 0.10\right) + \left(TWHC \cdot 0.20\right) \quad (3.50)$$

where *HS* refers to the hillshade calculated using the hillshade tool in ArcGIS operating on a DEM and with increasing scores representing increasing moisture content, *FA* refers to the flow accumulation calculated using the flow direction and accumulation tools in ArcGIS and with increasing scores contributing to increasing IMI, *C* refers to the curvature calculated with the

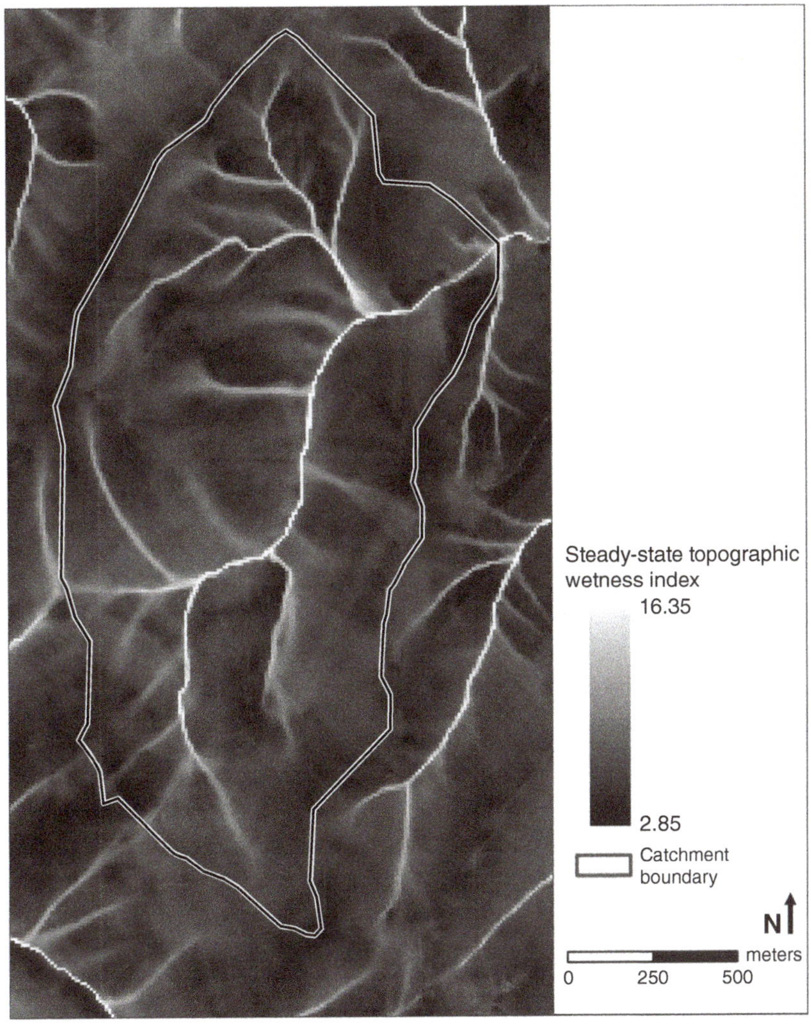


**Figure 3.31** Steady-state topographic wetness index derived for the Cottonwood Creek, MT study site using Equation 3.46, with the catchment boundary overlaid.

curvature tools in ArcGIS and with increasing positive scores representing increasing concavity (i.e. depressions) that will accumulate moisture and thus contribute positively to IMI, and *TWHC* refers to the total water-holding capacity which was derived from the SSURGO soils database by multiplying the sum of the depths of the A and B horizons and the available water-holding capacity per unit depth to estimate the total amount of water available to plants in the A and B horizons. The values for each of these four variables were standardized to scores ranging from 0 to 100 and then multipled using the weights shown in Equation 3.50 to produce IMI estimates that ranged from 0 to 100, with increasing values indicating higher relative moisture for those sites.

The choice of input variables and weights for the IMI summarized in Equation 3.50 was guided by pragmatic considerations and the forest landscapes of Ohio. The four variables included in Equation 3.50 were chosen because they could be calculated using existing data and tools and because they were related to site productivity (i.e. the site index, which refers to the average height of individual dominant trees in a stand of some specified age) and species composition in Ohio, as shown by the statistical analysis and results presented by Iverson et al. (1997). Similarly, the weights were chosen based on local experience, which indicated that the moisture availability on forest sites in Ohio is controlled by the intensity and duration of radiation (hillshade), micro- and macro-topography related to water infiltration and runoff (flow accumulation and curvature) and the capacity of soil horizons to store and release moisture over time (total water-holding capacity) (Iverson et al., 1997).

That said, Iverson et al. (1997) suggested the need to calibrate this approach to the specific location and application at hand (for example, the forest sites in Ohio show a maximum relief change of ~100 m and therefore there was no need to incorprorate elevation effects per se) and with this one caveat, concluded that the index would perform well in those applications where moisture is a key driving variable with respect to the ecological pattern or process of interest. More recently, Iverson, Prasad and Rebbeck (2004) and Davies et al. (2010) used a three-variable IMI that dropped the total water-holding capacity term in Equation 3.50 and weights of 0.50, 0.15, and 0.35 for the hillshade, curvature and flow accumulation variables, respectively. Iverson et al. (2004) have also compared both the IMI and the steady-state TWI summarized in Equation 3.46 to 2 years of soil moisture monitoring in Zaleski State Forest, Ohio. They found low correlations, notwithstanding the general (expected) trend by which soil moisture and the two land surface parameters increased in tandem, which they attributed to problems with the design and execution of the soil moisture field monitoring campaign. Davies et al. (2010) used the IMI with several other land surface parameters to evaluate the relationships between potential juniper cover and environmental variables (as explained in Section 3.2.2.2).

#### 3.2.1.3 Multi-resolution Valley Bottom Flatness Index

We turn our attention next to the MRVBF that was used by Wilson et al. (2005) as one of seven land surface parameters for predicting soil moisture (see Section 3.2.1.1 for additional details). The MRVBF was proposed by Gallant and Dowling (2003) to classify varying degress of valley bottom flatness, which may help to predict the depth of sediment deposits, the presence of groundwater constrictions, and the delineation of hydrologic and geomorphologic units.

The MRVBF algorithm identifies valley bottoms based on three guiding assumptions: (i) that valley bottoms are low to flat relative to their surroundings; (ii) valley bottoms occur at a range of scales; and (iii) large valley bottoms are flatter than small ones. The algorithm uses a slope classification constrained to convergent areas which is applied at multiple scales by progressive generalization of the DEM coupled with progressive reduction of the slope class thresholds (the DEM cell size increases by a factor of three starting with the second iteration and the latter is reduced by a factor of two at each iteration). Two land surface parameters – slope, calculated with the finite difference techniques in ArcGIS, and elevation percentile, computed with a stand-alone program (PCTL) that was part of the ElevResidGrid program suite described by Gallant and Wilson (2000) – are used with a series of iterative equations to measure the flatness and lowness characteristics of valley bottoms across a range of scales. Flatness is measured by the inverse of slope and lowness is measured by a ranking of elevation with respect to a circular surrounding area. The goal is to delineate the largest valley bottom area that is sufficiently low and flat at a specific scale and that is sufficiently flat (but not necessarily low) at all finer scales. Gallant and Dowling (2003) also proposed a multi-resolution ridgetop flatness index (MRRTF) that is derived in a similar way using the upper parts of the landscape that are identified with the same PCTL program used for the MRVBF index.

Gallant and Dowling (2003) used the results from two Australian sites – the Illalong Creek catchment located about 100 km northwest of Canberra, ACT and the Kyeamba Creek catchment located about 50 km southwest of Wagga Wagga, NSW – to demonstrate the efficacy of this pair of indices, notwithstanding the inclusion of numerous arbitrary choices in the approach itself. The authors thought that the various thresholds, resolution steps, and the other arbitrary parameters used in their method could be adjusted to match the characteristics of different landscapes but that ultimately a single approach was needed that could be applied across multiple landscapes to differentiate uplands (hilltops, ridges, and slopes) and lowlands (convergent areas and valley bottoms) as well as degrees of ridgetop and valley bottom flatness. Variants of the original method have been implemented in the GRASS and SAGA GIS platforms for calculating both the MRVBF and MRRTF indices thus far.

#### 3.2.1.4 Stream-power Index

Several terrain-based stream power and sediment transport indices have also been proposed over the years (Table 3.4), as outlined in this and the two subsections which follow this one.

Stream power is the time rate of energy expenditure and has been used extensively in studies of soil erosion, sediment transport and geomorphology as a measure of the erosive power of flowing water (I.D. Moore et al., 1991). It can be computed as:

$$d = \rho g q \tan\beta \qquad (3.51)$$

where $\rho g$ is the unit weight of water, $q$ is the discharge per unit width, and β is the slope gradient (in degrees). The compound topographic index, $A_s(\tan \beta)$, is therefore a measure of stream power, since $\rho g$ is essentially constant and $q$ is often assumed to be proportional to $A_s$.

Numerous researchers have used this index or variants of it to predict the locations of ephemeral gullies. Thorne et al. (1986), for example, multiplied the SPI by plan curvature and predicted both the locations of ephemeral gullies and the cross-sectional areas of the gullies after one year of development. Moore et al. (1988a) showed that ephemeral gullies formed where $W > 6.8$ and $A_s(\tan \beta) > 18$ for a small semi-arid catchment in Australia; however, Srivastava and Moore (1989) found that ephemeral gullies formed where $W > 8.3$ and $A_s(\tan \beta) > 18$ for a small catchment on the island of Antigua. These and similar results led I.D. Moore et al. (1991) to conclude that the threshold values for these indices are likely to vary from place to place because of variations in geology and soils. Moore and Nieber (1989) used the SPI to identify places where soil conservation measures which reduce the erosive effects of concentrated flow, such as grassed waterways, should be installed. Montgomery and Dietrich (1989, 1992) and Montgomery and Foufoula-Georgiou (1993) used a variation of the SPI, $A_s(\tan \beta)^2$, to predict the headwaters of first-order streams (i.e. locations of channel initiation).

This index is still used extensively. A recent project funded by the Minnesota Board of Soil and Water Resources and the University of Minnesota, for example, has used the SPI to measure the erosive power of overland flow (Natural Resources Research Institute, University of Minnesota Duluth, 2016). The index was calculated for the whole state and the results summarized by terrain zones, which represent regions of Minnesota with similar physiographic characteristics. Large SPI values from each of the five terrain zones were then used to create a critical area layer where erosion is likely to occur. In a similar application, the Wisconsin Department of Natural Resources (2016) has produced a GIS-based Erosion Vulnerability Assessment for Agricultural Land (EVAAL) tool to locate where best management practice assessments should be prioritized. The results are provided as a series of maps that show the erosion vulnerability index and the three components (soil loss, SPI, and internally drained areas) at their base resolution or aggregated to the level of an agricultural field or some other boundary dataset. In another project and completely different landscape setting, Jacoby, Peterson and Dogwiler (2011) used the SPI to estimate the stream erosion potential of cave levels in Carter Cave State Resort Park, a fluviokarst system in northeastern Kentucky containing multiple cave levels.

#### 3.2.1.5 Sediment Transport Indices

A second compound index was derived by Moore and Burch (1986a–c) from unit stream-power theory and a variant used in place of the original LS factor specified in the RUSLE (Renard et al., 1991) for slope lengths <100 m and slopes <14° as follows:

$$LS = (m+1)\left(\frac{A_s}{22.13}\right)^m (\sin\beta/0.0896)^2 \tag{3.52}$$

where $m = 0.4$ and $n = 1.3$ (Moore & Wilson 1992, 1994). Both this and the next equation are non-linear functions of slope and specific catchment area.

This first index calculates a spatially distributed sediment transport capacity and may be better suited to landscape assessments than the original empirical equation because it is easy to use (Wilson, 1986) and it explicitly accounts for flow convergence and divergence (Moore & Wilson, 1992).

The terrain-based sediment transport capacity index captures the change in the sediment transport capacity index along flow paths and can be used to differentiate net erosion and net deposition areas:

$$\Delta T_{cj} = \phi\, A_{sj-}^{m} \left(\sin\beta_{j-}\right)^{n} - A_{sj-}^{m} \left(\sin\beta_{j}\right)^{n} \tag{3.53}$$

where $\phi$ is a constant, subscript $j$ signifies the outlet of cell $j$, and subscript $j-$ signifies the inlet to cell $j$ (Moore & Wilson 1992, 1994). This index will predict erosion in areas experiencing an increase in sediment transport capacity and deposition in areas experiencing a decrease in sediment transport capacity. Mitášová et al. (1996) implemented variants of these two equations in the GRASS GIS (Neteler & Mitášová, 2008) and showed how net erosion areas coincided with areas of profile convexity and tangential concavity (flow acceleration and convergence) and net deposition areas coincided with areas of profile concavity (decreasing flow velocity).

The use of Equation 3.52 in place of the original slope gradient and length terms in RUSLE and its predecessors was criticized at first (see Foster & Wischmeier, 1994; Moore & Wilson, 1994; Wilson & Lorang, 1999 for additional details). However, the ease of implementing this new approach in a GIS has also seen several variants proposed during the past two decades. Desmet and Govers (1996a), for example, proposed a smiliar equation that used the QMFD1 (Quinn et al., 1991) multiple-flow direction algorithm to calculate upslope contributing area and a method for delineating slope units and, as a consequence, slope lengths in catchments with varying land cover. This last innovation was implemented to handle a series of special situations, such as forested upland areas which are unlikely to contribute to runoff and therefore can be ignored when calculating the slope length for RUSLE. On the other hand, Hickey (2000) and Van Remortel et al. (2001) attempted to delineate areas of deposition and their impact on slope length calculations, and all three of the aforementioned works as well as others (e.g. Yitayew, Pokrzywka & Renard, 1999) have used a series of comparisons to compare the LS values calculated with these approaches and the values calculated with the original manual approach proposed by Foster and Wischmeier (1974) and the RUSLE equations proposed by McCool et al. (1987, 1989) and McCool, Foster and Weesies (1997).

More recently, Winchell et al. (2008) proposed a new three-step GIS-based approach for calculating high-resolution, spatially distributed LS factor values for large watersheds. The first step uses the RUSLE method for estimating the slope factor:

$$S = 10.8 \cdot \sin\beta + 0.03 \quad \left(\text{where slope} < 9\%\right) \tag{3.54}$$

$$S = 16.8 \cdot \sin\beta - 0.50 \quad \left(\text{where slope} \geq 9\%\right) \tag{3.55}$$

where β is the slope angle in radians. The second step uses several custom GIS-based methods and datasets, including road and urban land-use datasets, to constrain slope lengths (relative to the original approaches proposed by Foster & Wischmeier, 1974 and McCool et al., 1989) and the third step uses the D∞ (Tarboton, 1997) single-flow direction algorithm which permits flow to one or two downslope cells to calculate upslope contributing areas. However, this new method did not include an algorithm like the one summarized in Equation 3.53 for predicting topographic zones of soil deposition.

Winchell et al. (2008) used the new method to calculate the LS factor for agricultural subwatersheds covering two-thirds of the Mississippi River basin. The results showed that this new GIS-based approach produced statistical distributions of the LS factor that were very similar to those described by the National Resources Inventory database (US Department of Agriculture, Natural Resources Conservation Service, 2000) at the eight-digit watershed level. Their GIS-based method built on the earlier work of Moore and Wilson (1992, 1994) and Desmet and Govers (1996a) and provides strong support for using GIS-based methods to represent the spatial heterogeneity and magnitude of the LS factor over large areas.

The two sediment transport indices summarized in Equations 3.52 and 3.53 have proved helpful in characterizing the environmental conditions for other applications as well. Iqbal et al. (2005), for example, set out to quantify the relationships between cotton lint yields and several measured soil properties and derived topographic attributes in a 42-ha field and showed that elevation, flow direction, sediment transport index (STI), percentage sand content, and volumetric water content ($\theta_v$) at −0.001 MPa pressure explained most of the lint yield variation. And in another study of this kind, Norouzi et al. (2010) set out to predict rain-fed wheat quality and quantity using artificial neural networks and soil and terrain characteristics. The overall results showed that: (i) the artificial neural network models could explain 89–95% of the total variability in wheat biomass, grain yield, and grain protein content; and (ii) the STI was identified as the most important topographic attribute influencing the wheat biomass, while grain yield and grain protein were affected the most by the total nitrogen.

#### 3.2.1.6 Mass Transport and Deposition Index

The development of a secondary land surface parameter to model erosion and deposition was taken a step further in the mass transport and deposition (MTD) index, which adds a deposition function to create a self-depleting flow that conserves mass between input and deposition (Gruber, 2007). The key idea embedded in this approach is the pre-definition of the maximum deposition for each cell based on slope, curvature, land cover, and possibly other characteristics (Gruber & Peckham, 2009). The flow through each cell can be defined using any of the single- and multiple-flow direction methods described in Sections 3.1.3.1 and 3.1.3.2 and the local deposition is substracted as follows:

$$A_i = \sum_{j=0}^{8} \left( A_{NBj} \cdot r_{NBj} \right) + I_i - D_i \tag{3.56}$$

where $A_i$ is the amount of mass (or volume, area or any other property) $A$ that is accumulated in cell $i$ as given by the sum of $A$ in each neighboring cell ($NB$) multipled by the respective receiving fraction $r$ plus the mass (or other quantity) input $I$ in cell $i$ itself, and $D_i$ is the deposition in cell $i$ (Gruber & Peckham, 2009).

The approach summarized in Equation 3.56 means that the flow passed through each cell $A_i$ is equal to the sum of the flow received from its neighbors plus its own source term $I_i$, minus the deposition $D_i$ in this cell. The deposition is limited by the amount of mass available (which is represented by the first two terms in Equation 3.56) and the maximum deposition $D_{max}$, which was described by Gruber et al. (2009) as a function of slope using one or other forms of the following equation:

$$D_{max} = \begin{cases} \left(1 - \dfrac{\beta}{\beta_{lim}}\right)^{\gamma} \cdot D_{lim} & \text{if } \beta < \beta_{lim} \\ 0 & \text{if } \beta \geq \beta_{lim} \end{cases} \tag{3.57}$$

where $D_{lim}$ is the limiting deposition (i.e. the maximum deposition that would occur on flat terrain), the limiting slope $\beta_{lim}$ denotes the maximum steepness at which mass is deposited, and the exponent $\gamma$ is used to control the relative importance of steep and gentle slopes. Gruber and Peckham (2009, p. 192) included a diagram showing how the maximum deposition varies as a function of slope. The various tools and functions required to implement the concepts and ideas summarized in Equations 3.56 and 3.57 are also available as IDL source codes (Gruber et al., 2009).

This approach has proved popular. It has been used to model the redistribution of eroded soil (Mitášová et al., 1996), the redistribution of snow by avalanches (Machguth et al., 2006), and debris flows and landslides in steep terrain (Gruber et al., 2009) for example, and similar concepts have been adopted and used to delineate lahar inundation zones (Iverson, Schilling & Vallance, 2003) and soil erosion and landscape evolution in the LAPSUS multi-module dynamic landscape evolution model (Schoorl, Veldkamp & Bouma, 2002; Claessens et al., 2006).

### 3.2.2 *Energy and Thermal Regimes*

This next group of secondary land surface parameters can be used to estimate the spatial and temporal distribution of the energy and thermal regimes at the Earth's surface. Variations in elevation, slope, aspect, and local topographic horizon (i.e. shadowing) can create very strong local gradients in solar radiation (Gates, 1980; Linacre, 1992), and thereby exert a large influence on the photosynthesis and evapotranspiration processes and ensuing vegetation diversity and biomass production at specific locations on the land surface; see Austin et al. (1984), Tajchman and Lacey (1986), Moore et al. (1993d) and Franklin (1995) for early examples documenting these kinds of relationships.

The interactions between the land surface and atmosphere typically occur over some combination of the meso- and topo-scales depicted in Figure 1.1. For example, the elevation and occasionally the aspect have been used to map a variety of precipitation and temperature variables measured at climate stations over large areas such as Australia and the USA (i.e. the meso-scale). Hutchinson (1995, 2008), Thornton et al. (1997, 2000), Thornton and Running (1999), Jarvis and Stuart (2001), Daly et al. (2002) and Lloyd (2005), among others, have proposed terrain-based approaches that can generate satisfactory results over large areas given a regular distribution of input data and proper representation of the topo-climatic settings present in each area of interest. The focus here is on the topo-scale (i.e. small to medium-sized catchments) and the four subsections which follow review the land surface parameters that have been proposed and used to assess the variability of the energy and thermal regimes across these land surfaces. The discussion starts with two relatively simple and straightforward indices and then moves to two groups of radiation and temperature indices which require more inputs and computational resources.

#### 3.2.2.1 Site Exposure Index

In the first of this group of four secondary land surface parameters, Balice et al. (2000) proposed a simple site exposure index (SEI) which can be calculated quickly and easily using slope and aspect. The SEI rescales aspect to a north–south axis and weights it by the steepness of the slope as follows:

$$\mathrm{SEI} = \beta \times \cos\left(\pi \cdot \frac{\varphi - 180}{\varphi}\right) \tag{3.58}$$

where $\beta$ is the slope and $\varphi$ is the aspect (Balice et al., 2000). The SEI creates a dataset of relative exposure ranging from −100 (coolest) to +100 (warmest) locations (in the northern hemisphere).

#### 3.2.2.2 Heat Load Index

In the second land surface parameter in this group, McCune and Keon (2002) proposed the heat load index (HLI) which can be calculated quickly and easily from slope, aspect, and latitude as follows:

$$\mathrm{HLI} = 0.039 + (0.808 \cdot \cos l \cdot \cos s) - (0.196 \cdot \cos l \cdot \sin s) - (0.482 \cdot \cos a \cdot \sin s) \tag{3.59}$$

where $l$ is the latitude, $s$ is the slope (in radians), and $a$ is the folded aspect calculated with Equation 3.60 as follows:

$$a = \left| \pi - \left| \text{aspect} - \frac{5\pi}{4} \right| \right| \tag{3.60}$$

The HLI values range from 0 to 1, with 0 being the coolest and 1 being the warmest (Davies et al., 2010).

Two recent studies by Davies et al. (2007, 2010) show how HLI, IMI, and SEI can be used to evaluate the relationships between potential land cover and one or more environmental variables. The latter study, for example, demonstrated how western juniper cover (*Juniperus occidentalis* subsp. *occidentalis* Hook.) could be accurately estimated from National Agriculture Imagery Program (NAIP) images and then used slope, northness (i.e. the cosine of aspect), and elevation along with HLI, IMI, and SEI to evaluate the relationships between potential cover and selected environmental variables. The three secondary land surface parameters were either not correlated (HLI, IMI) or weakly correlated (SEI) with juniper cover at stand closure unlike the first three primary land surface parameters that explained 40% of the variability in juniper cover at stand closure. However, in the first of the two studies, Davies et al. (2007) reported that HLI explained some of the variation in vegetation cover in several plant functional groups in sagebrush plant communities.

The variability of the results from these two studies may point to the failure of one or more of the three aforementioned secondary land surface parameters to represent the key environmental processes operating in these landscapes and/or variations in the plant communities themselves. The final two subsections discuss two sets of secondary land surface parameters that represent radiation fluxes and temperatures, and which can be expected to capture the key processes operating across multiple landscapes.

#### 3.2.2.3 Topographic Radiation Indices

The best methods for calculating the terrain-based radiation and temperature land surface parameters described in this and the following section will account for the three major causes of spatial variability of radiation at the land surface: (i) the orientation of the Earth relative to the sun; (ii) the presence of clouds and other atmospheric effects; and (iii) the topographic effects. In addition, it is important to note at the outset that the various software programs which have been proposed for calculating the radiation fluxes and the accompanying land surface parameters over the years vary tremendously in terms of both precision and completeness. Some provide partial solutions – the solar radiation tools in the ArcGIS and SAGA GIS platforms, for example, calculate the potential incoming solar radiation (without estimating the effect of overcast skies and clouds) in contrast to the solar radiation tools included in SRAD (Moore et al., 1993d; Wilson & Gallant, 2000c) and the r.sun model, which has been implemented in the

GRASS GIS, that can calculate estimates for all the aforementioned fluxes using grid spacings and time steps specified by the user. The ArcGIS solar radiation tools are based on work by Fu and Rich (2002) and the SAGA solar radiation tools are modeled after a subset of the SRAD solar radiation tools. The current r.sun solar radiation model (Šúri & Hofierka, 2004) has found widespread use in Europe and represents a substantial improvement over the original model (Hofierka, 1997), whose application was limited to small areas and clear-sky solar radiation as well. The SRAD model, described in more detail later, was recently used to build a series of mean monthly radiation surfaces for Australia at a 1-arcsecond resolution (Austin et al., 2013). In addition, the SRAD model used by Austin et al. (2013) was implemented in IDL (the orginal model was written in FORTRAN) and was modified to accept fractional vegetation cover, cloud fraction and albedo as surfaces for each month rather than single value parameters. It also used morning and afternoon sunshine fractions in place of the single value used for the whole day in the original version and the code was streamlined so that the slope, aspect and horizon angle calculations, which depend only on the DEM, were calculated just once for each model run.

That said, the best methods will use a two-step approach to (i) calculate the potential solar radiation as a function of latitude, slope, aspect, topographic shading and time of year; and (ii) modify these estimates using information about average monthly cloudiness and sunshine hours. This approach was used for SRAD and the following account is loosely modeled after the workflow used for the SRAD solar radiation modeling software package.

The basic approach adopted and used by SRAD can be summarized as follows. The short-wave irradiance is computed at both flat and sloping sites using a three-step method. The potential or extraterrestrial irradiance on a horizontal surface just outside the Earth's atmosphere is calculated first. Next, a series of instantaneous clear-sky, short-wave radiation fluxes are calculated for each of the DEM grid points at 12-minute intervals from sunrise to sunset, and direct beam and diffuse fluxes are calculated for flat sites and direct beam, circumsolar diffuse, isotropic diffuse and reflected fluxes are calculated for sloping sites. These instantaneous values are then summed to obtain daily totals and these values are adjusted to account for the effects of cloudiness. Daily temperature is extrapolated across the surface using a method that corrects for elevation via a lapse rate, slope-aspect effects via a short-wave radiation ratio, and vegetation effects via a leaf area index (see Hungerford et al., 1989; Running, 1991; Running & Thornton, 1996; Running, Nemani & Hungerford, 1987 for additional details) and the daily outgoing and incoming long-wave irradiances are calculated from the surface temperatures in the first instance and the air temperatures and fraction of sky visible at each grid point in the second case. The aforementioned short- and long-wave radiation fluxes are then used to estimate the surface energy budget at each grid point for a user-specified period ranging from one day to a year in length.

The description of the estimation methods which follows starts with the general approach used to estimate short-wave radiation on both flat and inclined sites which incorporates four steps.

The first task is to calculate the potential or extraterrestrial irradiance on a horizontal surface just outside the Earth's atmosphere. This step combines a series of fundamental angles, as defined in astronomy and related to one another by means of spherical trigonometry, with the solar constant to estimate daily amounts of direct sunlight incident on a horizontal surface just outside the Earth's atmosphere. The amount of sunlight incident at a point just outside the atmosphere, $R_{oh}$, depends on the time of year, time of day, and latitude as follows:

$$R_{oh} = \frac{I}{r^2} \cos z \tag{3.61}$$

where $I$ is the solar constant, $r$ is the ratio of the Earth–Sun distance to its mean, and $z$ is the zenith angle (Gates, 1980; Fleming, 1987). The magnitude of $r^2$ varies continuously throughout the year, from 1.0344 on January 3 to −0.9674 on July 5, but never deviates by more than 3.5% from 1.0 (Gates, 1980).

The zenith angle is the angle between the solar beam and the normal to the surface and can be calculated from the following equation:

$$\cos z = \sin\phi \sin\delta + \cos\phi \cos\delta \cos h \tag{3.62}$$

where $\phi$ is the latitude of the observer (degrees, negative in the southern hemisphere), $\delta$ is the declination of the Sun, and $h$ is the hour angle of the Sun from the solar noon (i.e. the angular distance from the meridian of the observer (Lee, 1978). The solar declination ($\delta$) measures the seasonally varying latitude of the Sun's path across the sky, north and south of the Equator. It varies from −23.5° at the northern winter solstice (December 22) to +23.5° at the northern summer solstice (June 22). Declination is independent of calendar year and the latitude of the observer and is a function only of time of year (Gates, 1980). The hour angle ($h$) measures the difference in time from solar noon expressed as 15° per hour of difference. Some authors substitute the solar elevation or altitude ($a$) in place of the zenith angle because this variable represents the height of the Sun above the horizon for an observer at a specific location and is the complement of the zenith angle (i.e. $\sin a = \cos z$). Both the zenith angle and solar altitude are functions of latitude, time of year, and time of day (Gates, 1980). SRAD calculates instantaneous values of $R_{oh}$ (and the other short-wave radiation fluxes described later) using 12-minute time steps organized symmetrically around noon and sums these to obtain daily totals (Wilson & Gallant, 2000c).

For the second step, a series of instantaneous clear-sky, short-wave radiation fluxes can be calculated for each of the grid points at 12-minute intervals from sunrise to sunset. Direct beam and diffuse fluxes are calculated for flat sites under clear skies, and the direct beam, circumpolar diffuse, isotropic diffuse, and reflected fluxes are calculated for sloping sites under clear skies, as described here.

The tasks accompanying this second step will usually start with the estimation of the direct-beam and diffuse radiation at horizontal sites under clear skies. The amount of solar energy reaching the ground is reduced because the atmosphere is semi-transparent to solar radiation. The molecular constituents of the atmosphere, along with water droplets, dust, and other particulate matter, scatter the direct solar beam and create a hemispherical (diffuse) source of radiant energy (Lee, 1978). In addition, the direct and diffuse irradiance are both reduced as a result of direct absorption and reflection to space along the light beam's path through the atmosphere to the ground (Linacre, 1992). We therefore need to know the transmission properties of the atmosphere to estimate the amount of extraterrestrial or potential direct solar radiation that traverses the Earth's atmosphere and is incident at the ground (Gates, 1980). One or other of two approaches is commonly relied on to estimate these fluxes: the first uses a lumped transmittance approach and the second calculates individual transmittance components.

The lumped sum approach assumes that the attenuation of the direct solar beam in passing through a homogeneous, cloudless atmosphere can be described by the following formula, often named after Beer, though first formulated by Pierre Bouguer in 1760:

$$R_{\mathrm{dirh}} = R_{\mathrm{oh}} \tau^{m} \tag{3.63}$$

where $R_{\mathrm{dirh}}$ is the direct-beam short-wave irradiance incident on flat surfaces under clear skies, $\tau$ is the transmission coefficient or fraction of radiation incident at the top of the atmosphere, which reaches the ground along the vertical (or zenith) path (i.e. the shortest path length between outer space and the ground surface), and $m$ is the ratio of the path length in the direction of the Sun at zenith angle $z$ to the path length in the vertical direction (Gates, 1980; Linacre, 1992).

The local transmission coefficient for each grid cell $\tau$ can be calculated from the elevation, monthly transmission coefficient at sea level, and transmissivity lapse rate as follows:

$$\tau = \tau_{\mathrm{sl}} + t_{\mathrm{lapse}} \cdot elev \tag{3.64}$$

where $\tau_{\mathrm{sl}}$ is the transmission coefficient at sea level, $t_{\mathrm{lapse}}$ is the transmissivity lapse rate, and *elev* is the elevation at a specific grid point. This equation mimics the common situation in which transmittance is greater at higher elevations because of the thinner layer of atmosphere that occurs above these locations (Linacre, 1992). The quantity $m$ is often called the relative air mass and is given by:

$$m = \sec z = \frac{1}{\cos z} \tag{3.65}$$

where $z$ is the zenith angle that was first used in Equation 3.61. However, this equation only works when the zenith angle is less than ~60°. A different approach is required when the Sun is low in the sky, because the curvature of the Earth reduces the length of the Sun's slant rays compared with the depth of the atmosphere in the zenith direction (Robinson, 1966; Gates 1980). SRAD relies on Equation 3.65 when the zenith angle is <60° and reverts to a table obtained from List (1968, p. 422) that summarizes optical air masses in 1° increments for sites with high zenith angles. These values are then corrected to account for the reduction in atmospheric pressure encountered at higher elevations by the factor $p/p_0$, where $p$ is the atmospheric pressure at the grid cell, and $p_0$ is the standard sea-level pressure of 1013.25 mbar (Gates, 1980; Fleming, 1987).

Equations 3.63, 3.64, and 3.65 are required to estimate the attenuation of the instantaneous direct-beam, short-wave irradiance incident at the ground surface. Some of the direct-beam radiation is transformed into diffuse radiation and the relationship for instantaneous transmittance to diffuse skylight derived by Liu and Jordan (1960) can be used to estimate the instantaneous diffuse irradiance, $R_{\text{difh}}$, as follows:

$$R_{\text{difh}} = \left(0.271 - 0.294\tau^{m}\right) R_{\text{oh}} \tag{3.66}$$

Equation 3.66 shows that the transmittance to scattered skylight decreases as the direct solar beam transmittance increases (Gates, 1980). The distinction between direct-beam and diffuse short-wave irradiance is an important one and we will see shortly how it affects the amount of short-wave irradiance striking a sloping surface (Linacre, 1992).

The second approach for estimating direct-beam and diffuse short-wave irradiance on horizontal sites under clear skies treats each of the transmittance components separately. The effects of water vapor, dust, and a clear atmosphere can be estimated as follows:

$$R_{\text{dirh}} = R_{\text{oh}} \left(AW \cdot TW \cdot TD \cdot TDC\right) \tag{3.67}$$

where $AW$ accounts for the absorption by water vapor, $TW$ the scattering by water vapor, $TD$ the scattering by dust, and $TDC$ the scattering by air molecules and density fluctuations in a clear-sky atmosphere (Gates, 1980). Four equations specified by Fleming (1987) and based on work by Monteith (no reference cited) and Idso (1969) are used to estimate each of these transmittance components as follows:

$$AW = 1 - 0.077 \cdot \left[u \cdot m\left(\frac{p}{p_0}\right)\right]^{0.3} \tag{3.68}$$

$$TW = 0.975^{um(p/p_0)} \tag{3.69}$$

$$TD = 0.95^{m(p/p_0)D} \tag{3.70}$$

$$TDC = 0.9^{m(p/p_0)} + 0.026 \cdot \left[m\left(\frac{p}{p_0}\right) - 1\right] \tag{3.71}$$

where $m$, $p$, and $p_0$ were defined earlier, $u$ is the water content of a vertical slice of atmosphere in centimeters, and $D$ is an empirically derived dust factor.

The above component atmospheric transmittance model, which is implemented in SRAD, assumes that absorption takes place first followed by scattering, and that the scattering is evenly divided between forward and back scattering. This approach tries to strike a balance between accuracy and input data requirements and, as a result, several effects are ignored (i.e. absorption effects of carbon dioxide and ozone) and each of the components that is considered is treated as acting uniformly over the whole solar spectrum when we know that most of these effects are strongly wavelength dependent (Gates, 1980). This particular approach also assumes that the scattered radiation gives rise to the diffuse radiation component without further absorption so that:

$$R_{\text{difh}} = 0.5\left(R_{\text{oh}} \cdot AW - R_{\text{dirh}}\right) \tag{3.72}$$

The next set of tasks to be completed under this second step focuses on the estimation of the circumsolar and isotropic diffuse radiation. Linacre (1992, p. 152) distinguished "skylight" diffuse radiant energy, which is isotropic (i.e. comes more or less equally from all directions in the sky), and "circumsolar" diffuse radiation, which comes from within approximately 5° of the direct solar beam. The circumsolar diffuse component moves across the sky with the Sun and can be separated and added back into the direct-beam component for the calculation of the radiation incident on sloping sites (Fleming, 1987). Wilson and Gallant (2000c, pp. 95–96) described how monthly average values of the fraction or diffuse radiation that is close to the solar disk were obtained from a site-parameter file and are used by SRAD to adjust the direct-beam and diffuse radiation fluxes as follows:

$$R_{\text{dirh}} = R_{\text{dirh}} + R_{\text{difh}} \cdot CIRC \tag{3.73}$$

$$R_{\text{difh}} = R_{\text{dirh}} + R_{\text{difh}} \cdot CIRC \tag{3.74}$$

where $CIRC$ is the fraction of diffuse radiation derived from within 5° of the direct solar beam (i.e. the circumsolar coefficient). Many of the solar radiation models that have been proposed to date do not separate these components and this omission may generate errors as large as ~10% when these models are used to estimate irradiance on sloping sites (Linacre, 1992).

The third and final set of tasks to be completed under this second step are required to estimate the direct-beam, diffuse, and reflected radiation at sloping sites under clear skies. The flux density of short-wave radiation on a sloping site differs from that on a horizontal surface primarily because the direct-beam radiation is modified. Minor differences also occur because the flux of diffuse sky radiation is affected and there is an added flux of short-wave radiation reflected from adjacent parts of the landscape (Lee, 1978; Fleming, 1987).

Four site attributes (slope, aspect, horizon, and skyview) are required to estimate the direct-beam and circumsolar irradiance with and without shading, the isotropic diffuse irradiance, and the reflected irradiance on sloping sites. The first three terms can be calculated with the standard tools incorporated in many GIS platforms and the one-dimensional horizon algorithm of Dozier, Bruno and Downey (1981) can be used to estimate the fraction of the sky hemisphere, $v$, visible at each grid point. This algorithm constructs profiles across the DEM and determines the horizon angle, $H_{\phi}$, for each grid point in a discrete number of directions, $\phi$ (i.e. 16). This algorithm is also computationally attractive because its computational efficiency is proportional to the number of grid points in a square-grid DEM. The skyview fraction, $v$, is computed from the horizon angles, $H_{\phi}$, by averaging the cosine of the horizon angles:

$$v = \frac{1}{n}\sum_{\phi=1}^{n} \cos H_{\phi} \tag{3.75}$$

The direct beam and circumsolar diffuse radiation on sloping surfaces depend on the solar elevation and the slope's angle to the horizontal (Linacre, 1992). Tilting a surface in the meridian plane north or south from the horizontal is the equivalent of going north or south in latitude by the same number of degrees (Gates, 1980). In addition, the slope and aspect effects are greatest during winter in middle latitudes and they tend to become negligible toward the equator and the poles (Lee, 1978).

The direct irradiance on a sloping surface without shading can be calculated using the following equations:

$$R_{\text{dirs}} = R_{\text{dirh}} \cos i \tag{3.76}$$

$$\cos i = A + B\cos h + C\sin h \tag{3.77}$$

$$A = \sin\delta\sin\phi\cos\beta + \sin\beta\cos\psi\cos\phi \tag{3.78}$$

$$B = \cos\delta\left(\cos\phi\cos\beta - \sin\phi\cos\psi\sin\beta\right) \tag{3.79}$$

$$C = \sin\beta\cos\delta\sin\psi \tag{3.80}$$

where $R_{\text{dirh}}$ in this instance is the direct-beam and circumsolar diffuse radiation on a horizontal surface under clear skies (i.e. as estimated with Equations 3.63 or 3.67 and 3.74), $i$ is the angle between the beam and the normal to the slope, $\beta$ is the slope angle, and $\psi$ is the aspect (Lee, 1978; Linacre, 1992). Instantaneous values can then be calculated in 12-minute time steps (or something similar) and these values multiplied by 12 and summed to estimate daily direct-beam radiation on sloping surfaces without shading ($R_{\text{dirsns}}$). This part of the analysis also needs to check whether the Sun is obstructed at 12-minute intervals and then calculates the direct-beam and circumsolar diffuse radiation taking shading into account ($R_{\text{dirss}}$). This approach means, of course, that $R_{\text{dirss}} \le R_{\text{dirsns}}$ at each sloping site.

The "isotropic" diffuse radiation on a sloping site is typically lower than that on a horizontal surface because part of the sky is obscured:

$$R_{\text{difs}} = R_{\text{difh}}\, \nu \tag{3.81}$$

where $R_{\text{difh}}$ in this instance is the isotropic diffuse irradiance estimated with Equations 3.65 or 3.71 and 3.72, and $\nu$ is the skyview factor (i.e. fraction of sky visible at a specific grid point).

However, the reductions in direct-beam and diffuse irradiance on sloping sites (compared to horizontal sites) may be partially offset by reflected radiation ($R_{\text{ref}}$) from other surfaces:

$$R_{\text{ref}} = \left(R_{\text{dirh}} + R_{\text{difh}}\right)\left(1-\nu\right)\alpha \tag{3.82}$$

where $R_{\text{dirh}}$ is the direct-beam radiation on a horizontal surface under clear skies estimated with Equation 3.63 or 3.67, $R_{\text{difh}}$ is the diffuse radiation on a horizontal surface under clear skies estimated with Equation 3.66 or 3.72, $\nu$ is the skyview factor, and $\alpha$ is the albedo (i.e. the fraction of sunlight reflected by the surface). For the sake of simplicity, this particular equation uses horizontal radiation values rather than attempting to sum the contributions of the spatially varying radiation within the ground-view area. The reflected radiation is calculated at 12-minute intervals and summed over the day to estimate reflected radiation from foreground surfaces facing the site. An upward-facing area will see less sky and more ground (from which it receives reflected light) as it is tilted upward from the horizontal (Lee, 1978). These inputs are important on many sloping sites and they can vary markedly with the type and status of the surface (Fleming, 1987).

For the third step, these instantaneous values are summed to obtain daily totals and these values are adjusted to account for the effects of cloudiness. The flux of direct and diffuse short-wave solar radiation incident at the Earth's surface is highly variable and difficult to predict when the sky is partially or totally overcast (Linacre, 1992). Clouds are highly variable in form, size, density, height, and duration. Very thin transparent cirrus clouds have little influence on global radiation, whereas thick, dark thunderstorm clouds may reduce radiation to 1% or less of its clear-sky value (Gates, 1980). There are several ways to execute this important step and the approach used by SRAD, which combines daily short-wave radiation estimates for clear skies with statistical averages of observational data collected over long periods of time (Wilson & Gallant, 2000c), is described next.

The accumulated radiation values at the end of each simulated day can be combined with the sunshine fraction and cloud transmittance for the month in question to estimate daily incoming short-wave solar radiation on horizontal sites $R_{\text{th}}$ as follows:

$$R_{\text{th}} = \left(R_{\text{dirh}} + R_{\text{difh}}\right)\left[\frac{n}{N} + \left(1-\frac{n}{N}\right)\beta\right] \tag{3.83}$$

where $n/N$ is the sunshine fraction (i.e. the observed duration of sunshine out of the maximum possible for that place and date), and $\beta$ is the cloud transmittance (i.e. the fraction of clear-sky radiation received when the sky is overcast).

For sloping sites, the cloud transmittance in SRAD can be adjusted downward to account for the reduction in sky view and upward to account for the enhanced flux of downward and upward diffuse radiation caused by reflection between ground and sky. Gates (1980) reports measurements by Kondratyev (1969) illustrating these effects and they are approximated in SRAD with the following equation:

$$\beta_s = \beta \nu \left( \frac{R_{\text{tsns}}}{R_{\text{tss}}} \right) \tag{3.84}$$

where $R_{\text{tsns}}$ is the total daily clear-sky, short-wave radiation without shading and $R_{\text{tss}}$ is the total daily clear-sky, short-wave radiation with shading on sloping sites. The daily incoming short-wave solar radiation on sloping sites, $R_{\text{ts}}$, is then estimated as follows:

$$R_{\text{ts}} = \left( R_{\text{dirss}} + R_{\text{difs}} \right) \left[ \frac{n}{N} + \left( 1 - \frac{n}{N} \right) \beta \right] + R_{\text{ref}} \tag{3.85}$$

Long-wave radiation estimates are also required to predict the surface energy budget. Long-wave radiation is emitted continuously by the Earth's atmosphere and surface. The incoming atmospheric flux, $L_{\text{in}}$, is almost always less than the outgoing surface flux, $L_{\text{out}}$, and this means that the average net daily longwave irradiance ($L_{\text{net}}$) represents a net loss of energy from the biosphere (Lee, 1978).

The daily incoming long-wave irradiance can be computed from air temperature taking skyview into account:

$$L_{\text{in}} = \varepsilon_{\text{a}} \sigma T_{\text{a}}^4 \nu + (1 - \nu) L_{\text{out}} \tag{3.86}$$

where $\varepsilon_{\text{a}}$ is the atmospheric emissivity (a function of air temperature, vapor pressure and cloudiness), $\sigma$ is the Stefan–Boltzmann constant, $T_{\text{a}}$ is the mean air temperature, and $\nu$ is the skyview factor discussed earlier (Lee, 1978). The inclusion of the skyview factor means that this equation handles both sloping and horizontal sites. Sloping sites see both the sky and adjacent terrain and $\nu$ accounts for the fraction of the sky that is visible and the $(1 - \nu)\, L_{\text{out}}$ term shows how a fraction of the outgoing long-wave radiation is added to the incoming component to account for radiation from foreground obstructions.

The daily outgoing long-wave irradiance can be calculated from surface temperature (see Equation 3.91 for additional details) as follows:

$$L_{\text{out}} = \varepsilon_{\text{a}} \sigma T_{\text{s}}^4 \tag{3.87}$$

where $\varepsilon_{\text{a}}$ is the atmospheric emissivity (>0.95 for most natural surfaces) and $T_{\text{s}}$ is the mean surface temperature (Lee, 1978).

The overarching goal of this approach is to estimate the net solar radiation, which is the quantity of energy available at the ground surface to drive air and soil heating, evaporation, and photosynthesis (Dubayah, 1992). Given the previous steps, the net radiation at each grid cell can be estimated by adding the incoming and outgoing fluxes for some user-defined time period as follows:

$$R_n = (1-\alpha) R_{th} + \varepsilon_s L_{in} - L_{out} \quad \text{(for horizontal sites)} \tag{3.88}$$

$$R_n = (1-\alpha) R_{ts} + \varepsilon_s L_{in} - L_{out} \quad \text{(for sloping sites)} \tag{3.89}$$

where $R_n$ is the net radiation, α is the surface albedo, $R_{th}$ is the global short-wave irradiance on a horizontal surface, $R_{ts}$ is the global short-wave irradiance on a sloping surface, $\varepsilon_s$ is the surface emissivity, $L_{in}$ is the incoming or atmospheric long-wave irradiance, and $L_{out}$ is the outgoing or surface long-wave irradiance (Wilson & Gallant, 2000c). These equations take the usual approach in that $R_n$ takes a positive sign when energy is transferred to the surface and a negative sign when the direction is reversed (Lee, 1978). This approach means, for example, that $R_n$ will normally be greater than zero during the summer months and might be negative in the winter months, particularly in the higher latitudes and at locations facing away from the sun that receive little or no direct beam radiation.

The topographic radiation and temperature indices described in this and the following section should be read with two caveats in mind. The first is to reiterate that the radiation indices that can be calculated with different software packages will likely vary in terms of the methods, data sources, and assumptions used to estimate the various components. The SRAD approach described here provides an approximate method for estimating each of the aforementioned fluxes at any location in a topographically heterogeneous landscape. The variation in potential solar radiation can be estimated over a catchment as a function of slope, aspect, topographic shading and time of year, and then adjusted for cloud, atmospheric, and land cover effects with this method. The variables that serve as model inputs, such as albedo, cloudiness, emissivity, sunshine fraction, mean air and surface temperatures and clear-sky transmittances, can be varied on a monthly or annual basis (Moore et al., 1993d; Wilson & Gallant, 2000c). Many of the competing models offer partial solutions. SRAD and the r.sun solar radiation model implemented in the GRASS GIS (Šúri & Hofierka, 2004) share many similarities and probably present the most complete solutions available at this time.

The second and final caveat concerns the difficulty of verifying the parameter estimates because most radiation stations are located on flat terrain (i.e. horizontal surfaces). One possible way around this problem is to use satellite data for estimating incoming solar radiation and some of the other inputs. Böhner and Antonić (2009, pp. 199–207), for example, described one such study by Dubayah and Loechel (1997) which combined the coarse spatial resolution data of Geostationary Satellite Server imagery with the fine spatial resolution DEM-based topography using the direct–diffuse partitioning algorithm of Erbs, Klein and Duffie (1982), the elevation correction formulations of Dubayah and van Katwijk (1992), and the various equations describing the topographic effects on direct, diffuse and reflected radiation noted earlier. Similarly, Austin et al. (2013) recently noted how the validation of the SRAD radiation surfaces which they developed for Australia was constrained by observations gathered on horizontal sites and how many of the components of SRAD could be improved by substituting satellite observations for parameters such as albedo and cloud cover that are currently estimated from ground observations.

#### 3.2.2.4 Topographic Temperature Indices

Wilson and Gallant (2000c) also described how the temperature at each grid cell can be estimated as a function of the mean minimum and maximum air temperatures, the surface temperature, the minimum, maximum and average temperature lapse rates, and elevation for a reference station. The topographic effect on temperature is introduced via the short-wave radiation ratio, $S$, at each grid point:

$$S = \frac{R_{\mathrm{rs}}}{R_{\mathrm{th}}} \tag{3.90}$$

where $R_{\mathrm{th}}$ and $R_{\mathrm{ts}}$ are the daily total or global short-wave irradiance on horizontal and sloping sites, respectively.

The minimum air, maximum air, and surface temperature, $T$, for each grid point can then be calculated using:

$$T = T_{\mathrm{b}} - \frac{T_{\mathrm{lapse}}\left(z - z_{\mathrm{b}}\right)}{1000} + C\left(S - \frac{1}{S}\right)\left(1 - \frac{LAI}{LAI_{\mathrm{max}}}\right) \tag{3.91}$$

where $z$ is the elevation of the grid point, $z_{\mathrm{b}}$ is the elevation of the temperature reference station, $T_{\mathrm{b}}$ is the monthly minimum, maximum, or surface temperature at the reference station, $T_{\mathrm{lapse}}$ is the monthly temperature lapse rate (°C/1000 m), $C$ is a constant (which in past applications was set to 1.0), $LAI$ is the leaf area index at the grid cell, and $LAI_{\mathrm{max}}$ is the maximum leaf area index. In addition, the $LAI$ and short-wave radiation ratio corrections should not be applied to minimum temperature because these occur during the night, and the average air temperature is assumed to be the average of the minimum and maximum air temperature.

These temperatures may vary significantly over relatively short distances due to topographic effects. Wilson and Gallant (2000c, p. 127), for example, show maps of the short-wave solar radiation ratio and mean annual average temperature for the Cottonwood Creek catchment that was first introduced in Chapter 1 in which the short-wave ratio varies from ~0.7 to ~1.1 and the mean annual average air temperature in this relatively small catchment varied from ~3.6 to ~4.8 °C. These temperature indices were also used with numerous other indices by Moore et al. (1993d) to simulate the spatial distribution of radiation, thermal and hydrologic regimes in a series of topographically complex forested catchments within the Brindabella Range in southeastern Australia. The computed fluxes were used to characterize the fine-scale environmental heterogeneity and environmental domains of the five main subalpine forest types found in the study area. Two variables – the average minimum temperature in the coldest month (July) and the annual net radiation – were used to differentiate the occurrence of the three eucalypt species (*Eucalyptus pauciflora*, *E. fastigata*, and *E. delegatensis*) investigated.

## 3.3 Final Comments

This chapter has described more than 100 land surface parameters given that each of the 24 flow-direction algorithms summarized in Table 3.2 can be employed to estimate some or all of the flow-based primary and secondary land surface parameters listed in Tables 3.1 and 3.4, respectively. Their number has grown steadily with time along with the methods that can be used to calculate some of these land surface parameters. In addition, the survey of applications provided in Section 1.3 shows that these parameters have been adopted and used in a large number and variety of environmental applications and landscape settings during the past 40 years and that their popularity is likely to grow further in the future as we work to characterize and hopefully solve a series of environmental challenges such as global warming, non-point source pollution and ecosystem destruction, that express themselves across local, regional, and global scales.

That said, there are three sets of issues connected with the choice of scale, terrain modeling and statistical methods that will continue to confound the adoption and use of the primary and secondary land surface parameters described in this chapter.

The first issue concerns the role of scale with respect to the calculation of the primary land surface parameters. The local terrain shape, which is usually thought of as the continuous variation of elevation values over the terrain surface from point to point, has an enormous impact on both the local and regional land surface parameters described in Section 3.2, but this role is influenced by data and computational factors as well. Florinsky (1998) has gone so far as to suggest that local attributes, such as slope gradient, aspect, and curvatures, are mathematical variables rather than real-world values. This statement may be extended to all local terrain attributes for two reasons. First, local terrain shape may rely on different mathematical descriptions, so that calculated local attributes depend on algorithm selection. Second, the terrain shape portrayed by DEMs is a function of scale, combining the complexity of the terrain, scale or resolution of data, and spatial scale at which the terrain surface is observed (e.g. Deng, Wilson & Gallant, 2008). Thus it is possible to use the same local land surface parameter to describe terrain shape at different scales (i.e. grid spacings or resolutions). The special feature of non-local primary attributes is that they rely on the terrain shape of a larger, non-neighbor area and need to be defined with reference to other non-local points. Therefore, calculating non-local attributes is more difficult because it incurs additional efforts in constructing point-to-point connections over the landscape and involves more complex algorithms and scale considerations (e.g. Desmet & Govers, 1996a; Gallant & Wilson, 2000; Wilson & Gallant, 2000c; Winchell et al., 2008).

The second issue concerns the more or less continuous development and evolution of the methods for creating DEMs and calculating specific land surface parameters. The 30- and 100-m DEMs that were used extensively in the 1980s have given way to 3–10 m DEMs today and many of us may find ourselves working with 1–3 m DEMs in the next few years. Similarly, the number of single- and multiple-flow direction algorithms has tripled during the past 25 years and nearly all the methods listed in Table 3.2, for example, are likely to generate varying estimates of upslope contributing area, SCA, TWI and stream power index as well as the other flow-based land surface parameters. The rapidly evolving character of both the underlying data support and methods used to calculate specific land surface parameters means that many of the conclusions drawn from the terrain modeling work conducted in the past may not apply to the terrain modeling work which is executed nowadays.

The third and final problem is that search for the biophysical meaning of these land surface parameters may be exacerbated by the multi-collinearity of numerous pairs of attributes. Tables 3.1 and 3.4 list 68 and 15 primary and secondary land surface parameters, respectively, and these counts do not anticipate the possibility of using as many as 24 flow-direction algorithms to estimate several of the flow-based primary and secondary land surface parameters. In addition, many of the land surface parameters are empirical in nature and, as such, based on the first and second derivatives of the Earth's surface. This state of affairs means that several of the parameters may yield some of the same information (the classic example is relief and slope) and they should therefore be used and interpreted carefully.

The significance of this last observation can be illustrated by looking forward to the extraction and classification of landforms and other kinds of land surface objects in Chapter 4. The use of primary and secondary land surface parameters with a fuzzy *k*-means classification to delineate landform classes should start, for example, with correlation analysis to confirm that the candidate inputs are not highly correlated with one another (see Burrough et al., 2000, 2001; Deng & Wilson, 2006, 2008; Deng, Wilson & Sheng, 2006; Deng, 2007; Deng et al., 2007, 2008 for examples of good practice) and to ensure that the inclusion of land surface parameters that are correlated with one another does not introduce unwanted weights or bias. This advice can be extended to cover all forms of terrain modeling and is discussed in considerably more detail in both of the next two chapters.

# 4 Delineating Land Surface Objects and Landforms

This chapter marks the shift from general to specific geomorphometry, where general geomorphometry describes the continuous surface and specific geomorphometry describes a series of discrete landforms such as alluvial fans, drumlins, eskers, mountains, and valleys (Evans, 1972).

MacMillan and Shary (2009, pp. 229–233) described landforms as widely recognized natural objects that differ in terms of the physical processes that contributed to their formation and that continue to operate on them at the present time. They also noted how landscapes and landforms occur over a hierarchy of scales and sizes, and how many of the applications thus far have treated the variation in the land surface that occurs below a given scale as random noise to be ignored (or worse, removed) and the variation at the preferred scale or over a range of scales as the signal that is to be recognized and interpreted. This is a big challenge because there is probably not an optimum resolution at which to calculate land surface parameters in order to portray and classify terrain (Fisher, 1997; Hengl, 2006). The initial task for every study, therefore, involves the selection of an operational scale (i.e. grid spacing or resolution assuming we start from a square-grid DEM) that is appropriate to capture and describe the surface features of interest for the specific application at hand.

Looking beyond these scale issues, the automated extraction and classification of land surface objects and landforms can be viewed in a number of ways. Dehn, Gärtner and Dikau (2001) proposed a semantic-based approach to describe landforms that considered geometric form, topology, and semantic definitions. Mackay et al. (2003), for example, advocated the use of higher-order geographic objects, rather than cells, for topographic characterization and they incorporated inexact reasoning to indicate the fuzzy certainty level of the resultant delineations. Romstad and Etzelmüller (2009, 2012) proposed a simple method to segment a DEM into terrain units based on curvature. MacMillan and Shary (2009) offered the

*Environmental Applications of Digital Terrain Modeling*, First Edition. John P. Wilson.

opinion that most automated landform classification approaches seek to replicate some previously conceived system of manual landform classification and mapping. The works by Dikau, Brabb and Mark (1991), Dikau et al. (1995), Gallant, Douglas and Hoffer (2005), Hrvatin and Perko (2009), Drăguţ and Eisank (2012), and Karagulle et al. (2017) described in the following sections represent examples that fit this description. And finally, Tomer and Anderson (1995), Clarke (1988) and Shary et al. (2002), among others, have argued that the land surface is composed of deterministic and noisy components and that geomorphometry is concerned with the task of extracting deterministic landforms from the noisy land surface (MacMillan & Shary, 2009).

MacMillan and Shary (2009) also noted the five steps that need to be completed to extract land surface objects and/or landforms from the land surface parameters that were described in the previous chapter.

1. The specification of the spatial objects that will be classified and in the case of a map showing the spatial pattern, the preparation of the map legend.
2. The specification and calculation of the input variables (i.e. square grids representing one or more land surface parameters, as described in Chapter 3).
3. The specification of the extraction and classification rules using one or a combination of unsupervised, supervised and knowledge-based (heuristic) approaches.
4. The extraction of the land surface objects and landforms using the aforementioned classification rules.
5. An assessment of the accuracy of the classification.

However, the transition from square-grid cells (i.e. pixels) to land surface objects and landforms is not straightforward, as Fisher (1997, p. 679) so eloquently stated nearly 20 years ago:

> … it is my contention that the pixel, the elementary unit of analysis in remote sensing and the usual vehicle for integrating data between GIS and remote sensing, is a delusion which may become a snare for the unwary given the way in which it is treated in modern software. It has the potential to be a powerful modeling device, but is usually misused, and so becomes a handicap to analysis.

That said, the remainder of this chapter is divided into five parts. The first four sections describe some of the ways in which land surface elements, land surface objects based on flow variables as well as other kinds of fuzzy specific landforms, and repeating landform types have been extracted and classified thus far, while the fifth and final section draws on some recent work by Minár and Evans (2008) and Drăguţ and Eisank (2011), among others, to revisit the transition from general to specific geomorphometry and reflect on what has and has not been accomplished thus far, and what will be needed to advance this work in the years ahead.

## 4.1 Extracting and Classifying Specific Landform Elements

MacMillan and Shary (2009) defined landscape elements as those parts of the landscape (i.e. landscape types) that are relatively homogeneous in terms of shape (profile/plan curvature), steepness (slope gradient), orientation or exposure (aspect or solar radiation), moisture regime, and relative landform position. The landform elements therefore act at a more detailed level of abstraction and a larger scale than the classification of specific landforms and landform types. Not surprisingly, many studies have been conducted with the goal of extracting and classifying specific landform elements during the past 30–40 years.

The use of land surface parameters to segment the landscape into landform elements can be traced to the pioneering work of Speight (1968) and Dikau (1989). Recent developments have included the use of automated fuzzy classification algorithms to detect landform elements (e.g. Burrough et al. 2000, 2001; Schmidt & Hewitt 2004). The focus of these fuzzy applications spans the identification of specific landforms (e.g. mountains, valleys, river basins and channel networks), landform elements (i.e. geometric shapes that constitute part or all of a hillslope or some other specific landform), and repeating landforms types (e.g. a series of rolling hills and valleys). The primary focus here is on the use of these fuzzy techniques to identify landscape elements.

Nearly all of the classification approaches for identifying and characterizing landform elements have relied on various representations of landscape position and/or the shape of the land surface itself (e.g. Pennock et al., 1987; Dikau, 1989; Skidmore et al., 1991; Irvin, Ventura & Slater, 1997; Schmidt, Merz & Dikau, 1998; Schmidt & Dikau, 1999; Shary et al., 2002). However, perhaps the two most widely cited approaches have used curvatures to classify landform elements, as described next.

For example, Dikau (1989) divided the landscape into combinations of concave, straight, and convex plan curvatures on the one hand and concave, straight and convex profile curvatures on the other hand. This approach (like many subsequent ones) relies on the inferred relationship between surface shape (i.e. local curvature) and the accumulation of surface flow and consequently that of surface deposition through two accumulation mechanisms. The first mechanism reflects the divergence and convergence of flow across a hillslope, whereas the second reflects the relative deceleration of flow in the downslope direction as influenced by changes in profile curvature (I.D. Moore et al., 1991; MacMillan & Shary, 2009).

Shary and colleagues have criticized Dikau's (1989) original approach on two levels, and proposed a more robust and predictable classification based on curvatures. Their first suggestion was to use tangential curvature in place of plan curvature in the classification of basic form elements (Figure 4.1), because tangential and profile curvature are both curvatures of normal

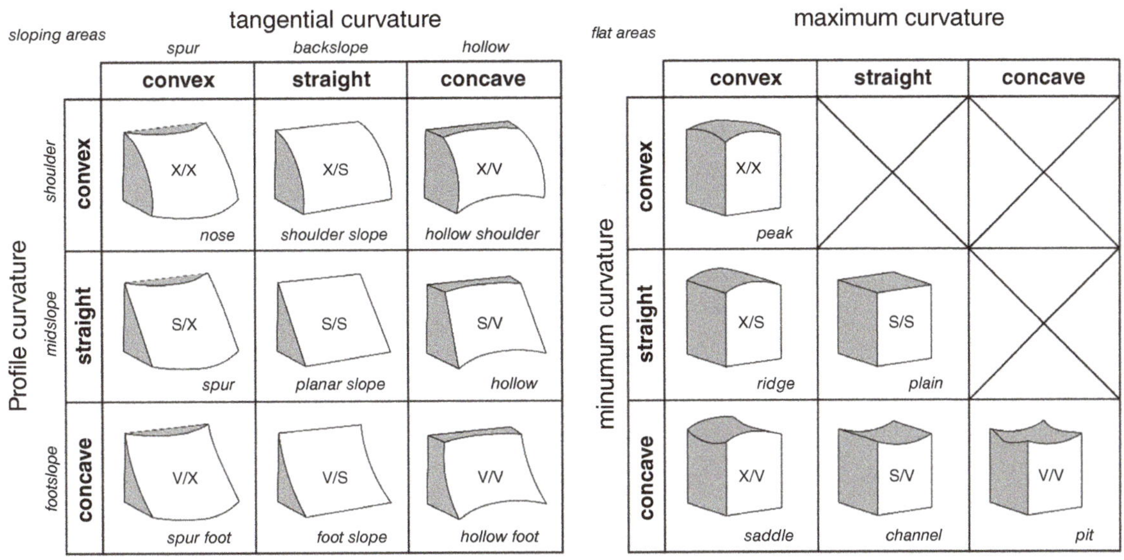


Figure 4.1 The modified Dikau (1989) classification of form elements based on the profile and tangential curvatures. The elements have been further classified as positive or negative based on the radius of curvatures (>600 or <600 m) and the planform curvature in the original classification was replaced by tangential curvature based on Shary and Stepanov (1991). *Source*: Gerçek et al. (2011, p. 1015). Reproduced with permission of Taylor & Francis.

sections and they both exhibit similar statistics of distribution, unlike the case with plan curvature (MacMillan & Shary, 2009, p. 238). The second criticism concerned the contradiction in Dikau's (1989) original premise that this approach differentiated form elements with homogeneous plan and profile curvature because these facets invariably contain even more homogeneous form facets with similar gradients, aspects, and curvatures. Shary (1995) and Shary et al. (2005), in turn, have proposed an objective, local, scale-specific classification of elemental landform features based solely on consideration of the signs the tangential, profile, mean, difference, and total Gaussian curvatures as a way of avoiding these kinds of problems (Figure 4.2).

However, neither of the aforementioned approaches incorporates the contextual position of specific facets that make up the landscape. Most of the automated classifications that incorporate context build on the conceptual classifications of hillslopes like that of Ruhl (1960) which divided hillslopes into five units (summits, shoulders, backslopes, footslopes, and toeslopes) along a toposequence from a ridge crest to a valley bottom following concepts proposed by Milne (1935) and elaborated by Ruhl and Walker (1968) and Huggett (1975) (Figure 4.3).

Many other such classifications have been proposed: Skidmore et al. (1991), for example, calculated the absolute and relative horizontal and vertical distance to ridgelines and channels; Schmidt and Dikau (1999) calculated the position of the landscape relative to the order of the nearest stream channel into which a hillslope drained; and Conacher and

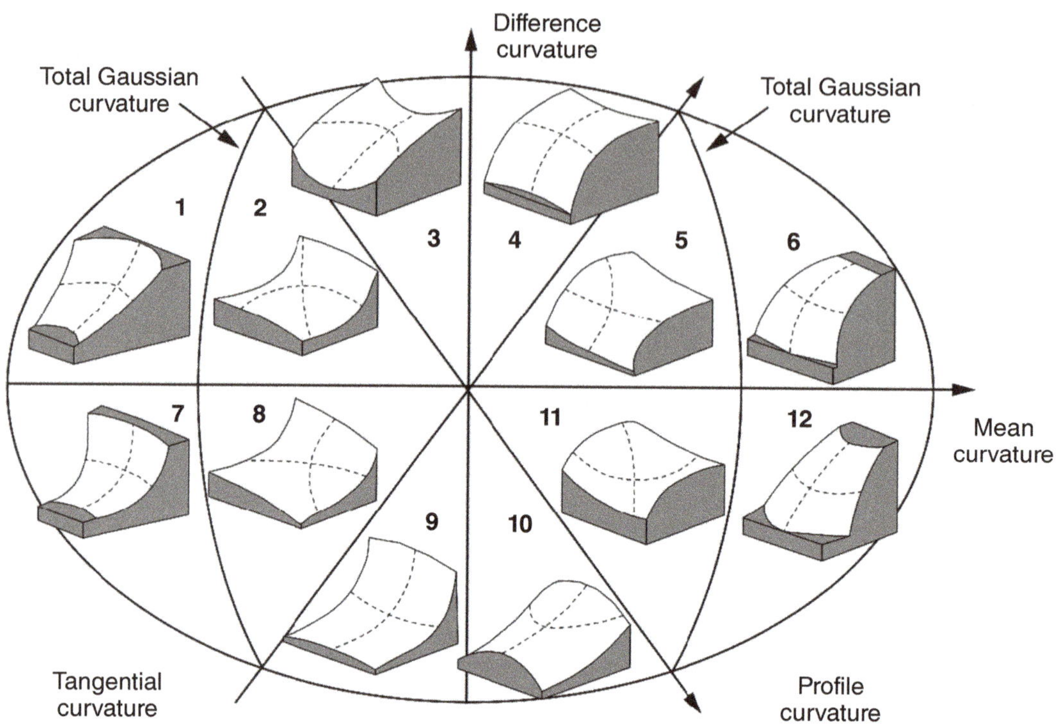


**Figure 4.2** Shary's complete system of classification of landform elements based on signs of tangential, profile, mean difference, and total Gaussian curvatures. *Source*: Shary et al. (2005). Reproduced with permission of the Italian Glaciological Committee.

Dalrymple (1977) and Speight (1990) divided hillslopes into nine and ten units, respectively, by delineating finer resolution hillslope facets and/or by extending the hillslope to include the channel or various parts of the channel at the bottom of the hillslope (Tables 4.1 and 4.2).

Taken as a whole, these conceptual classes consider slope gradient and relative slope position along a toposequence from divide to channel in addition to curvatures and the automated classification efforts that have followed have included absolute and relative horizontal and vertical distance to ridgelines or channels (e.g. Skidmore et al., 1991) and position in the landscape relative to the order of the nearest stream channel below a hillslope (e.g. Schmidt & Dikau, 1999; Schmidt, Hennrich & Dikau, 2000).

### *4.1.1 Fuzzy Concepts and Fuzzy Classification Methods*

The adoption and use of fuzzy classification algorithms to detect landform elements marked an important step forward. The importance of this innovation can be traced to the fact that each of the aforementioned approaches will work in some instances and not others, and they will seldom produce satisfactory answers to questions linked to the locations of a specific

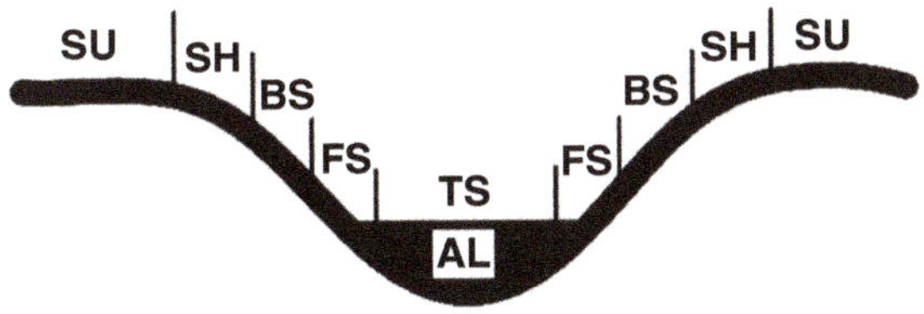


| | |
|---|---|
| **SU Summit** | An upland surface with an inclination which differs distinctly from the hillslope which ascends to it. |
| **SH Shoulder** | The convex component between the summit and the backslope. |
| **BS Backslope** | The typically linear, inclined part of the hillslope. |
| **FS Footslope** | The concave part of the hillslope that welds the linear segments to the lower terrain and is both an erosional and depositional surface. |
| **TS Toeslope** | The region which extends away from the base of the hillslope and is composed of depositional debris. |
| **AL Alluvium** | Unconsolidated depositional debris. |

**Figure 4.3** Landscape elements on a hillslope profile between two interfluves as delineated by Ruhl (1960) and Ruhl and Walker (1968). *Source*: Ventura & Irvin (2000, p. 269). Reproduced with permission of John Wiley & Sons, Inc.

**Table 4.1** Conceptual landform units defined by Conacher and Dalrymple (1977).

| Land surface units | Characteristics |
|---|---|
| Interfluve | Interfluve with predominant pedomorphometric processes caused by vertical (i.e. up and down) soil-water movements and 0–1° slope gradients |
| Seepage slope | Upland area where responses to mechanical and chemical eluviation by lateral subsurface soil-water movements predominate |
| Convex creep slope | Convex slope element where soil creep is the predominant process, producing lateral movement of soil materials |
| Fall face | Areas with slope gradients >45° characterized by the process of fall and rockslide |
| Transportational midslope | Inclined surfaces with 1–45° slope gradients and responses to transport of large amounts of material downslope by flow, slump, slide, erosion and cultivation |
| Colluvial footslope | Concave areas with responses to colluvial redeposition from upslope |
| Alluvial toeslope | Areas with responses to redeposition from upvalley alluvial materials and 0–4° slope gradients |
| Channel wall | A channel wall distinguished by lateral corrosion by stream action |
| Channel bed | A stream channel bed with transportation of material downvalley by stream action as the predominant process |

*Source*: MacMillan & Shary (2009, p. 240). Reproduced with permission of Elsevier.

mountain or valley for example (Fisher et al., 2004). Indeed, we can think of many phenomena that are difficult to locate or delineate because their meaning is not well defined and/or because of the subjectivity, vagueness and ambiguity that have often characterized the ways in which we describe

**Table 4.2** Morphologic type (i.e. topographic position) classes of Speight (1990).

| Morphologic type classes | Definitions |
|---|---|
| Crest | Area high in the landscape, having a positive plan and/or profile curvature |
| Depresson (open, closed) | Area low in the landscape, having a negative plan and/or profile curvature; closed: local elevation minimum; open: extents at same or lower level |
| Flat | Areas having slope <3% |
| Slope | Planar element with an average slope >1%, subclassed by relative position |
| Simple slope | Adjacent below a crest or flat and adjacent above a flat or depression |
| Upper slope | Adjacent below a crest or flat but not adjacent above a flat or depression |
| Midslope | Not adjacent below a crest or flat and not adjacent above a flat or depression |
| Lower slope | Not adjacent below a crest or flat but adjacent above a flat or depression |
| Hillock | Compound element where short slope elements meet at a narrow crest <40 m |
| Ridge | Compound element where short slope elements meet at a narrow crest >40 m |

*Source*: MacMillan & Shary (2009, p. 241). Reproduced with permission of Elsevier.

the world around us (Burrough, 1996; Wilson & Burrough, 1999). Fuzzy set theory represents an alternative approach to classic set theory (Burrough & McDonnell, 1998; Robinson, 2003) and has been employed in many environmental domains to solve these kinds of problems. Hence, one or more forms of fuzzy classification have been used to describe the spatial transitions or gradations between slope positions, for example ridge, shoulder slope, backslope, footslope, and valley (Qin et al., 2009); soil variability (e.g. McBratney & Odeh, 1997; Zhu, 1997a,b, 1999; Ahn, Baumgartner & Biehl, 1999); land cover (e.g. Fisher & Pathirana 1994; Foody, 1996; Brown, 1998; DeBruin, 2000); site selection and multi-criteria evaluation (Jiang & Eastman, 1996; Charnpratheep, Zhou & Garner, 1997); and the parameterization of land surface models (e.g. Mackay et al., 2003).

The aforementioned examples indicate how many authors have used fuzzy sets and fuzzy logic operators with land surface parameters to generate partial and multiple memberships of spatial objects of various kinds during the past 20 years. The various membership functions that have been described in considerable detail by Robinson (2003) represent the core of the method because they allow us to express irreducible observation and measurement uncertainties in their various manifestations, and to make these uncertainties intrinsic to the classification (i.e. using grades of membership). In this way, the fuzzy logic approach will associate a fuzzy likelihood of each output class with each value or class on each input map (Figure 4.4). This means that when fuzzy data are processed, their intrinsic uncertainties are processed as well and their results are more meaningful than their counterparts obtained by processing the usual crisp data (Klir & Yuan, 1995; Robinson, 2003). The advantages of using such an approach are evident in the landform elements extracted by MacMillan et al. (2000) based

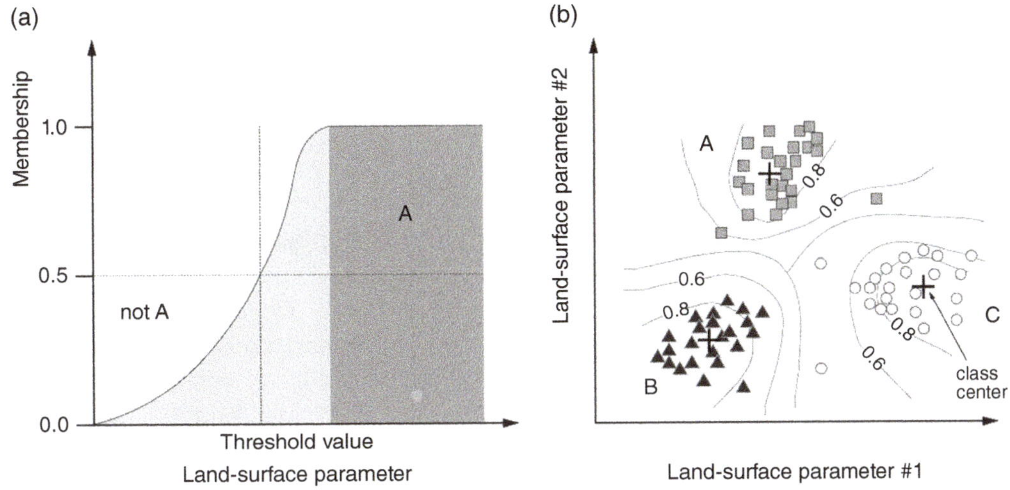


**Figure 4.4** Schematic showing the derivation of fuzzy memberships using (a) the definition of thresholds and (b) the definition of class centers. *Source*: Hengl & MacMillan (2009, p. 453). Reproduced with permission of Elsevier.

on a combination of measures of local surface shape (convexity/concavity) plus relative slope position for a 64-ha site in Alberta, Canada (three-dimensional views of the landform classification for this site are reproduced in black and white in MacMillan & Shary, 2009, p. 243 and in color in Hengl & Reuter, 2009, p. 718).

There are two basic approaches. The first relies on expert knowledge and is often referred to as the semantic import (SI) model, whereas the second, which is referred to as the similarity relation (SR) model approach, relies on the identification and characterization of data clusters (Bezdek, Ehlich & Full, 1984). The fuzzy sets generated in both of these instances will often provide tremendous expressive power (Fisher, 2000a,b; Robinson, 2003). For example, Zhu (1997a,b, 1999) clearly demonstrates the utility of the fuzzy paradigm for capturing and representing spatially explicit soil knowledge from human experts. Zhu and his colleagues solicited expert knowledge of observable environmental inputs, including several land surface parameters, and targeted outputs (soil map units), and adopted a limiting factor approach for computing the overall similarity. The value for the overall similarity between an unclassified and a reference entity was obtained by simply selecting the smallest similarity value from among all similarity values computed for all attributes for an unclassified entity in this application.

Both the expert-driven (SI) and data-driven (SR) fuzzy classification approaches will result in multiple maps of membership, where each class will be represented on a separate map. The advantages of using memberships are threefold since we are able to: (i) determine which classes are connected with which land surface parameters; (ii) assess the confusion between the classes; and thereby (iii) detect the areas where the confusion between two or more classes is high (Burrough & McDonnell, 1998; Hengl et al., 2004; Shi, Li & Zhu, 2005; Evans et al., 2009).

However, there are many subtleties connected with these fuzzy classification methods. The knowledge in the aforementioned approach could have been acquired via several methods, and each would have varied in the degree to which it is theoretically, empirically or statistically valid (see Qi & Zhu, 2003; Qi, 2004; Qi et al., 2006 for an extended discussion of these kinds of issues). Similarly, the results can be expected to vary with the method used to compute overall similarity scores. MacMillan et al. (2000), in their implementation of the SI model, relied on a weighted average method to calculate the overall similarity of an unclassified site to a reference entity, based on the assumption that all input variables should be included in calculating the similarity of a site to a reference entity. However, a case can also be made that some inputs may deserve to be afforded a greater importance or weight than others (Hengl & MacMillan, 2009). The sensitivity of the FCM approach to the choice of input variables and weights assigned to them when calculating overall similarity values has been explored extensively by Deng and colleagues (e.g. Deng et al., 2006; Deng & Wilson, 2006).

There is also the need to choose an appropriate scale for both the SI model and SR-FCM classification approaches. The fact that there is no single true or fixed value for local land surface parameters such as slope or curvature at a point but rather a whole range of values that are dependent upon the horizontal and vertical resolution has already been noted. Not unexpectedly, there can be no single best resolution at which to calculate local land surface parameters to portray and classify terrain (Hengl, 2006; Smith et al., 2006; Deng et al., 2007) and the final scale or scales that are chosen should be appropriate for capturing and describing the surface features of interest for a particular application (Deng et al., 2008). The size or extent of the study area needs to be added to this list of sensitive variables since some land surface parameters will vary in systematic ways across the landscape and may generate locally specific results when the fuzzy classification is implemented for a limited area (Evans et al., 2009).

## 4.2 Extraction and Classification of Land Surface Objects Based on Flow Variables

The extraction and classification of land surface objects from DEMs based on flow variables has a relatively long and varied history (e.g. Mark, Dozier & Frew, 1984; O'Callaghan & Mark, 1984; Band, 1986; Douglas, 1986; Jenson & Domingue, 1988; Morris & Heerdegen, 1988; Tarboton et al., 1991; Chorowicz et al., 1992; Martz & Garbrecht, 1992, 1993; Tribe, 1992). That said, the typical workflow starts with the calculation of flow directions, flow accumulation (i.e. upslope contributing areas), and possibly one or more other land surface parameters, and finishes with the delineation of the drainage networks along with the basin boundaries and suites of channel and basin attributes.

### *4.2.1 Drainage Networks and Channel Attributes*

The two groups of methods that have been used to delineate drainage networks are often referred to as the valley recognition (VR) and similarity relation (SR) methods, and this same terminology is used here.

The VR methods typically use a simple two-step process to extract the drainage network from a DEM: (i) the calculation of a flow-direction grid (e.g. Figure 4.5) using one of the 24 single- and multiple-flow direction algorithms described in Section 3.1.3; and (ii) the selection and use of an appropriate method to prune the dense drainage tree so that the flow vectors on the hillslopes are excluded. However, several problems or challenges may arise at one or both of the aforementioned steps.

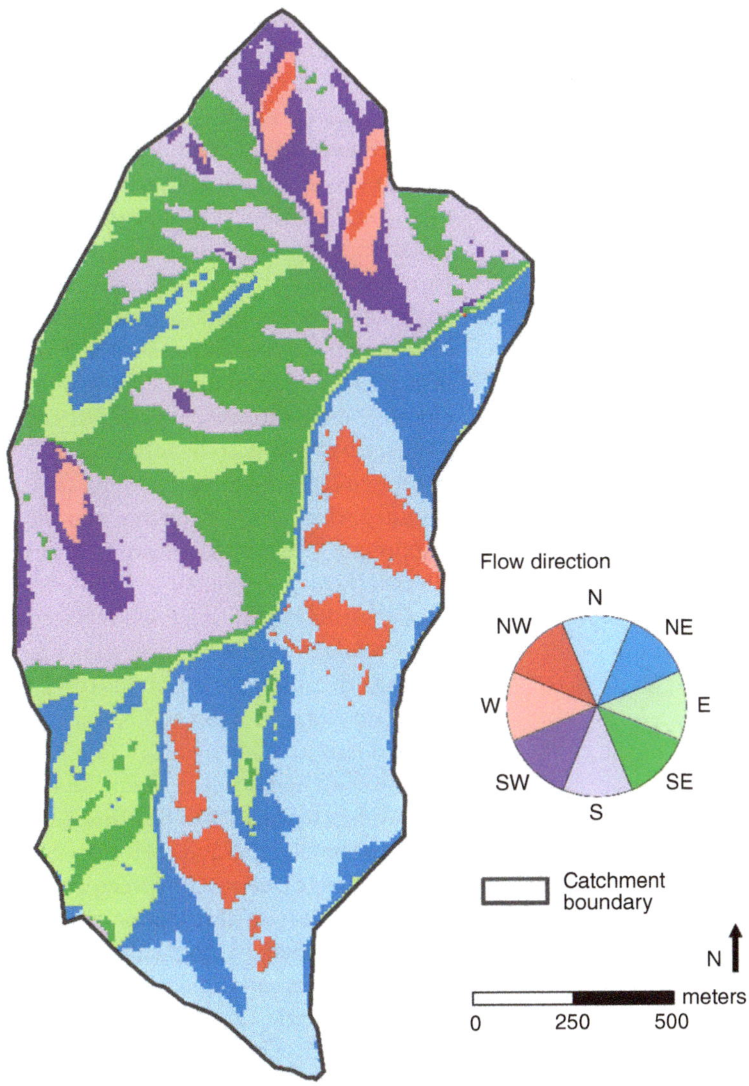


**Figure 4.5** D8 (O'Callaghan & Mark, 1984) flow direction derived for the Cottonwood Creek, MT study site with the catchment boundary overlaid.

The flow-direction grid that is produced at the first step is usually calculated with the D8 (O'Callaghan & Mark, 1984) single-flow direction algorithm when the goal is the extraction of the drainage network(s). See Band (1986, 1989), Montgomery and Dietrich (1989, 1992) and Peckham (1998) for examples that successfully used the D8 single-flow direction algorithm to delineate drainage networks. This algorithm works best in mountainous topography, and numerous studies have examined the performance of this (and other flow-direction algorithms) in large flat plains, lakes, wetlands, and other relatively flat areas with noisy DEM source data (e.g. Smith et al., 1990; Mackay & Band, 1998; Liang & Mackay, 2000; Lindsay, 2006; Hengl, Heuvelink & van Loon, 2010).

Turning next to the second step, Gruber and Peckham (2009) described two popular pruning methods. In the first and simpler of the two methods, a flow accumulation grid is calculated and the flow vectors for grid cells with upslope contributing areas less than some specified threshold are removed. Tarboton et al. (1991) suggested using a break in slope on a scatterplot of slope versus area to identify this threshold. However, Gruber and Peckham (2009) have argued that this threshold may not be apparent in every scatterplot and even if it was, this simple threshold method may not capture the natural variability because the drainage density will vary with geology, elevation, and other factors. More recently, Ariza-Villaverde, Jiménez-Hornero and Gutiérrez de Ravé (2013) have successfully used multifractal analysis to find appropriate flow accumulation threshold values that avoided some if not all of the aforementioned problems by comparing the Rényi multifractal spectra from one catchment to the next. Ariza-Villaverde, Jiménez-Hornero and Gutiérrez de Ravé (2015) used three DEM resolutions and four drainage basins to test the influence of DEM resolution on drainage network extraction by comparing the Rényi spectra of the drainage network that was obtained with the D8 algorithm and those determined by photogrammetric restitution. The results showed that: (i) DEM resolution influences the selected flow accumulation threshold and the simulated channel morphology; and (ii) the suitable flow accumulation threshold increases as the DEM resolution increases and shows greater variability for basins with low drainage densities. Qin et al. (2016) went one step further by describing a case-based formalization and reasoning method for knowledge discovery in a digital terrain modeling workflow. The authors illustrated this approach by using it with 124 cases of drainage network extraction drawn from peer-reviewd journal articles to determine the catchment area threshold.

For the second method, a Horton–Strahler order grid is calculated for the dense drainage tree and flow vectors of grid cells with orders less than some threshold value (e.g. three) are removed. Gruber and Peckham (2009) noted how: (i) this method adapts to the variability of the landscape (unlike the first method, assuming a single threshold value is chosen); and (ii) that one of the single-flow direction algorithms will work best with both of these approaches because grid cells must increase along every streamline or a disconnected network will result. The D∞ (Tarboton, 1997) single-flow direction algorithm should be counted as one of the multiple-flow direction algorithms in this instance because it permits flow to one or two downstream neighbors in certain situations.

Several SR drainage network extraction methods have been proposed that focus on source identification (i.e. channel initiation) as well. Montgomery and Dietrich (1988, 1989, 1992), for example, calculated slope and upslope contributing area grids and derived a series of threshold values for network initiation. This approach works well in rugged landscapes but is likely to produce spurious channels in flat areas (Tribe, 1992; Gruber & Peckham, 2009).

Lindsay (2006) compared the performance of four VR and two SR methods for extracting channels in a small upland catchement in the Peak District National Park of the southern Pennines, UK and found that: (i) the channel network extracted using all six of the tested algorithms will likely contain significant artifacts due to error where the magnitude of this error is moderate to large and the degree of spatial autocorrelation in the error surface is low; and (ii) the VR algorithms are more sensitive to topographic uncertainty (both error magnitude and degree of spatial autocorrelation) than the SR algorithms.

Several attempts have been made recently to combine elements from both the VR and SR approaches. Pelletier (2013), for example, proposed a new two-step, two-parameter method for extracting drainage networks from high-resolution LiDAR-based DEMs. This method first identifies the valley heads, which differ from hillslopes because they are zones of positive topographic (i.e. tangential) curvature where flow transitions from unconfined sheetflow to localized, confined flow. The method next identifies the valley bottoms downstream from the valley heads, where the valley bottoms are those parts of the landscape that are located downstream from the valley heads in which localized confined flow is maintained. Two user-defined parameters, one a threshold for the tangential curvature required to define a valley head and another the flow confinement threshold required for a valley to be maintained downstream from the valley head, are necessary. The method is implemented in six steps: (i) optimal Wiener filtering to remove microtopographic noise; (ii) mapping of the tangential curvature; (iii) identification of valley heads using the user-defined contour-curvature threshold criterion; (iv) routing of a unit discharge of water from each valley head using the FMFD (Freeman, 1991) multiple-flow direction algorithm; (v) removal of discontinuous reaches from the drainage network using a user-defined discharge-per-upstream-valley-head-threshold criterion; and (vi) the thinning of the valley network to a single grid cell width. Pelletier (2013) used a series of synthetic valley networks that mimicked the complexities of real-world landscapes and for which the true drainage network was known, to demonstrate the efficacy of this new joint VR/SR approach.

Clubb et al. (2014), on the other hand, proposed a new channel initiation method which identifies the change from channel to hillslope topography along a profile using a transformed longitudinal coordinate system, and tested the performance of this and three existing methods against field-mapped channel head data from multiple field sites. Of the three existing methods used, two were based on geometric techniques – Pelletier's (2013) tangential curvature method, which requires two user-defined parameters, and the method of Passalacqua et al. (2010) which requires five user-defined parameters and also relies on a tangential curvature threshold to specify the channel head – and the third was a process-based, slope–area scaling relationship proposed by Montgomery and Dietrich (1988, 1992), Istanbulluoghu et al. (2002), and Jefferson and McGee (2013). The field data (167 channel heads) were drawn from Feather River in California's Sierra Nevada Mountains, the Mid Bailey Run and Indian Creek in Ohio's Wayne National Forest, and Virginia's Piedmont.

Their new approach is based on the transformation of river long profiles and is called the Drainage Extraction by Identifying Channel Heads (DrEICH) method (Clubb et al., 2014). This algorithm has two principal steps: (i) the basins in which the channel heads are to be identified are selected based on the tangential curvature (i.e. the geometry of the valley); and (ii) the exact position of the channel heads within these basins is calculated using the longitudinal profiles of the channels and hillslopes. The second step uses chi transformations to compare the steepness of channels, normalized for drainage area. The transformed coordinate, $\chi$, can be calculated from any topographic dataset, suffers less from errors and uncertainties in topographic data than slope–area analysis (Perron & Royden, 2013), and is best thought of in the context of the stream power equation that proposes the fluvial incision rate is proportional to the stream power, which represents the energy expenditure of the flow (Howard, 1994; Sklar & Dietrich, 1998). The steady-state stream power equation is integrated in the upstream direction from some base level to the observation point following the river profile. The chi-plots are used to predict the channel head locations with an algorithm that assumes that the chi-plot is composed of channel and hillslope segments and that the channel head occurs at the transition point between these two segments. The concave portions of the landscape were identified in order to locate basins for channel head identification based on the drainage extraction algorithm of Peucker and Douglas (1975) that was described by Band (1986). Valleys were identified if the stretch of landscape is at least 10 m long (a 1-m LiDAR-based DEM was used for this study) and has a tangential curvature of $\geq 0.1\,m^{-1}$, which is the same threshold used by Pelletier (2013). The chi segment fitting algorithm is then run from each of the first-order valley outlets identified by the tangential curvature to the hilltop and a series of statistical parameters are used to predict the position of the channel head, from which the flow routing generates the channel network.

A comparison of the results from three of the four methods (the slope–area analysis method was not considered further because it performed poorly across all four study sites) showed that Pelletier's (2013) tangential curvature method and the DrEICH method were the most successful at locating the channel heads as mapped in the field (Figure 4.6). The distribution of errors in both of these cases were similar, both in terms of magnitude and direction,

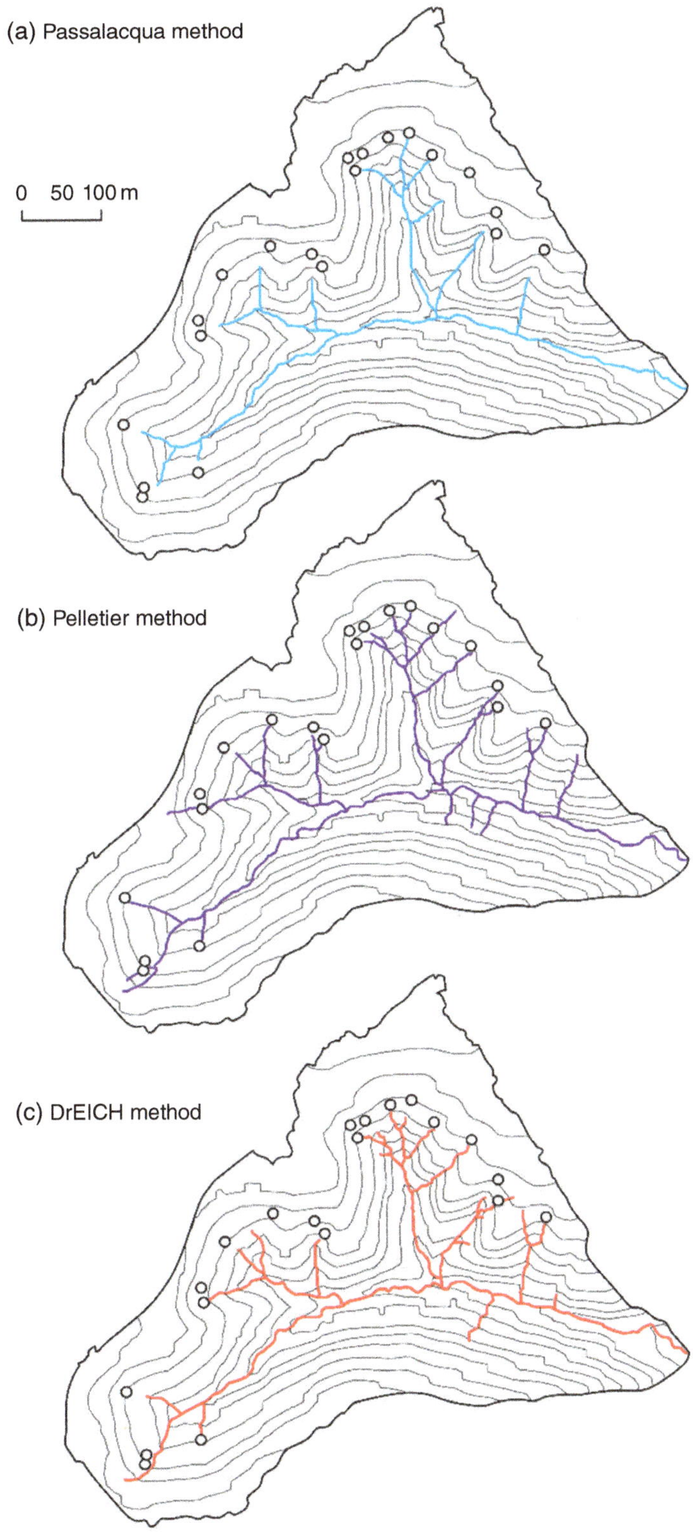


**Figure 4.6** Contour maps showing the results of using three methods to predict channel head locations for a catchment in Indian Creek, Ohio. The circles indicate mapped channel heads and the contour intervals are 10 m. The stream networks resulting from the (a) Passalacqua et al. (2010) method, (b) Pelletier (2013) method, and (c) DrEICH (Clubb et al., 2014) methods are shown in the three maps as well. *Source*: modified from Clubb et al. (2014, p. 4294). Reproduced with permission of John Wiley & Sons, Inc.

Table 4.3 List of channel attributes and their significance.

| Channel attribute | Type | Description |
|---|---|---|
| Absolute sinuosity | Regional | Ratio of along-channel and straight-line length |
| Along channel length | Regional | Along channel length of all upstream channels |
| Along channel slope | Regional | Along channel slope of all upstream channels |
| Contributing area (above downslope end) | Regional | The upslope contributing area above the downslope end of each grid cell |
| Downslope end pixel ID | Local | The identity of the grid cell that marks the end of the channel (and discharge of water to another tributary, a lake or the ocean) |
| Drainage density | Regional | The ratio of the total length of drainage lines and the area drained by them (Horton, 1932; Tarboton, Bras & Rodriguez-Iturbe, 1992; Dobos, Daroussin & Montanarella, 2005) |
| Elevation drop | Regional | Elevation drop of all upstream channels |
| Length of longest channel | Regional | Length of the longest channel among all upstream channels |
| Network diameter | Regional | The maximum number of grid cells between any upstream source |
| Relief | Regional | The difference between the elevation of a grid cell and the highest elevation in the catchement that drains into it |
| Shreve magnitude | Regional | Total number of sources (i.e. tributaries) located upstream of the grid cell |
| Source density | Regional | The number of sources above the grid cell divided by the total catchment area |
| Straight-line length | Regional | Straight-line length of all upstream channels |
| Straight-line slope | Regional | Straight-line slope of all upstream channels |
| Stream order | Regional | An integer valued measure of stream hierarchy (Horton, 1932; Strahler, 1957; Peckham & Gupta, 1999) |
| Total length | Regional | Total length of all upstream channels |
| Upslope end pixel ID | Local | The identity of the grid cell that marks the start of the channel |

*Source*: modified from Gruber & Peckham (2009, p. 190). Reproduced with permission of Elsevier.

and suggests these two approaches identify similar features as channel heads despite completely different methodologies (Clubb et al., 2014).

Once a drainage network has been delineated using any of the aforementioned approaches, then it can be stored as an array of channel segments or Horton–Strahler streams, along with the network topology or connectedness and numerous attributes (Table 4.3) (Gruber & Peckham, 2009). One potential advantage of the Horton–Strahler stream order approach is that attributes for the ensembles of sub-basins with the same Horton–Strahler stream order exhibit topologic and statistical self-similarity such that measurements at one scale can be extrapolated to other scales (Peckham, 1995; Peckham & Gupta, 1999).

### 4.2.2 Basin Boundaries and Attributes

The D8 (O'Callaghan & Mark, 1984) flow-direction grid can also be used to extract basin boundaries as polygons with the accompanying attributes (Table 4.4) (Gruber & Peckham, 2009). The boundary for the Cottonwood

Table 4.4 List of basin attributes and their significance.

| Basin attribute | Type | Description |
|---|---|---|
| Area | Regional | Surface area of the basin |
| Centroid | Local | The identity of the grid cell that marks the center of the basin |
| Diameter | Regional | The maximum distance between any two points on the boundary |
| Mean elevation | Regional | The average elevation of the grid cells that are part of the basin |
| Mean slope | Regional | The average slope of the grid cells that are part of the basin |
| Perimeter | Regional | The length of the line marking the boundary of the basin |

*Source*: modified from Gruber & Peckham (2009, p. 190). Reproduced with permission of Elsevier.

Creek study site can be calculated by identifying the downstream end pixel and then tracing the boundary around all of the grid cells that drain to this outlet to delineate the basin boundary (see Figure 4.5 for additional details).

This same approach can also be used to partition the catchment or watershed into hydrologic subunits (i.e. subwatersheds), where each subwatershed represents the set of grid cells that drain to a particular stream segment or reach that marks the tributary or main channel outlet.

## 4.3 Extracting and Classifying Specific (Fuzzy) Landforms

There are numerous authors who have attempted to extract and classify other types of discrete landscape features or units, such as alluvial fans, cirques, drumlins, dunes, mountains, ridgelines and valleys, as well. Many of these works have used one or other of the two basic fuzzy classification methods described in Section 4.1.1.

Several works have adopted the SI model and used a priori knowledge to assign a value of fuzzy membership to a landscape feature with a particular metric property such as height. For example, Usery (1996) used the height above a certain elevation as a membership function with membership increasing with height to calculate a fuzzy extent for Stone Mountain, Georgia, and Cheng and Molenaar (1999a,b) used height to determine membership functions of separate elements of dynamic beach landforms. The SR model, which uses surface derivatures such as slope and curvature as input to a multivariate classification which yields the membership values has also proved popular (e.g. Irvin et al., 1997; MacMillan et al., 2000; Burrough et al., 2000, 2001). In addition, several studies have documented how the results of using these methods will vary with the resolution of the DEM (e.g. Arrell et al., 2007; Deng et al., 2007), so that what may be a channel at one scale may be another morphometric class at another scale (Figure 4.7). One of the distinguishing characteristics of the three cases

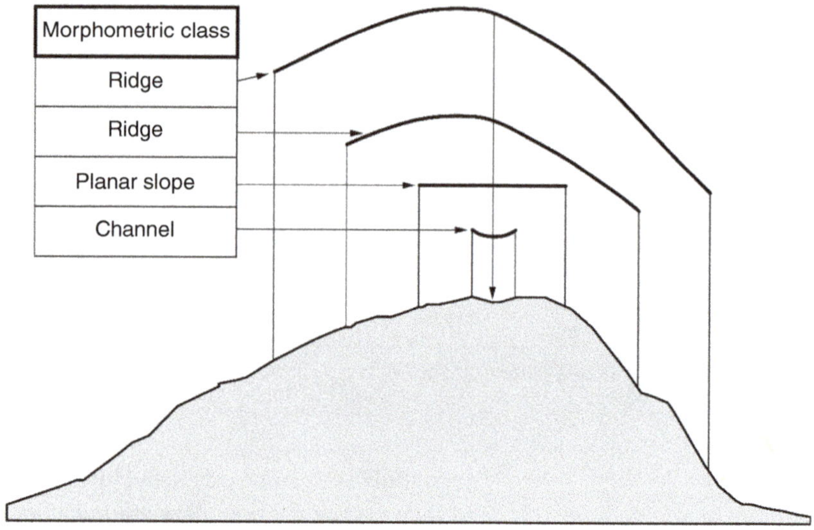


**Figure 4.7** Schematic showing how the morphometric class at the point indicated by the vertical arrow varies as shown with the scale over which it is measured. *Source*: Fisher et al. (2004, p. 109). Reproduced with permission of John Wiley & Sons, Inc.

described below is the way in which they have sought to use this information to strengthen the extraction and classification of the landforms themselves.

The first case focuses on the MRVBF index proposed by Gallant and Dowling (2003) and described in Section 3.2.1.3. A short summary of the core concepts and how this index can be used to delineate valley landforms, among several possible outcomes, is provided here. The MRVBF index uses two inputs: (i) slope, which is calculated on DEMs with coarsening resolutions and decreasing slope thresholds; and (ii) elevation percentile that is calculated with an expanding window. Slope and elevation percentile are inverted in flatness and lowness indices, respectively, and these indices are combined by multiplication at each scale to generate a valley bottom flatness (VBF) score at that scale. The VBF scores are then summed across a range of scales to generate the MRVBF which represents the multiscale membership of valley bottoms and, with an appropriate threshold, can be used to define the presence and spatial extent of valleys. Gallant and Dowling (2003) also proposed the MRRTF index which could be used in the same way to define the presence and spatial extent of ridgelines and ridgetops.

In the second case study, Fisher et al. (2004) adopted a multi-resolution approach to define the fuzzy set membership of six morphometric landscape classes – pits (all neighbors higher), peaks (all neighbors lower), passes (neighbors on two opposite sides higher and on the orthogonal sides lower), channels (neighbors on two opposite sides higher), ridges (neighbors on two opposite sides lower), and plains (no prominent curvatures defining distinct shapes) – and used the results to identify a series of culturally recognized landscape features (i.e. mountains and passes) in the English Lake District.

Fisher et al. (2004) started by explaining why landscape features such as mountains, ridges and valleys are fiat objects (Smith & Varzi, 2000), which only exist in the human understanding and division of the landscape. A mountain, for example, is not a real (bona fide) feature because it can only be interpreted as a region in the continuum of variation of the surface of the Earth. Some mountains are sufficiently noteworthy that people have named them (i.e. the Helvellyn mountain used to motivate this particular journal article), but it is much more difficult to describe or understand the spatial extent of or region associated with these kinds of features. There may be full agreement about the location of the core concept (i.e. the mountain peak in this example) but that core concept fails to capture their full identity, and they have a spatial extent beyond the core area, where some (most) people would to some extent say they exist (Usery, 1996; Wood, 1996a,b; Fisher & Wood, 1998). And to make matters worse, the class to which a location is assigned will likely vary due to the scale of observation or measurement giving rise to further ambiguity as to the correct classification and so vagueness.

To execute their multiscale analysis, Fisher et al. (2004) modeled the surface as a square-grid DEM measuring 50 m on a side and then generalized it to a series of alternative scales by fitting a local quadratic surface centered on an expanding window or filter. Morphometric analysis was then performed on the generalized surface over a range of window sizes – this study used filter sizes ranging from 3 × 3 to 75 × 75 grid cells to represent features with planimetric sizes ranging from 100 to 3700 m. The resulting morphometric classifications were executed for progressively generalized versions of the DEM and the results, which were saved at the original scale, were then combined and used with a slope threshold (Wood, 1996b, 1998) to define fuzzy memberships for each of the six morphometric classes over an area of the central Lake District. The fuzzy image of peaks was next converted to a vector theme and the vector polygons associated with each peak were isolated and used to evaluate the landscape that could be seen from specific peaks and which of the mountains and their peaks could be seen from specific locations.

The results were impressive and showed, among other things, that the class "peak" corresponds to many known mountains and hills. However, Fisher et al. (2004) also noted several problems with their method, including the difficulty of parameterizing the morphometric extraction function, and the lack of knowledge to guide the decisions about the weighting of the results from different scales and the choice of scales over which to perform the analysis.

The third and final case study illustrates how mountain peaks can be conceptualized and mapped as multiscale and multi-criteria fuzzy entities (Deng & Wilson, 2008). Four semantic meanings were imported and quantified to first characterize peaks at a range of spatial scales and, second, to evaluate the multi-criteria peakness at each scale. The individual-scale peakness was then summarized across spatial scales into a univariate membership surface to allow the mapping of vague peak entities as non-homogeneous peak regions whose boundaries depend on user-specified thresholds. The approach is illustrated using a 17.4 × 9.3 km$^2$ study area characterized by steep slopes, high relief and rugged, deep-cutting fluvial topography in the Santa Monica Mountains, California.

The case study used three circular moving windows with 5-, 20- and 80-cell radii and four criteria to define prototypical peaks. These peaks have: (i) high downslope relief in a local area; (ii) are steep in the local area; (iii) have high elevation compared with a large surrounding area; and (iv) do not have any competing peaks in this larger neighborhood. These criteria were operationalized and used to calculate peak prototypicality using equal weights for each of the four peak properties but not the scales (the larger circular windows were assigned higher weights). A threshold was next used to select the most typical summits (i.e. peak prototypes) at each spatial scale and then across all three scales, and the peak properties were then used to calculate the similarlity of each point to a scale-specific peak class center by: (i) converting an attribute distance between each point and a peak center class; and (ii) converting the attribute distance into a relative similarity score ranging from 0 to 1. And finally, four thresholds ranging from 0.65 to 0.80 were applied to the grids with the aforementioned similarity scores to identify four sets of peak entities in the study area.

## 4.4 Extracting and Classifying Repeating Landform Types

MacMillan and Shary (2009) used the term "repeating landform types" to refer to distinctive patterns of terrain (i.e. plains, hills, mountains and valleys) that exhibit variation in the size, scale, and shape of the landscape features and often occur in a recognizable contextual position relative to the adjacent geomorphc features. Different authors over the years have used different terms to describe these types: Speight (1974), for example, used the term "landscape patterns" and Dikau (1989) used the term "relief forms" for these repeating landform types.

The classification of landform types typically occurs at a higher level of absraction and at a broader scale than the classification of the landform elements described in Section 4.1. These classifications have a long history and the earliest examples relied on manual, subjective systems to analyze contour maps and generate results at two or more specific levels in multilevel hierarchies of landform types (e.g. Veatch, 1935; Fenneman, 1938; Hammond, 1954, 1964, 1965). Many of the more recent examples, in contrast, have incorporated varying levels of automation but most have focused on relatively small geographic scopes and some stopped short of specifying landscape types (e.g. Troeh, 1964, 1965; Speight, 1974, 1990; Pike, 1988; Dikau et al., 1991, 1995; Guzzetti & Reichenbach, 1994; van Engelen & Ting-Tiang, 1995; Brabyn, 1998).

The remainder of this discussion of these classification methods focuses on the Hammond method because it has been automated and applied in a variety of regional settings over the years (e.g. Dikau et al., 1991, 1995; Brabyn, 1998; Gallant et al., 2005; Morgan & Lesh, 2005; Hrvatin & Perko, 2009) and just

Table 4.5 Landform classification criteria used by Dikau et al. (1991).

| Distribution of gentle slopes | Local relief | Profile type |
|---|---|---|
| (A) More than 80% of the area is gently sloping | (1) 0–30 m | (a) More than 75% gentle slope is lowland |
| (B) 50–80% of the area is gently sloping | (2) 30–91 m | (b) 50–75% of gentle slope is lowland |
| (C) 20–50% of the area is gently sloping | (3) 91–152 m | (c) 50–75% of gentle slope is upland |
| (D) Less than 20% of the area is gently sloping | (4) 152–305 m<br>(5) 305–915 m<br>(6) >915 m | (d) More than 75% gentle slope is upland |

*Source*: MacMillan & Shary (2009, p. 246). Reproduced with permission of Elsevier.

recently it was modified and used to propose a new global classification of landform types (Karagulle et al., 2017).

Hammond's (1954) extensive field experience provided the rationale for the empirical definition of slope gradient, local relief, and profile type as the essential parameters for classifying repeating landform types. This method incorporated four classes of gentle slopes, six classes of relative relief, and four classes of profile type and provided 96 possible landform subclasses (Brabyn, 1998). These 96 subclasses are commonly regrouped into 24 landform classes and five landform types (Bayramin, 2000). Hammond (1964) later assessed these parameters through visual analysis of 1:250,000-scale topographic maps and produced a pioneering map entitled *Classes of Land Surface Form of the Forty-eight States, USA* (Karagulle et al., 2017).

Dikau and colleagues were the first to automate Hammond's approach to generate a digital landform map of the State of New Mexico (Dikau et al., 1991, 1995). These authors used the same inputs (i.e. four slope, six local relief and four profile type classes; Table 4.5) and produced the same outputs (five landform types, 12 landform classes, and 96 landform subclasses; Table 4.6) as Hammond (1954, 1964). However, Dikau and his colleagues used a 200-m DEM and a 9.65-km neighborhood analysis window (NAW) within which the three parameters were calculated in place of the 1:250,000-scale topographic maps used by Hammond. The slope gradient was calculated in a $3 \times 3$ moving window and the distribution of slopes for the cells included in the NAW was used to estimate the proportion of cells that were flat (i.e. <8%) versus sloping (see column 1 in Table 4.5 for additional details). The relative relief was calculated for each grid cell as the difference between the maximum and minimum cell elevations within each NAW (column 2, Table 4.5). The third profile type input required by the Dikau method (column 3, Table 4.5) was defined as the proportion of flat or gently sloping terrain that occurs in upland versus lowland areas. To generate these inputs, the center cell in the NAW was examined first to determine whether

**Table 4.6** Landform classes and subclasses used by the Dikau method.

| Landform type | Landform class | Landform subclass code |
|---|---|---|
| Plains | Flat or nearly flat | A1a, A1b, A1c, A1d |
| | Smooth plains with some local relief | A2a, A2b, A2c, A2d |
| | Irregular plains with low relief | B1a, B1b, B1c, B1d |
| | Irregular plains with moderate relief | B2a, B2b, B2c, B2d |
| Tablelands | Tablelands with moderate relief | A3c, A3d, B3c, B4c |
| | Tablelands with considerable relief | A4c, A4d, B4c, B4d |
| | Tablelands with high relief | A5c, A5d, B5c, B5d |
| | Tablelands with very high relief | A6c, A6d, B6c, B6d |
| Plains with hills or mountains | Plains with hills | A3a, A3b, B3a, B3b |
| | Plains with high hills | A4a, A4b, B4a, B4b |
| | Plains with low mountains | A5a, A5b, B5a, B5b |
| | Plains with high mountains | A6a, A6b, B6a, B6b |
| Open hills and mountains | Open very low hills | C1a, C1b, C1c, C1d |
| | Open low hills | C2a, C2b, C2c, C2d |
| | Open moderate hills | C3a, C3b, C3c, C4c |
| | Open high hills | C4a, C4b, C4c, C4d |
| | Open low mountains | C5a, C5b, C5c, C5d |
| | Open high mountains | C6a, C6b, C6c, C6d |
| Hills and mountains | Very low hills | D1a, D1b, D1c, D1d |
| | Low hills | D2a, D2b, D2c, D2d |
| | Moderate hills | D3a, D3b, D3c, D3d |
| | High hills | D4a, D4b, D4c, D4d |
| | Low mountains | D5a, D5b, D5c, D5d |
| | High mountains | D6a, D6b, D6c, D6d |

*Source*: Dikau et al. (1991, p. 9).

it lay in the upper or lower half of the NAW and the proportion of flat cells in upland and lowland areas within the NAW was then calculated and used to differentiate areas with flat tablelands from areas with flat bottomlands (Brabyn, 1998; MacMillan & Shary, 2009). Both Guzzetti and Reichenbach (1994) and Brabyn (1997, 1998) noted various concerns with misclassification (e.g. the tendency for the Dikau/Hammond method to classify areas with different landform classes into the same class) and imprecise boundaries, particularly in transition zones between areas of high and low relief.

Several similar methods have been developed and deployed during the past 20 years. True (2002), for example, used a variant of the Hammond/Dikau method that omitted the profile type parameter and used a 1-km NAW and 30-m NED source DEM (Gesch et al., 2002). This approach, which is also referred to as the MoRAP (Missouri Resource Assessment Partnership), was subsequently adopted by the USGS and used to map Hammond's landform types in South America (Sayre et al., 2008), the conterminous USA (Sayre et al., 2009), Africa (Sayre et al., 2013), and the world (Sayre et al., 2014). The global product was developed on top of a 250-m global DEM (Danielson & Gesch, 2011) and used a 1-km NAW. Karagulle et al. (2017) have subsequently reviewed the classified landforms in the global product against a multidirectional shaded relief map where they expected to see mountain, hill, and plains landforms visually aligned with matching terrain features in the hillshade product. These authors found that: (i) landforms had a faceted and fragmented pattern rather than a regionally homogeneous appearance; (ii) tablelands were conspicuously absent; and (iii) certain areas commonly regarded as hills were classified as plains. They attributed some of these problems to the omission of Hammond's profile parameter in the MoRAP method and the choice of a 1-km NAW in the Sayre et al. (2014) global application.

The choice of the size of the NAW has been frequently examined during the past 25 years. The key challenge is how to establish the extent of the search window within which the various parameters are calculated and the general consensus is that the extent of these NAWs should be allowed to vary to match the spatial texture of the land surface. MacMillan and Shary (2009, pp. 247–251) recently reviewed some of the proposals for how this broad-scale land surface texture might be captured and used to vary NAWs before concluding that much work needs to done before automated classification of repeating landform types could make effective use of the various strategies and measures that have been proposed thus far.

That said, Karagulle et al. (2017) recently published a new characterization of global Hammond landform regions. They began by implementing the approach of Morgan and Lesh (2005) as an ArcGIS geoprocessing script tool written in Python. Morgan and Lesh (2005) had previously adapted and used the approach of Dikau et al. (1991, 1995) in ArcGIS to classify Hammond landforms in Maryland using a 30-m DEM. They chose the Morgan and Lesh (2005) approach to start with because it used the preferred computational platform (i.e. ArcGIS) and it included the profile parameter that some of the more recent work had skipped altogether. Karagulle et al. (2017) also chose the 250-m GMTED (Danielson & Gesch, 2011) source data, which was also used by Sayre et al. (2014) for their global product, and made four modifications to the Morgan and Lesh (2005) approach that allowed for: (i) identification of optimal NAW sizes; (ii) inclusion of negative (i.e. below sea level) elevation values; (iii) increased separation of low undulating hills and flat areas; and (iv) reduction of the fragmented appearance of the landforms generated with the MoRAP approach. The GMTED was projected from the original World Geodetic System (WGS) 1984 geographic coordinate system to World Equidistant Cylindrical to ensure that the circular NAW would use the same number of cells and a consistent distance from the center of the NAW to the edge, anywhere on Earth.

The three parameters were calculated as follows. Percent slope was calculated using a 12.5-cell (3-km) NAW. Each cell in the NAW was assigned the average of the slopes between that cell and its eight neighbors and these slope values were then averaged within the NAW and interpreted using the four rules listed in the first column of Table 4.5. Local relief represents the amount of elevation change in the NAW and was calculated as the difference between the maximum and minimum elevations in the NAW and interpreted using the six rules listed in the second column of Table 4.5. The profile parameter was calculated by first assigning every cell in a 25-cell (6-km) NAW into an upland or lowland class relative to the midpoint of the cell elevations observed in the NAW and then calculating the amount of gentle slope within the NAW. Karagulle et al. (2017) added a fifth profile class labeled "Less than 50% of upland or lowland in gentle slopes" to the four listed in the third column of Table 4.5 before acknowledging that the affected areas were already implicitly classified correctly using the slope and local relief parameters.

Karagulle et al. (2017) next calculated Hammond landform classes and types by summing the slope, relief, and profile outputs and visually compared this output with a shaded relief model for topographic context. This yielded numerous mismatches and so they extracted plains features separately using a 1-km NAW in a single global implementation. The plains-centric adjustment was deployed as an override and eliminated four dual-landform classes from the original Hammond/Dikau classification model (Table 4.7). The majority filter tool was also run four times to remove or smooth the "salt-and-pepper" effect caused by the one- and two-cell landform features that resulted from the initial integration of the three profile outputs.

Overall, this new method used a 3-km NAW for the slope parameter and a 6-km NAW for both the local relief and profile parameters, introduced a new plains-centric override adjustment (that incorporated a 1-km NAW), and used data filtering to remove numerous small inclusions. Table 4.7 shows how six of the landscape classes included in the Hammond/Dikau method were dropped with this new method and how the assignment of landform classes to landform types was changed when the number of landform types was reduced from five to four with this new method of Karagulle et al. (2017). Table 4.8, on the other hand, shows how the seven landform classes used in the global application of Sayre et al. (2014) were increased to 16 in this new application and how both the Sayre et al. (2014) and Karagulle et al. (2017) landscape classes were assigned to the four landform types used in the first of these two applications.

And finally, Figure 4.8 shows the comparison of the total area of the four landform types (plains, hills, low mountains, and high mountains) for the Sayre et al. (2014) and Karagulle et al. (2017) applications. This bar chart shows that the new method identified considerably less area in plains and a corresponding increase in low and high mountains. This result thereby addressed one of the shortcomings (the over-prediction of plains and under-prediction of low and high mountains) noted by Sayre et al. (2014) when they had published their global landform classification. The new global Hammond landform classification and map produced by Karagulle et al. (2017) and the underlying data can be accessed via ArcGIS Online and the Living Atlas at http://esriurl.com/hammond or http://rmgsc.cr.usgs.gov/outgoing/ecosystems/Global/.

**Table 4.7** Comparison of landform classes used by the Dikau and Karagulle methods and their assignment to landform types.

| Landform class | Dikau landform type | Karagulle landform type |
|---|---|---|
| Flat or nearly flat plains | Plains | Plains |
| Smooth plains with some local relief | | Plains |
| Irregular plains with low relief | | Hills |
| Irregular plains with moderate relief | | Hills |
| Tablelands with moderate relief | Tablelands | Hills |
| Tablelands with considerable relief | | Hills |
| Tablelands with high relief | | Hills |
| Tablelands with very high relief | | Hills |
| **Plains with hills** | Plains with hills or mountains | |
| **Plains with high hills** | | |
| **Plains with low mountains** | | |
| **Plains with high mountains** | | |
| **Open very low hills** | Open hills and mountains | Hills |
| **Open low hills** | | Hills |
| Open moderate hills | | Low mountains |
| Open high hills | | High mountains |
| Open low mountains | | |
| Open high mountains | | |
| **Very low hills** | Hills and mountains | Hills |
| **Low hills** | | Hills |
| Moderate hills | | Low mountains |
| High hills | | High mountains |
| Low mountains | | |
| High mountains | | |

*Source*: Karagulle et al. (2017). Reproduced with permission of John Wiley & Sons, Inc.

**Table 4.8** Comparison of global Hammond landform classes and types modeled by Sayre et al. (2014) and Karagulle et al. (2017).

| Landform types | Sayre landform classes | Karagulle landform classes |
|---|---|---|
| Plains | Flat or nearly flat plains | Flat or nearly flat plains |
| | Smooth plains with some local relief | Smooth plains with some local relief |
| Hills | High hills | High hills |
| | Moderate hills | Moderate hills |
| | Low hills | Scattered (i.e. open) high hills |
| | | Scattered (i.e. open) moderate hills |
| | | Irregular plains with low relief |
| | | Irregular plains with moderate relief |
| | | Tablelands with moderate relief |
| | | Tablelands with considerable relief |
| | | Tablelands with high relief |
| | | Tablelands with very high relief |
| Low mountains | Low mountains | Low mountains |
| | | Scattered (i.e. open) low mountains |
| High mountains | High mountains | High mountains |
| | | Scattered (i.e. open) high mountains |

*Source*: Karagulle et al. (2017). Reproduced with permission of John Wiley & Sons, Inc.

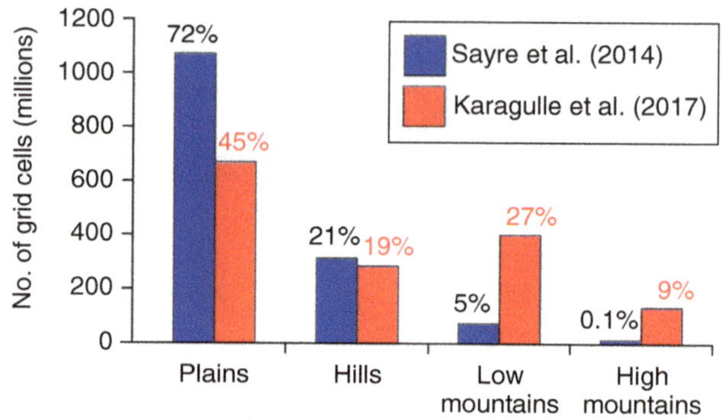


**Figure 4.8** Comparison of major landform types between the Sayre et al. (2014) and Karagulle et al. (2017) maps. *Source*: modified from Karagulle et al. (2017). Reproduced with permission of John Wiley & Sons.

## 4.5 Discrete Geomorphometry: Coupling Multiscale Pattern Analysis and Object Delineation

The applications described thus far give some sense of the great progress that has been made with automated landform classification during the past quarter century. However, the successful deployment of these techniques requires considerable knowledge and experience with the science of geomorphology, the techniques themselves, and the study area to which they are to be applied (Minár & Evans, 2008; Evans, 2012). In addition, most of the applications that have been described thus far are scale-specific and constrained by the availability of gridded elevation datasets and/or the author's knowledge of the landscapes to which the techniques were applied. Some of the applications used multiscale analysis to address the scale constraints (e.g. Gallant & Dowling, 2003; Deng & Wilson, 2008) and smaller numbers have used field observations (e.g. Clubb et al. 2014) and/or digital reconnaissance (e.g. Karagulle et al., 2017) to acquire and utilize additional knowledge of the study area in the implementation of the final classification method.

However, a series of recent papers has questioned whether or not the approaches described thus far are the appropriate ones for the task at hand and the title used for this section is taken from perhaps the most influential of these papers authored by Drăguţ and Eisank (2011). The discussion which follows summarizes the various arguments and describes one potential pathway to better connect general and specific geomorphometry. The term "discrete geomorphometry" was coined and used by Drăguţ and Eisank (2011) to describe this general approach.

Most of the landform classification systems implemented in the applications described thus far started with the classification of cells and finished up with the clustering of these results to delineate the extent of the final objects, instead of delineating the objects first and then classifying those objects (Drăguţ & Eisank, 2011). The approach of classifying cells directly brings five sets of problems.

1. The cells are treated as spatially independent from one another and thus adjacent cells are frequently assigned to different classes resulting in a highly scattered spatial representation of landforms (Burrough et al., 2001), creating the so-called "salt-and-pepper" effect that Karagulle et al. (2017) noted and used filtering to minimize in their global Hammond landform classification.
2. The scale of the analysis is tied to the footprints of cells and the NAW which will likely have nothing to do with the size and shape of real-world entities such as landforms (Fisher, 1997; Goodchild, Yuan & Cova, 2007). Making matters worse, the land surface parameters calculated for specific cells may be assigned different values depending on the DEM resolution and the size of the NAW.
3. The question about which DEM resolution and NAW are best suited to a given purpose is more often than not decided by data availability (Schmidt & Andrew, 2005) and only occasionally decided by expert knowledge (e.g. Gustavsson & Kolstrup, 2009) or an exploratory approach to the selection and use of land surface parameters (Deng, 2007; Deng et al., 2007).
4. The difficulty of including topological relationships, such as those represented by the single- or multiple-flow directions (i.e. pathways) connecting upslope and downslope cells in specific catchments and those represented by the tributaries that coalesce and connect subcatchments into larger hydrologic basins, in the classification.
5. The difficulty of developing hierarchies of landforms that incorporate all of the important levels, including the various examples of landscape elements, specific landforms, and repeating landform types that were described in Sections 4.1–4.4.

Drăguţ and Eisank (2011) proposed discrete geomorphometry as a way to connect general and specific geomorphometry and to thereby avoid one or more of the aforementioned limitations.

The approach articulated in Drăguţ and Eisank (2011) incorporated three components: (i) the objects were defined as homogeneous areas delineated by discontinuities in one or more land surface parameters; (ii) these objects were used to reveal the land surface patterns at a given scale or across multiple scales; and (iii) the character of these objects was assigned after or along with the segmentation methods used to delineate homogeneous areas. These authors saw these objects as intermediate building blocks in the transition from grid cells to landform divisions and that possible candidates for such objects have been called *land components* by Dymond, Derose, and Harmsworth (1995), *terrain facets* by Rowbotham and Dudycha (1998), *elementary forms* by Minár and Evans (2008), *pattern elements* by Drăguţ et al. (2009), and *morphometric primitives* by Gessler et al. (2009) in past work. Once the objects have been delineated, they can be used with further statistical, relational, and semantic rules to map each object to the closest landform concept and assigned to classes of elemental form or given interpretations such as "alluvial fan," "mountain" or "valley."

This new approach starts with the definition of objects and delineation of homogeneous areas and in both instances the paper by Minár and Evans (2008) is particularly instructive here. For the first task, these authors specified elementary forms as landform elements with a constant value of altitude or of two or more readily interpretable morphometric variables (i.e. local land surface parameters), bounded by lines of discontinuity. Each of the forms was represented by a specific mathematical function (i.e. fitted function) and the lines of discontuinuity were represented by a sudden change of value of one (simple discontinuity line) or more (compound discontinuity line) properties. Their paper includes a series of instructive diagrams and the form-defining equations for the first-, second- and third-order derivatives. Turning to the second task, Minár and Evans (2008) also thought that segmentation of the land surface could provide the transition from the field to the object model (Brändli, 1996; Cova & Goodchild, 2002) and from general to specific geomorphometry (Evans, 1972, 1987).

A number of segmentation algorithms that were originally developed for the analysis of remotely sensed images have been successfully applied for geomorphologic mapping or the delineation of homogeneous areas from DEMs during the past 10–15 years (e.g. Lucieer & Stein, 2005; Drăguţ & Blaschke, 2006, 2008; Möller et al., 2008; Schneevoight et al., 2008; Anders, Seijmonsbergen & Bouten, 2009; Jellema et al., 2009; Romstad & Etzelmüller, 2009; Stepinski & Bagaria, 2009; Gerçek, Toprak & Strobl, 2011). The optimum scale (or scales) needs to be determined in all these segmentation algorithms and, here again, a variety of approaches has been described for selecting the appropriate scale parameters (Ming, Wang & Zhang, 2015).

Drăguţ and Eisank (2011) favored the multi-resolution segmentation (MRS) algorithm (Baatz & Schäpe, 2000), as implemented in the eCognition software, and the local variance (LV) data-driven scale detection method. These are described in more detail later because these same authors have adopted and used these methods to support an automated object-based classification of topography from SRTM data that covers most of the Earth's land surface (Drăguţ & Eisank, 2012).

The MRS algorithm is a multiscalar region-merging technique to create objects from pixels (i.e. grid cells) through an optimization process that minimizes the internal weighted heterogeneity of each object at a given scale. The objects are then merged or split to create objects at consecutive scales, either higher, created in a bottom-up approach, or lower, created in a top-down approach. In this way, each decision of merging or splitting is based on the attributes of homogeneous structures of a recent scale and on a user-defined heterogeneity threshold, called the scale parameter. This particular segmentation approach has proved popular for delineating landform elements and specific landforms in recent years (e.g. Drăguţ & Blaschke, 2006, 2008; Schneevoight et al., 2008; Anders et al., 2009; d'Oleire-Oltmanns et al., 2013), even though there was no tool available to objectively guide the selection of appropriate scales for segmentation in the earliest applications.

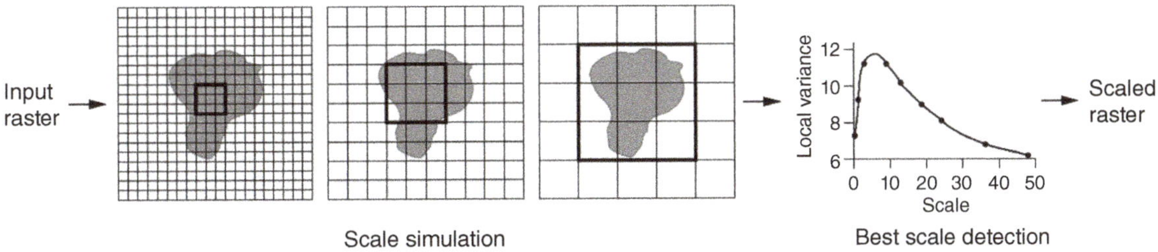


**Figure 4.9** Schematic showing local variance (LV) method applied to the grid cells in a DEM. *Source*: Drăguţ, Eisank & Strasser (2011, p. 164). Reproduced with permission of Elsevier.

Li (2008) and Drăguţ et al. (2009) have both advocated using LV to guide the selection of appropriate scales for multiscale analysis in geomorphometry. This method was originally developed in image analysis to help with scale detection (Woodcock & Strahler, 1987). The approach examines the relationship between the size of objects in the real world and pixel (i.e. grid cell) resolution, as expressed by the spatial structure of the images. The typical pattern is captured in Figure 4.9. The image on the left shows a high-resolution scene where the one object covers multiple grid cells. The spatial autocorrelation is high and the local variance, which is calculated as the average value of standard deviation measured in a small $3 \times 3$ moving window is therefore small. The next two scenes (moving left to right) show coarser scale images, which were produced from the initial dataset through resampling, where the grid cells are similar in extent to the object in this scene. The maximum value of LV will be recorded with the third scene because the likelihood of neighbors being similar is minimized. At coarser grid cell sizes, the LV will decrease again as a consequence of including more objects within a grid cell and thereby reducing the difference between neighboring grid cells (Drăguţ & Eisank, 2011).

Drăguţ et al. (2011) set out to show that the LV method can help in detecting characteristic scales in geomorphometric analysis, as is the case in remote sensing applications using LiDAR DEMs for low relief and mountainous areas. The procedure included four steps: (i) three land surface parameters (slope gradient, plan and profile curvatures) were generated at constant increments with either resampling (cell-based) or image segmentation (object-based); (ii) LV was calculated as the average standard deviation with a $3 \times 3$ moving window for each scale level; (iii) the rate of change of LV (ROC-LV) was calculated from one level to another; and (iv) the values so obtained were plotted against scale. Visual assessment demonstrated the efficacy of this metric and showed that the LV method performed better on scale levels generated through segmentation rather than upscaling through resampling.

The aforementioned procedure was then implemented in a tool called ESP (Estimation of Scale Parameter; Drăguţ, Tiede & Levick, 2010) for application in Trimble's eCognition Developer software (http://www.ecognition.com). This new tool iteratively generates image objects at multiple scale levels in a bottom-up approach and calculates the LV for each scale. The thresholds in the ROC-LV are then used to indicate the scale levels at which the image can be segmented in the most appropriate manner, relative to the data characteristics at the scene level. Drăguţ et al. (2010) conducted a series of tests on different types of imagery to show how this new tool enabled fast and objective parameterization when performing image segmentation.

The efficacy of this new approach was demonstrated in two recent applications. In the first application, Drăguţ and Eisank (2011) automatically segmented and classified an SRTM elevation layer at three scale levels by using the self-adaptive, data-driven techniques described above. This application used one layer (elevation) as input, which did not need any preprocessing, and the segmentation and classification required just two parameters, elevation and standard deviation of elevation. The results resembled the patterns of existing global and regional classifications, displaying a level of detail close to manually drawn maps and statistical analysis showed that most classes maximized the internal and minimized the external homogeneity. The methodology was implemented as a customized process for the eCognition software.

In another application, d'Oleire-Oltmanns et al. (2013) used the aforementioned object-based workflow to extract landforms at multiple scales from two distinct data types. Optical data (i.e. contrast information and shape properties of segments gathered from aerial photographs) were used for gully mapping on a very local scale and terrain data (i.e. contextual and shape information gathered from DEMs) were used for drumlin mapping at a slightly broader scale. An accuracy assessment was performed by comparing classification results with independent reference datasets that were delineated manually from the input data and the results showed that this method achieved satisfactory agreement between mapped and reference landforms.

The various examples highlighted throughout this chapter have demonstrated steady but relentless growth in the number and sophistication of the methods and data sources used to classify and map land surface objects and landforms. The knowledge-based methods highlighted in this last section offer new opportunities to improve both the accuracy and transferability of these classification systems, and can be further used to develop hierarchical classifications of landform elements, landforms, and repeating landform types.

That said, the next chapter explores three topics – the identification and treatment of error and uncertainty, the concept of fitness-for-use, and the role of multiscale analysis and cross-scale inference – that influence nearly every aspect of the typical digital terrain workflow that has been the overarching focus of Chapters 2 through 4.

# 5 Measuring Error and Uncertainty

The previous chapters have shown how: (i) DEMs can be constructed in a variety of ways nowadays; (ii) there are many subtleties associated with the various elevation data sources and terrain modeling methods; and (iii) errors can be introduced at every stage of the production process. In addition, there are many challenges connected with these errors and the accompanying uncertainty. Some can be attributed to the fact that these errors may vary with the choice of sensor, the specific application (i.e. the method of deployment and use of elevation data), and/or the landscape (i.e. study area) of interest. Another set of challenges concerns the propagation of the elevation errors in the primary and secondary land surface parameters and the considerable effort that will usually be required to identify them. The usual approach for propagating errors relies on statistical modeling of the error in the DEM (which is often only partially known) and then running a Monte Carlo analysis. The best work will utilize these techniques to check for and hopefully remove some or all of the errors; however, the errors in the source data cannot always be eliminated and those interested in using the land surface parameters and/or objects calculated from DEMs must be cognizant of these errors and how they may affect their own workflows and the efficacy and significance of the results.

That said, this chapter takes stock of what we know about the accuracy of DEM elevation values and the land surface parameters and objects that are calculated from these elevations. The discussion starts with a review of the various ways in which researchers have identified and treated errors and uncertainty in digital terrain modeling (Section 5.1) and then moves on to consider some of the ways in which error and uncertainty intersect with two concepts that were introduced in earlier chapters – fitness-for-use and scale – in Sections 5.2 and 5.3, respectively. The chapter concludes in Section 5.4 with a description of the US National Water Model that was recently launched to show how useful products and services can be constructed notwithstanding the presence of error and uncertainty.

*Environmental Applications of Digital Terrain Modeling*, First Edition. John P. Wilson.

## 5.1 Identification and Treatment of Error and Uncertainty

Many papers have been published with the goal of describing DEM errors since the landmark paper on the general and specific modes of geomorphometric analysis authored by Evans in 1972 (Figure 5.1). Some of these papers also proposed methods for reducing or eliminating these errors and an even smaller number described ways to characterize the resulting uncertainty and then incorporate it into the subsequent modeling workflows.

Wise (2000a) described several easy-to-follow examples which show why the details matter when working to understand and minimize error. Similarly, Hebeler and Purves (2009, pp. 4–6) did a good job of introducing the concepts of error and uncertainty and why both matter. There are many possible sources of error and Hebeler and Purves (2009, p. 5) treated this error, whatever its source, as "the deviation of a measurement from its true value, implying that the elevation error can only be determined if a set of more accurate reference data is available," as was the case in Rexer and Hirt (2014) for example.

This definition represents a good start but it falls short in a couple of ways because it ignores both the presence of systematic bias and the spatial pattern of errors which may be critical for those land surface parameters that are heavily influenced by the shape of the land surface (Hutchinson & Gallant, 2000; Deng et al., 2008; Wise, 2010) on the one hand and the possibility of introducing additional errors at subsequent stages of a typical terrain

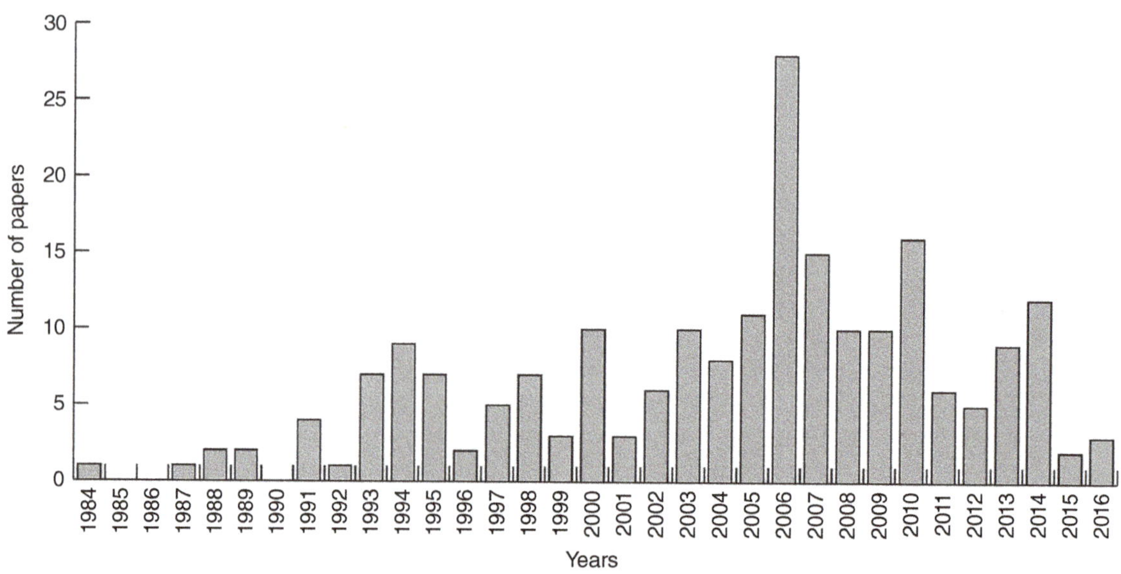


**Figure 5.1** Number of papers focused on DEM error and uncertainty cited in Section 5.1 by year of publication.

modeling workflow (i.e when calculating flow-based primary and secondary land surface parameters for example) on the other hand.

The key observation is that no matter what the source of the error, neither the magnitude nor spatial distribution of the error will usually be known at any particular location and this creates uncertainty. Hebeler and Purves (2009, p. 5) therefore used the term "uncertainty" to refer to the situation "where a (modeled) value is expected to deviate from its true value, but it is uncertain to what extent, and is often associated with confidence intervals (e.g. Kyriakidis, Shortridge & Goodchild, 1999; Endreny & Wood, 2001; Shortridge, 2001)." This uncertainty can be approximated through the use of geostatistics or uncertainty models, as illustrated by several of the papers highlighted in Section 5.1.2.

Several authors have also described the types of error which might be present. Two examples are introduced here because the various types of error may help one to assess the magnitude or extent of the error and the various ways in which it might be introduced to a typical terrain modeling workflow.

In the first example, Wise (2000b) distinguished blunders, and systematic and random errors. Blunders are gross errors which may occur because of equipment failure or human error, as might happen, for example, when an inappropriate map projection is chosen. Systematic errors show a common trend or dependency and usually reflect some problem with the data capture or processing workflow, such as the striping that plagued the early efforts to develop USGS 7.5-minute DEMs (e.g. Brown & Bara, 1994; Oimoen, 2000), the systematic biases that have plagued some LiDAR and SAR missions (e.g. Filin, 2003; Katzenbeisser, 2003; Lalonde et al., 2010), and the tendency for streams that have been encased in pipes or covered over in urban areas to be omitted from vector hydrography datasets (e.g. Sheng et al., 2007). Random error may originate from many sources and generally no trend can be observed.

This classification is helpful because the general expectation is that the systematic error can be computationally reduced or eliminated by some appropriate method and/or with the help of one or more ancillary data sources once it has been identified. However, this will only work sometimes. Hence, Sheng et al. (2007) were able to use ANUDEM (Hutchinson, 1989) and historical aerial photography to locate, model, and add the "missing" streams to the NHD streamlines in Los Angeles, California. However, Oksanen and Sarjakoski (2006) compared a series of fine toposcale DEMs with 5–50 m grid spacings derived from contours with high-quality reference data derived from airborne laser scanning and found that: (i) the spatial autocorrelation of the fine toposcale DEM error was the result of a complex combination of random and systematic-like components; (ii) appropriate modeling by geostatistical methods was problematic because of the small extent of the areas in which the assumption of stationarity was valid; and (iii) the description of the shape of the DEM error distribution was impossible with a single parameter of dispersion because of the large number of outliers. The results from this last study point to the immense challenges we sometimes face and the need for more robust descriptors of the error than the conventional error statistics in some applications.

The second classification of error offered by Fisher and Tate (2006) identified three main sources of DEM error:

1. the error associated with the measurement and generation of the source data used to create the DEM(s);
2. the error generated with the data processing (i.e. conditioning) that preceded the generation of the DEM from the source data; and
3. the error related to the properties of the terrain surface being modeled with respect to their representation in the DEM.

The third source of error listed is particularly important for a couple of reasons. First and as noted by Hebeler and Purves (2009), this source highlights a fundamental and often neglected consideration when working with DEMs, namely that the resolution and representation of the land surface and the derivation of primary and secondary land surface parameters from that terrain surface can introduce further ambiguity and contribute additional uncertainty. The second reason is to repeat once again that much of the work on error and uncertainty thus far has focused on the DEMs themselves, rather than the terrain properties in the landscape (i.e. study area) of interest, when both should command our attention (Hutchinson & Gallant, 2000; Liu et al., 2015).

That said, the next two subsections review the work to date, starting with a description of the errors and then moving to uncertainty and finishing up with a case study showing how uncertainty can be modeled and used with primary and secondary land surface parameters in various kinds of environmental process modeling workflows.

### 5.1.1 Error

The following review of the 200-plus terrain modeling papers that have tackled some aspect of error and/or uncertainty is organized around the terrain modeling workflow that was summarized in Figure 2.1. This means that the errors associated with: (i) the choice of data model; (ii) the discretization of space, the choice of elevation source data and the interpolation and/or filtering approach that was used; (iii) the preprocessing of the source data to remove unwanted depressions, resolve flow directions in flat terrain and merge the DEM with streamlines acquired from some other vector hydrography source; and (iv) the methods used to calculate the many primary and secondary land surface parameters which have been proposed, including (v) the flow-based parameters which must process activity in the correct order, are described in turn. The goal is not to describe everything that has been learnt, since this might double the length of this book, but rather to describe the kinds of topics that individual works have covered so the reader is provided with a roadmap of sorts to help them follow topics of specific interest.

#### 5.1.1.1 Choice of Data Model

This review starts with the error that may accompany the choice of data model. The review of the strengths and weaknesses of the three elevation data networks in Section 2.1 is based primarily on I.D. Moore et al. (1991). There were several papers around this same time that explored the effects of different land surface or terrain representation schemes (e.g. Band, 1993; Kumler, 1994; Band et al., 1995), but the ways in which the data sources and processing methods have evolved during the past 25 years – from ground surveys and topographic map conversion to passive methods of remote sensing and more recently to active sensing with LiDAR and radar interferometry – have helped to consolidate the role of the square-grid DEM as the predominant data model. However, the TIN has made a comeback of sorts in recent years and many of the new flow direction algorithms that have been proposed during the past decade, for example, have divided the individual grid cells into triangular facets as the initial step when calculating (i.e. estimating) the magnitude of the flows from one grid cell to the next (e.g. Seibert & McGlynn, 2007; Zhou et al., 2011; Pilesjö & Hasan, 2014).

#### 5.1.1.2 Discretization of Space, Data Source, Interpolation and Filtering Approaches

Many, many papers have focused on the error accompanying the ways in which space was discretized, the choice of data source, and the interpolation and/or filtering approaches that were used. Some of these works were introduced in earlier chapters. For example, the role and importance of scale (i.e. resolution) were introduced in Section 1.2 and the argument was made there that the choice of DEM resolution should be guided by the spatial resolution of the processes and/or patterns of interest. The grid spacing was discussed again in Section 2.4, which focused on data preprocessing and DEM construction, and here the argument was expanded such that the goal should be to choose an operational scale for terrain modeling work that not only suited the work at hand but also one that could be supported by the source data. A lengthy review of some of the studies that have examined the sensitivity of selected land surface parameters to various combinations of data source, data model, and grid resolution was also provided in Section 1.2 and there will inevitably be some overlap with the summary provided here. This is because the potential sources of error frequently blend together with one another when constructing DEMs, and my goal is to describe the breadth of the work describing the error associated with the discretization of space, the ways in which the source data were acquired, and the interpolation and/or filtering techniques that were used to construct the DEMs.

The early works focused on the error associated with the elevation datasets that were used for the first terrain modeling applications and new interpolation methods for reducing or eliminating this errror. Issacson and Ripple (1991), for example, compared the quality of the 7.5-minute and 1° DEMs produced by the USGS that were popular elevation data sources around this time. A few years later, Brown and Bara (1994) and Oimoen (2000) described ways to recognize systematic error and proposed methods to either reduce or eliminate this error in elevations derived from 7.5-minute DEMs produced by the USGS. Day and Miller (1988) evaluated the quality of DEMs derived from the automatic stereo matching of SPOT image pairs.

Turning next to interpolation methods, Hutchinson (1988, 1989) first proposed his ANUDEM software for building hydrologically sound DEMs around this same time, and Lee (1991) compared a series of existing methods for constructing TINs from gridded DEMs. A few years later, Mitášová and Hofierka (1993) and Mitášová and Mitas (1993) proposed using regularized splines with tension for terrain modeling and surface geometry analysis.

The continued popularity of DEM error and uncertainty studies illustrated in Figure 5.1 can be traced, in part, to the emergence of new GPS, airborne laser scanning, and InSAR sources that have enabled the development and sharing of high-resolution DEMs that cover continents and, in some instances, the whole globe during the past 20 years.

GPS has been used in a couple of different but complementary ways. The most popular method of deployment has been as a form of high-resolution, high-accuracy data to evaluate the airborne laser scanning and SAR sources that cover large geographic areas. Lee et al. (2005), for example, described how kinematic GPS campaigns could be stood up and used to validate DEMs produced with satellite-based SAR data such as the ASTER and SRTM datasets, and several of the studies discussed later illustrate this kind of deployment. The second group of GPS studies have used digital photogrammetry and/or kinematic GPS to build very precise DEMs for small areas. In one such study, Baldi et al. (2000, 2002) compared the use of digital photogrammetry and kinematic GPS for deriving fine-scale DEMs that would support volcanic monitoring. Nico et al. (2005b), on the other hand, examined the influence of the operational GPS precision and density of GPS samples on the quality of the DEMs derived from kinemtic GPS surveying (see Section 2.2.2 for additional details). Staying with photogrammetry for a moment, Tadono et al. (2009, 2014) recently examined the accuracy of the DEMs and ORIs derived from archived data collected with the ALOS-PRISM system from 2006 to 2011.

Turning next to the LiDAR-based DEMs that were introduced in Section 2.2.5, this increasingly popular source of capturing elevation data has spawned many empirical assessments that have used elevation data drawn from one or more other sources for calibration and/or validation.

Webster and Dias (2006), for example, proposed an automatic method for comparing GPS measurements with proximal LiDAR elevations. The various works that have examined the error associated with LiDAR-based DEMs include those of Harding et al. (1994), Raber et al. (2002, 2007), Ahokas, Kaartinen and Hyyppä (2003), Filin (2003), Hodgson et al. (2003, 2005), Katzenbeisser (2003), Zhang et al. (2003), Hodgson and Bresnahan (2004), Glenn et al. (2006), Lloyd and Atkinson (2006), Su and Bork (2006), Aguilar and Mills (2008), Cavalli et al. (2008), Liu (2008), Bater and Coops (2009), Heritage and Milan (2009), Notebaert et al. (2009), Aguilar et al. (2010), Estomell et al. (2011), and Hasan et al. (2013b), among others. In addition, the recent manual on airborne topographic LiDAR authored by Renslow (2012) is timely for those wanting more detail as the USGS continues to roll out new LiDAR products as part of its 3DEP initiative (see Section 2.5 for additional details).

The rollout of the SRTM products derived from the SAR data captured on the 11-day mission aboard the space shuttle *Endeavor* in February 2000 has generated a large body of work focused on the identification and/or treatment of error as well. The various works that have examined the error associated with the SRTM-based DEMs include those of Farr and Kobrick (2000), Rabus et al. (2003), Rodriguez et al. (2005, 2006), Carabajal and Harding (2006), Hofton et al. (2006), Selige et al. (2006), Shortridge (2006), Slater et al. (2006), Berry et al. (2007), Bhang et al. (2007), Farr et al. (2007), Reuter et al. (2007), Hirt et al. (2010), Lalonde et al. (2010) and Rexer and Hirt (2014), among others. Two studies in the aforementioned list – those by Reuter et al. (2007), which reviewed a number of interpolation algorithms and provided recommendations on the best algorithms for filling the 3.3 million voids in the SRTM-3 dataset and Rexer and Hirt (2014), which compared the ASTER GDEM2 and two DEMs based on the SRTM released by the USGS (USGS SRTM-3 Version 2.1) and the CSI (CGIAR-CSI Version 4.1) using 228,000 "accurate" station heights from the ANGD – are described in more detail in Sections 2.2.7 and 2.2.8, respectively.

The release of the first ASTER GDEM products a few years later has also produced a flurry of work focused on the identification and/or treatment of error (e.g. Hirano et al., 2003; Eckert et al., 2005; Slater et al., 2009; Hirt et al., 2010; Krieger et al., 2010; Gesch et al., 2011; Tachikawa et al., 2011a,b; Rexer & Hirt, 2014). Several of these studies, including Rexer and Hirt (2014), have compared the performance of both the ASTER and SRTM products because they both offer near-global coverage at similar resolutions, and for some applications the final choice of DEM may well come down to choosing between one or other of these products. In addition, Gesch et al. (2014) recently provided an accuracy assessment of the USGS NED and a comparison of the accuracy of this product with ASTER and SRTM.

There is also a large collection of papers that have looked at the impact of the choice of interpolation and/or filtering method on grid DEM accuracy.

Carara et al. (1997), for example, proposed five objective criteria for evaluating the quality of TINs and square-grid DEMs derived from digital contour lines as follows.

1 The DEM should have the same values as contours close to the contour lines.
2 The DEM values must be in the range given by the bounding contour lines.
3 The DEM values should vary almost linearly between the values of the bounding contour lines.
4 The DEM patterns must reflect realistic shapes in flat areas.
5 The artifacts must be limited to a small proportion of the dataset.

A contour map (see Figure 1.3) is a very effective vehicle for communicating information about the shape and character of the land surface and the abovementioned criteria will certainly help in efforts to differentiate high- and low-quality DEMs derived from contours, which one might use nowadays, along with DEMs derived from other sources, to assess landscape change over time (e.g. James et al., 2012; Beattie, 2014).

The large numbers of papers which follow incorporate a variety of perspectives. Wise (1998), for example, examined the effect of GIS interpolation errors on the use of DEMs in geomorphology. Desmet (1997) evaluated a suite of methods to interpolate a DEM from a ground survey with respect to precision and the ability of the DEM to maintain the shape of the original height data. The shape reliability was evaluated by comparing the spatial patterns of secondary land surface parameters derived from the interpolated elevation data and the best interpolation method for the specific study area was found to be a spline interpolation in this particular study. Hutchinson and Gallant (2000) suggested a larger and more diverse list of simple metrics for measuring the quality of DEMs constructed from surface-specific point elevation data and contour- and stream-line data that incorporated some of the same ideas as the list offered by Carara et al. (1997). This list included the visual assessment of: (i) spurious sinks and drainage analysis; (ii) views of shaded relief and other land surface parameters; (iii) derived elevation contours; and (iv) frequency histograms of slope and aspect. Albani et al. (2004) examined the effect of the choice of window size in approximating topographic surfaces from DEMs. Florinsky (2002) examined three classes of errors in DEMs related to: (i) the Gibbs phenomenon, which can cause overshoots and undershoots near escarpments and other jump discontinuities; (ii) the tendency for some interpolation methods to mimic the high-frequency components of the source data when building a DEM; and (iii) the displacement of a DEM grid. Wilson et al. (2000) examined the effects of the choice of elevation data source, grid spacing, and flow-direction algorithm on slope gradient, slope length, SCA, the RUSLE slope–length factor, and the STI.

More recently, Aguilar et al. (2005) examined the effects of terrain morphology, sampling density, and interpolation method on the accuracy of square-grid DEMs. Pain (2005) examined a range of DEM grid sizes to determine the largest grid spacing that could represent the shape of the land surface near Picton in New South Wales, Australia. Shi et al. (2005) examined the model error of several higher-order interpolation algorithms and how these same algorithms propagated the measurement error from the original nodes of the network. Hancock (2006) examined the impact of different gridding methods on catchment geomorphology and soil erosion over long timescales using a landscape evolution model. Hengl (2006) proposed methods for calculating the appriopriate grid spacing or pixel size based on the density of reflection points and contours for DEMs constructed from sampled points and contours, respectively. Shi and Tian (2006) proposed a hybrid approach that incorporated linear and non-linear methods for interpolating square-grid DEMs. Hu et al. (2009b) showed how approximation theory could be used to evaluate vertical accuracy. Liu et al. (2012) have shown how the error bands produced with this approach can be used to assess the overall accuracy of a DEM created by linear interpolation, and to highlight those areas where more effort might be expended to collect additional reference data in order to reduce this error further. Heritage et al. (2009) quantified the survey strategy and interpolation error for a gravel bar on the River Nent, Blagill, Cumbria, UK. Vaze et al. (2010) examined the impact of using 1-m versus 10- and 25-m resolution DEMs on hydrologically important land surface parameters and the loss of accuracy and reliability of the results as one moves from finer to coarser resolutions. Arun (2013) compared the performance of the IDW, kriging, ANUDEM, nearest neighbor, and spline interpolation approaches using a variety of source data which covered several different landscape settings. Shi, Wang and Tian (2014) provided an analytical approach to document the relationship between the interpolated DEM accuracy and the spatial resolution of the DEM and terrain slope, and proposed rules for directly correlating the horizontal resolution of DEM data with vertical accuracy.

There are also numerous papers that have tried to evaluate the impact of DEM accuracy on one or more land surface parameters. Walsh et al. (1987), for example, examined the effect of DEM error on slope and aspect calculations (along with the errors inherent in land-cover maps derived through classification of Landsat digital data and soil-type data acquired from USDA Soil Conservation Service soil survey reports). Bolstad and Stowe (1994) and Zhou and Liu (2004a) also estimated the effect of DEM errors on slope and aspect. Ruiz (1997) examined the viewshed errors derived from a 30-m USGS DEM using a specially commissioned 10-m DEM as a reference model. Holmes, Chadwick and Kyriakidis (2000) calculated the error in a 30-m USGS DEM by comparing it with 2652 differential GPS elevation measurements and showed how the spatial pattern of this error influenced the estimation of slope, aspect, plan and profile curvature, flow accumulation, upslope flow path length, roughness, and CTI. Kienzle (2004) examined the effect of DEM resolution on the estimation of slope, aspect, plan and profile curvature, and TWI. Zhou, Liu and Sun (2006) examined the impact of terrain complexity represented by terrain steepness and orientation on slope and aspect. Deng et al. (2007) examined the DEM resolution dependencies of five primary land surface parameters (slope, plan curvature, profile curvature, north–south and east–west slope orientation) and one secondary land surface parameter (TWI) across a landscape in the Santa Monica Mountains of southern California. Aryal and Bates (2008) examined the effects of catchment discretization on TWI distributions. Liu and Bian (2008) examined the effect of the spatial autocorrelation of DEM errors on slope calculations. Wu et al. (2008) examined the effect of DEM resolution on the estimation of four primary land surface parameters: slope, upslope contributing area, slope length, and watershed area. Chow and Hodgson (2009) examined the effects of LiDAR post spacing and DEM resolution on mean slope estimation. Ying et al. (2014) estimated the effect of the DEM resolution and underlying topography on the surface area estimated for individual grid cells (see Section 3.1.1 for additional details).

Some studies looked beyond the impact of these errors on the estimation of the primary and secondary land surface parameters themselves to other outcomes. Lagacherie et al. (1993), for example, examined the effects of DEM data source and sampling pattern on a series of land surface parameters and the outputs of a topography-based hydrological model. Garbrecht and Martz (1994) examined the effect of DEM grid size on the estimation of a series of channel network properties (i.e. number of channel links, total channel length, and drainage density). Wolock and Price (1994) also examined the effect of map scale and DEM resolution on the performance of a topography-based watershed model. Zhang and Montgomery (1994) examined the influence of DEM grid size on landscape representations and hydrologic simulations. Gyasi-Agyai et al. (1995) examined the effects of the map scale and vertical resolution of DEMs on the channel networks derived from square-grid DEMs. Florinsky and Kuryakova (2000) calculated slope, plan, profile and mean curvature over a series of grid sizes to determine the best grid size for landscape investigations focused on soil moisture distribution. Tang et al. (2002) examined the effect of grid spacing on the hydrologic information derived from DEMs. Zhou and Liu (2004b) examined the effects of grid resolution, grid orientation and terrain complexity on slope and aspect calculations. Lindsay and Evans (2006) examined the influence of elevation error on the morphometrics of stream networks extracted from DEMs and the implications of these effects on hydrologic modeling. Pirotti and Tarolli (2010) examined how 1-m DEMs derived from different ground LiDAR point densities and landscape curvatures calculated with different smoothing factors influence the extraction of stream networks from DEMs. Mukherjee et al. (2013) examined the effects of terrain roughness and DEM grid spacing on TWI calculations.

And last but not least, several other studies have examined the impact of these kinds of errors on a variety of land surface objects and process model outcomes. Vieux (1993) and Vieux and Needham (1993) examined the effects of DEM aggregation and smoothing on surface runoff and non-point source pollution modeling. Walker and Willgoose (1999) compared several published cartographic and photogrammetric DEMs of various grid spacings with a surveyed ground truth dataset to determine the implications of these differences on the delineation of stream networks and catchments and some related hydrologic statistics. Schoorl et al. (2000) examined how the erosion and sedimentation predictions in a simple landscape process model changed when the DEM grid size was increased. Hengl et al. (2004) used a variety of interpolation and filtering methods to reduce the error in a DEM and improve the mapping of landform facets (the classification accuracy increased from 51.3 to 72%) and solum thickness (the $R^2$ of the regression model for the prediction of the solum thickness increased from 0.27 to 0.40). Thompson et al. (2001) examined the effect of grid spacing on land surface parameter calculation and quantitative soil-landscape modeling. Claessens et al. (2005) examined the effects of DEM resolution on shallow landslide hazard and soil redistribution modeling and Hancock (2008) examined the impact of depression removal on catchment geomorphology, soil erosion, and landscape evolution. Lassueur et al. (2006) examined whether very high resolution DEMs improve models of plant species distribution and, of course, the work of Leempoel et al. (2015), featured in Section 1.2, suggests that considerable research is needed to answer this question for specific species and/or landscape settings. Van Niel et al. (2004) and Van Niel and Austin (2007) examined the effect of error in DEMs on a series of indirect (i.e. elevation, slope, cosine of aspect, topographic position) and direct environmental (i.e. net solar radiation, average air temperature, TWI) variables that are often used for predictive vegetation modeling, and showed that this error can affect the reliability of interpretations of model results and level of accuracy in predictions, as well as the spatial extent of the predictions. Hengl et al. (2010) fitted a variogram model to sampled elevations, generated 100 DEM realizations using conditional Gaussian simulation, extracted the stream network for each of these realizations, and analyzed this collection of stream networks to estimate error propagation. Martinez et al. (2010) used a hierarchical scaling approach with a 5-m DEM created from Differential Global Positioning System (DGPS) data and a 25-m DEM acquired from the New South Wales Land and Property Information Authority to investigate the effect of increasing DEM grid size on several geomorphic and hydrologic descriptors (i.e. soil moisture, area–slope relationship, cumulative area distribution, hypsometric curve, width function, Strahler stream order, and stream network statistics), and showed that a 5-m DEM was necessary to accurately capture catchment geomorphology and hydrology and to model the spatial distribution of soil moisture for their study site. Ariza-Villaverde et al. (2013, 2015) used multifractal analysis to examine the relationship between DEM grid spacing and the D8 flow accumulation threshold values that are frequently used to determine the grid cells that belong to drainage networks (see Section 4.2 for additional details).

#### 5.1.1.3 DEM Preprocessing and Conditioning

Many terrain analysis and modeling workflows also include one or more preprocessing or conditioning steps that modify the DEM before calculating primary and secondary land surface parameters or classifying landforms and other kinds of land surface objects. As described in Section 2.4, there are three major areas of concern: (i) how to deal with unwanted depressions; (ii) how to specify flow directions across flat terrain; and (iii) how to reconcile DEM-derived stream networks with those acquired from other sources like the NHD.

To eliminate unwanted depressions or spurious pits, Martz and Garbrecht (1999) proposed an outlet breaching algorithm for the treatment of closed depressions on DEMs. Planchon and Darboux (2001) later proposed a fast, simple and versatile but different algorithm to fill the depressions in DEMs. Soille (2004) proposed an optimal method that used both of the aforementioned carving and filling approaches by minimizing the elevation modifications that were needed to transform an input DEM into a pitless DEM. Temme, Schoorl and Veldkamp (2006) proposed a new algorithm for dealing with depressions in dynamic landscape evolution models (see Section 5.1.2 for additional details). Lindsay and Creed (2005a,b, 2006) proposed a minimum impact approach for the removal of unwanted depressions, documented the sensitivity of digital lansdscapes to artifact depressions in remotely sensed DEMs, and suggested methods to distinguish artifact and real depressions in digital elevation data. Wang and Liu (2006) proposed an efficient method for identifying and filling surface depressions in DEMs. Grimaldi et al. (2007) proposed a physically based method to remove depressions in DEMs. Reuter et al. (2007) reviewed a series of void-filling algorithms and recommended the best algorithms for filling the 3,319,913 voids in the SRTM data (see Section 2.2.7 for additional details). Arnold (2010) proposed another new approach for dealing with depressions in DEMs when calculating flow accumulation rates. Jiang and Tang (2015) proposed a new method for filling depressions that took into account the local microrelief. And finally, Lindsay (2016a) recently proposed an efficient hybrid method that uses breaching and/or filling to remove sinks and enforce flow paths in DEMs.

The determination of flow directions in flat areas constitutes a more difficult challenge in some ways because these problems may be exacerbated by the methods decribed above to eliminate unwanted depressions. We typically use ANUDEM (Hutchinson, 1989; Hutchinson et al., 2013) to tackle this (i.e. the need to resolve currently unresolved flow directions in flat terrain) and the other two problems (i.e. the need to fill unwanted depressions and the reconciliation of elevation and hydrography datasets in our own work). However, there are other options. Kenny, Matthews and Todd (2008), for example, recently proposed a new method for routing overland flow through both depressions and flat areas in gridded DEMs and Hasan, Pilesjö and Persson (2013a) have described a system in which the user is able to decide if sinks will be filled or passed, how water is supposed to flow over flat surfaces, and if/how artificial or natural walls in the landscape will be treated. This last system might work well for small study areas but it is difficult to imagine a way to scale up this kind of approach to work with large catchments.

The third problem noted at the start of this section arises because there are multiple representations of stream networks in many parts of the world nowadays and one may want or need to reconcile the DEM and the streamlines acquired from one or more of these other sources (Lindsay et al., 2008). Vector hydrography datasets have been produced in the UK (e.g. Regnauld & Mackaness, 2006), the USA (Moore & Dewald, 2016) and several other countries during the past 20 years, and one or other of two basic approaches is typically used to reconcile the DEM and vector hydrography datasets. The first approach relies on stream "burning" where the local topography is altered to provide consistency (Saunders & Maidment, 1996) and the second approach utilizes the stream network as part of the surface-fitting method used to generate the final square-grid DEMs (as has been implemented, for example, in ANUDEM; see Section 2.4 for additional details). The technique involves adjusting the elevations of grid cells that are coincident with the features of a vector hydrography layer. The first method has proved popular but it may cause problems because the method modifies the DEM and the values of many of the land surface parameters (i.e. slope, aspect, curvature) calculated in those parts of the landscape in which the elevation was altered. Lindsay (2016b) also described the topological errors that often result from the mismatched scales of the hydrography and DEM datasets and proposed a novel alternative stream burning method. This TopologicalBreachBurn method, which uses total upstream channel length (TUCL) to prune the vector hydrography layer to a level of detail that matches the raster DEM grid resolution, was tested by Lindsay (2016b) using five DEMs of varying resolutions (1–30 arcseconds) for a large area in southwestern Ontario, Canada.

#### 5.1.1.4 Choice of Methods to Calculate Land Surface Parameters

Another group of papers has examined the error associated with the methods used to derive some of the land surface parameters that were introduced in Chapter 3. Many of these studies focused on the methods used to calculate slope, aspect and curvature, and a smaller number on one or more of the flow-based primary and secondary land surface parameters, as illustrated with the various examples reported in this section.

Skidmore (1989), for example, compared several methods for calculating slope and aspect from gridded DEMs. Guth (1995) also compared several methods for calculating slope and aspect. Hodgson (1995) examined the cell sizes (i.e. areal extents) represented by various slope and aspect calculations on gridded DEMs. Dunn and Hickey (1998) compared several GIS-based methods for calculating slope, and a few years later Hickey (2000) compared several GIS-based methods for calculating slope gradients and lengths. Florinsky (1998) examined the effect of four different algorithms for calculating four local land surface parameters: slope, aspect, plan and profile curvature. Jones (1998) examined the effect of choosing one of eight different algorithms for calculating slope using both synthetic and real-world data as test surfaces. Schmidt et al. (2003) examined several methods for calculating land surface curvatures. Warren et al. (2004) compared several methods for calculating slope from square-grid DEMs. Raaflaub and Collins (2006) examined the effect of using several different algorithms and grid sizes for calculating slope and aspect. Tang, Pilesjö and Persson (2013) examined eight algorithms for estimating slope from gridded DEMs at varying resolutions on both flat and steep terrain.

Turning next to the flow-based land surface parameters, Pan et al. (2004) compared several GIS-based algorithms for calculating the CTI used in TOPMODEL. Erskine et al. (2006) examined the effects of DEM grid spacing and the choice of flow-direction algorithm for calculating upslope contributing area. Baker et al. (2006b) compared the effects of several automated methods for delineating watersheds based on a variety of metrics connecting land cover and nutrient discharge. Pilesjö et al. (2006) compared two methods for calculating potential wetness in rugged fields. Sørensen et al. (2006) examined several methods for calculating TWI based on field observations. Nguyen and Wilson (2010) examined the effects of DEM grid spacing, the choice of flow-direction algorithm, and the ways in which the soil properties were characterized on quasi-dynamic TWI. Persson et al. (2012) examined modeled TWI against site-specific wetness measurements in permafrost landscapes.

#### 5.1.1.5 Performing Calculations in the Correct Order

And finally, there are some additional errors that may occur because most of the flow-based land surface parameters and several of the other secondary land surface parameters rely on a series of calculations being performed in a specific order. This is particularly problematic for flow direction and flow width and the many other secondary land surface parameters described in Sections 3.2.1 and 3.2.2 that incorporate one or both of these parameters. The fundamental challenge is one of representing a continuous surface with a square-grid DEM and this is an active area of research. A total of nine single-flow and 15 multiple-flow direction algorithms were described in Sections 3.1.3.1 and 3.1.3.2 and each, in their own way, suggests a new approach that is designed to overcome one or more of the shortcomings of earlier algorithms. It is perhaps not surprising then that many of the more recent papers proposing new flow-direction algorithms have taken some time to compare and contrast the performance of the new algorithm with some of the earlier algorithms (e.g. Costa-Cabral & Burges, 1994; Holmgren, 1994; Quinn et al., 1995; Tarboton, 1997; Pilesjö et al., 1998; Orlandini et al., 2003; Qin et al., 2007; Seibert & McGlynn, 2007; Pilesjö et al., 1998; Zhou et al., 2011; Pilesjö & Hasan, 2014; Wright et al., 2014).

The 24 flow-direction algorithms described in Chapter 3 can be described as follows: (i) the D8 algorithm (O'Callaghan & Mark, 1984) remains very popular despite its well-documented flaws; (ii) the multiple-flow direction algorithms have increased in both number and popularity over time; (iii) several of the newest flow-routing algorithms divide grid cells into triangular facets as a way to avoid some of the problems encountered using multiple-flow direction algorithms that rely solely on square grids for the calculation of the flow direction and/or flow width land surface parameters; and (iv) there is no single- or multiple-flow routing algorithm that will perform better than the others in every application and/or landscape setting (i.e. study area).

This last claim has been substantiated by a series of papers that have set out to evaluate the strengths and weaknesses of two or more of the aforementioned single- and multiple-flow direction algorithms during the past 20 years. Some of these papers relied on visual comparisons, some others used a series of reference surfaces for which the upslope contributing areas are known, and several used field observations to examine the performance of these flow-direction algorithms. The examples which follow illustrate the breadth of the approaches that have been used to date and the interested reader can find more detailed descriptions of the best of these studies in Section 3.1.4.3.

Wolock and McCabe (1995) examined the effects of choosing a single- versus multiple-flow direction algorithm for calculating the CTI used in TOPMODEL. Zhou and Liu (2002) proposed four mathematical surfaces for which specific and total contributing area can be computed and used these surfaces to assess the performance of several previously published flow-direction algorithms (see Figure 3.14). Endreny and Wood (2003) examined the spatial congruence between the DEM-based overland flow networks delineated with four different flow-direction algorithms and observed overland flow from two agricultural hillslopes in New Jersey. Kenny and Matthews (2005) described a method to align the computed raster flow directions with photogrammtrically mapped hydrology. Desmet and Govers (1996b) examined how the choice of flow-routing algorithm influenced predictions of the locations of ephemeral gullies. Wilson et al. (2007, 2008) compared the performance of numerous flow-direction algorithms in a watershed that had been subdivided into a series of components with known characteristics (steep versus gentle slopes, north versus south facing slopes, etc.) using a fuzzy classification approach. Zhao et al. (2009) compared two different methods for calculating flow direction and deploying it in hydrologic models in the Xitiaoxi catchment, located in the Taihu Basin in southern China, which has both mountainous and flat terrain. Kopecký and Čížková (2010) used vegetation data from 521 georeferenced plots and showed how the choice of algorithm used to calculate TWI does matter in vegetation ecology. Orlandini et al. (2012) compared the performance of various flow-direction algorithms using detailed field observations of overland flow dispersion on small areas of two hillslopes in the European Alps of northern Italy. Peckham (2013) also described a series of mathematical surfaces for which the specific and total contributing areas can be calculated and used these surfaces to evaluate the performance of several flow-direction algorithms. Qin et al. (2013a) proposed four new artificial surfaces (a convex-centered slope, a concave-centered slope, a saddle-centered slope, and a straight-ridge slope) that are typical and widespread in real-world landscapes, and used them to evaluate the performance of four existing grid-based flow-direction algorithms. And finally, Tang et al. (2014) examined the influence of two different flow-direction algorithms on soil–water–vegetation interactions using the LPJ-GUESS dynamic ecosystem model.

#### 5.1.1.6 Summary

The results from all of the aforementioned papers that have focused on error, taken as a whole, suggest that the results obtained by terrain analysis and modeling need careful interpretation. DEMs can be constructed from a variety of sources and the precision and accuracy of the data acquired from these sources will vary depending on the instruments and sensors that are used together with the choice of data model(s), resolution, preprocessing, and the quality of the interpolation and/or filtering methods. These workflows may introduce error from multiple sources, and no matter what the source of the elevation data, a high-resolution DEM may still have greater uncertainty than a low-resolution DEM if we are less certain of its attribute values and the errors in DEMs may propagate to the land surface parameters and modeling results in ways that are not easily predicted.

The previous subsections provide a roadmap of sorts for someone new to digital terrain modeling to identify the errors embedded in their methods and data. The preceding descriptions of the potential sources of error also identified several strategies and methods that might be deployed to reduce or eliminate some of these errors. However, the knowledge that neither the magnitude nor spatial distribution and statistical stationarity of this error will usually be known at a particular location creates uncertainty. This state of affairs should motivate efforts to predict these errors and their likely impact on the final outcomes of the work at hand. The identification and treatment of uncertainty is therefore taken up in Section 5.1.2.

### *5.1.2 Uncertainty*

Numerous papers have tackled the uncertainty embedded in one or more outcomes of the various digital terrain modeling workflows described in the previous chapters. This work, irrespective of the approach that is taken, is nearly always complicated and laborious, as illustrated by many of the studies described here.

We can start with the work of Peter Fisher, who investigated several ways to characterize the uncertainty embedded in viewshed calculations in a series of papers he authored 18–25 years ago. In the first paper, Fisher (1991) used a Monte Carlo simulation (MCS) to add repeated error fields with varying parameters to the original DEM and determined the viewsheds in the resulting noisy DEMs. The results showed that the area of the viewshed calculated in the original DEM may significantly overestimate the viewshed area. In a companion paper, Fisher (1992) used multiple realizations of an error simulation algorithm to generate multiple Boolean viewsheds which were then combined to yield a fuzzy viewshed with all the properties of a fuzzy set. A third paper (Fisher, 1993) examined three major features of viewshed algorithms – how elevations in the DEM are inferred, how viewpoints and targets are represented, and the mathematical formulation of the comparison – and showed that the second of these elements produced the greatest variability in the viewable area (up to 50% over the mean viewable area), while the last produced the least. The next paper (Fisher, 1995) showed how probable viewsheds generated in one or other of three ways – by union, intersection, and weighted average – gave much more information about the nature of the viewable area and thereby enabled a more thorough approach to planning the use of viewable areas. In the final paper in this series, Fisher (1996) noted how current implementations of the viewshed function within GIS had profound shortcomings in their ability to yield useful and generalizable answers to simple queries related to the viewable area. He proposed three variants – the horizon, and local and global offset viewsheds – and showed how the images derived from these new versions of the viewshed can be used both to identify the binary-coded viewshed and to recover each of the other viewsheds.

This group of papers, viewed as a whole, illustrates the variety of ways in which we can think about and measure uncertainty, many of which could be applied and used with other primary and secondary land surface parameters as well. These papers also promoted the use of geostatistics as a viable approach for treating the propagation of DEM error, which we will see momentarily has come to dominate efforts to anticipate and manage the uncertainty accompanying DEM error. The next three papers were chosen because they illustrate how the choice of final method may depend on whether or not and, if so, how well the spatial structure of the error can be estimated.

The first paper by Hunter and Goodchild (1997), for example, had no knowledge of the spatial structure of the error. However, this pair of authors did show how the errors in slope and aspect calculations depend on the spatial structure of the DEM errors, and they proposed a general purpose model of DEM error in which a spatially autoregressive random field was added as a disturbance term to elevations. They also proposed a general procedure for generating such errors and showed how a series of worst-case assumptions (in the absence of explicit information on the spatial structure of DEM error) could be used to make general observations about the nature and severity of slope and aspect errors.

The second paper by Fisher (1998), in contrast, used actual elevations measured in ground and aerial survey at higher precision in place of the elevations in the DEM that were derived from standard paper maps to determine the statistical and spatial distribution of the error. He then substracted these high-precision elevations from values stored in the DEM for approximately the same locations, and showed how these values can be used with conditional stochastic simulation to derive alternative realizations of the error, the DEM itself and any derived products (i.e. land surface parameters).

The final paper, by Wechsler and Knoll (2006), incorporated partial knowledge of the spatial distribution of the error. This pair of authors used MCS with four methods for representing DEM uncertainty that utilized metadata and the spatial characteristics of a DEM to evaluate the effects of uncertainty on elevation, slope, upslope contributing area, and TWI. They then derived seven statistics from the simulation results to quantify the effect of DEM error and used these to show how the range of deviation was 5–8% for slope, 460–950% for upslope contributing area, and 4–9% for TWI across the four methods that were used to represent DEM uncertianty.

Overall, the number of papers focused on DEM uncertainty has grown rapidly since 2000. The examples reviewed briefly here show how these papers vary widely in terms of the level of prior knowledge of the magnitude and spatial distribution of the DEM error that is required and the level of sophistication used to evaluate uncertainty. Endreny and Wood (2001) proposed several novel ways to propagate errors and quantify how elevation uncertainty influenced runoff modeling and flow path mapping. Wechsler (2007) reviewed the uncertainty derived from DEM error, the land surface parameters frequently derived from DEMs and the associated algorithms used to derive these parameters, the influence of DEM scale as imposed by grid cell resolution, DEM interpolation, and the terrain surface modification used to generate hydrologically viable DEM surfaces, which may impact the use of DEMs for hydrologic applications. Darnell, Tate and Brunsdon (2008) presented a prototype tool to support uncertainty analysis in which the effect of correlated gridded DEM error can be investigated and stochastic conditional simulation then used to generate multiple equally likely representations of an actual terrain surface. Hebeler and Purves (2009), reviewed in more detail later, investigated the influence of elevation uncertainty on the derivation of several derived land surface parameters. O'Neil and Shortridge (2013) introduced and evaluated a new method for quantifying uncertainty in flow-direction rasters derived from DEMs which uses flow-direction values derived from finer resolution digital elevation data to estimate uncertainty, on a cell-by-cell basis, in flow directions derived from coarser digital elevation data. Wheaton et al. (2010) presented two new methods to support more robust and spatially variable estimation of DEM uncertainties and to propagate these forward to evaluate the consequences for estimates of geomorphic change. The first relies on a fuzzy inference system to estimate the spatial variability of elevation uncertainty in individual DEMs, and the second approach modifies this estimate based on the spatial coherence of erosion and deposition units. Both techniques allow for probabilistic representation of uncertainty on a cell-by-cell basis and thresholding of the sediment budget at some user-specified confidence interval. Gonga-Saholiariliva et al. (2011) demonstrated how local and global spatial autocorrelation can be analyzed conjointly and used to identify and map the locations where corrections to elevation are required in order to minimize error accumulation or propagation. Qin et al. (2013b) used MCS to compare uncertainties due to DEM error in two representative landslide susceptibility mapping approaches: (i) a recently developed expert knowledge and fuzzy logic-based approach to landslide susceptibility mapping, and (ii) a logistic regression approach that is representative of multivariate statistical approaches to landslide susceptibility mapping.

The final case study, by Hebeler and Purves (2009), is examined in considerably more detail to illustrate how statistical modeling can be used to propagate error in a MCS. This pair of authors set out to: (i) develop a robust model of the error and uncertainty in GLOBE data and its relationship (if any) to the underlying terrain surface; and (ii) illustrate the potential uses of this uncertainty model in geomorphometry through a simple case study that was conducted in the European Alps.

The GLOBE data were acquired using a variety of sources and methods, with major contributions being USGS DEM data for the USA and DTED and Digital Chart of the World (DCW) data in the remainder of the northern hemisphere. The GLOBE data give better accuracy than GTOPO30, even though it uses many of the same sources (Hastings & Dunbar, 1998), and the accuracy reported in metadata range from 18 m RMSE for GLOBE data derived from DTED to 97 m RMSE for DCW data. Hebeler and Purves (2009) next derived error surfaces for the GLOBE DEM using SRTM as ground truth in three study areas in the European Alps, the Pyrenees, and Turkey with elevations ranging from sea level to approximately 2500 m. The SRTM3 data provided by CGIAR-CIAT, which resembles the post-processed NASA SRTM data with voids filled using generated contours and auxiliary data, was used where available (Jarvis et al., 2004, 2006). The ~100-m SRTM data were downscaled to fit the extent and resolution of the corresponding ~1-km GLOBE data following the approach of Jarvis et al. (2004) which assumed that the GLOBE data represent the average altitude within each grid cell and that the SRTM data represent the maximum height of each posting, so that the SRTM could be averaged for the work at hand. Both the GLOBE and SRTM DEMs were clipped to eliminate water bodies and the error surface representing the derivation of GLOBE from the averaged SRTM dataset was calculated by subtraction. The SRTM data were assumed to be error-free in the subsequent steps described next.

Hebeler and Purves (2009) tested the 11 land surface parameters listed in Table 5.1 for correlation with GLOBE error. A simple linear regression model of the error with land surface parameters was developed through stepwise substitution of different combinations of parameters. The results showed that most of the variation in the surface was explained by just two parameters, roughness of altitude and minimum extremity, and that adding one or two additional parameters provided only a slight improvement. The

Table 5.1 Land surface parameters calculated and tested for correlation with GLOBE data.

| Land surface parameters | Descriptions |
|---|---|
| Altitude | Value of GLOBE cell |
| Error, absolute error | (Absolute) deviation of GLOBE from mean SRTM value |
| Aspect | Direction of first derivative of elevation |
| Slope | Magnitude of first derivative of elevation |
| Plan curvature | Second derivative orthogonal to direction of steepest slope |
| Profile curvature | Second derivative in direction of steepest slope |
| Total curvature | Compound curvature index |
| Maximum-mean-minimum extremity | Derivation of center cell from the maximum/mean/minimum of its 3 × 3 cell moving window following Carlisle (2000) |
| Roughness (slope) | Standard deviation of slope in a 3 × 3 cell moving window |

*Source*: Hebeler & Purves (2009, p. 8). Reproduced with permission of Elsevier.

final model meant that a substantial quantity of the error was not explained, and stochastic elements were added next to provide a suitable input for MCS and to ensure that every one of the uncertainty surfaces that was produced would be different.

Hebeler and Purves (2009) next built a final error model in several steps using a regression equation of the modeled magnitude of error, a second regression equation of the residuals from the modeled magnitude of the calculated error (the residuals from this second equation were essentially random and when added to the residuals from the first regression model, introduced the first stochastic component to the final model), and a third equation that calculated the sign of the error as a function of the mean extremity and aspect was used to add a second, constrained random component to the final model. The values of the error surface derived above were transformed to a normal distribution and a convolution filter (Oksanen & Sarjakoski, 2005), using a range that was empirically estimated in order to adjust the spatial correlation of the modeled uncertainty to that of the original measured error, before the uncertainty surface was transformed back to its original distribution. Forty realizations of this final uncertainty surface were then added to the original GLOBE DEM and used to examine the influence of this uncertainty on derived land surface parameters.

All sinks in the resulting DEMs were filled using ArcGIS and TauDEM (Tarboton, 2016) and the DEMs were then used to calculate a set of land surface parameters for each realization. The results were tabulated in three ways as: (i) global statistics, (ii) derived watersheds and statistics, and (iii) a comparison of two watersheds. The global statistics describing elevation, hypsometry, and mean relative catchment areas were relatively robust to uncertainty, though slope, particularly for lower-order streams showed large values of uncertainty. Most of the watershed parameters were also robust to uncertainty, although some statistically significant differences in mean watershed slope were found for Strahler sixth-order watersheds. The hypsometric curves, catchment area, elevation, and slope displayed considerable uncertainty as a function of uncertainty in elevation values when two large watersheds were examined. However, the need for an uncertainty model for GLOBE DEM data in Fennoscandia that could be used as input in MCSs for sensitivity testing of ice sheet models, which provided the primary motivation for this study, was not realized.

## 5.2 Fitness-for-Use Revisited

The study by Hebeler and Purves (2009) introduced in the previous section described a statistical approach for modeling the uncertainty associated with terrain modeling workflows, but stopped short of clarifying exactly how the results could be used with more complex process models to guide knowledge discovery in another domain, such as ecology, geomorphology, or hydrology. The five examples used in this section revisit the concept of fitness-for-use and show how the measurement of error and uncertainty accompanying terrain modeling workflows might be used to improve our understanding of predictive vegetation modeling (Van Niel & Austin, 2007) and soil redistribution resulting from water erosion (Temme et al., 2009); how catchment area calculations, slope estimates and numerical simulations of landscape development (Shelef & Hilley, 2013) and the soil–water–vegetation interactions in the LPJ-GUESS dynamic ecosystem model (Tang et al., 2014) are influenced by the choice of flow-direction algorithm; and how a new sub-grid TOPMODEL parameterization and the associated uncertainties influence the modeling of the spatiotemporal dynamics of global wetlands (Zhang et al., 2016).

### *5.2.1 Predictive Vegetation Modeling*

In the first of these examples, Van Niel and Austin (2007) examined the impact of error propagation from errors in a DEM on all stages in the fitting of statistical regression models for predicting the distribution of species and their potential habitats. The study was conducted in the Murramarang National Park and South Brooman State Forest on the south coast of New South Wales in Australia. The elevation at the site ranges from 8 to 260 m with an average elevation of 66 m. This particular area has been used for numerous vegetation prediction studies (e.g. Moore,

Lees & Davey, 1991; Fitzgerald & Lees, 1992, 1994) and there was an extensive GIS database and set of 424 field plots that were able to be reused for the work at hand.

The error model was developed from field-verified data and used to create a probability distribution function (PDF) to derive 10 random grids which simulated known error in the DEM. The DEM for the study area was developed from 10-m elevation contours, streamlines and spot heights that were digitized from a 1:25,000-scale topographic map and then interpolated to a 30-m DEM using Idrisi (Clark Labs, Worcester, MA), as described by Lees (1999). A total of 2097 points were captured along roads with differential GPS at a vertical accuracy of ≤3.8 m and then aggregated into grid cells taking the mean of all values. The DEM was assessed at a standard deviation of 5.3 m, normally distributed, with a spatial autocorrelation equivalent to a Moran's $I$ of 0.2776 (Van Niel et al., 2004) and these settings were used to develop the PDFs used for this particular study as well.

Van Niel and Austin (2007) developed predictive models for two species – *Corymbia maculata* (Hook.), a common sclerophyll tree (63% presence on plots) and *Acmena smithii* (Poir.), a less common rainforest tree (20% presence on plots) – and spatial predictions were developed for a region that included the field sites. A series of direct and indirect variables, as defined by Austin and Smith (1989), were used to develop separate models for each of the two species. The direct variables used were net solar radiation (Moore et al., 1993d; Leathwick, 1995), average air temperature (Austin & Meyers, 1996; Vayssières, Plant & Allen-Diaz, 2000), and TWI (Moore et al., 1993d; Barling et al., 1994), and the indirect variables were elevation (e.g. Guisan, Theurillat & Kienast, 1998; Vayssières et al., 2000), slope (Franklin, 1998; Guisan et al., 1998), the cosine of aspect (e.g. Lees & Ritman, 1991; Guisan et al., 1998), and topographic position. A geologic nutrient dataset, as described in D.M. Moore et al. (1991), was included and used as a proxy for soil characteristics.

The PDFs were used to create 10 realizations of the the original DEM and the predictor variables. These realizations were then used to create 10 separate statistical models, which were compared against one another and against the original models for the two species. One particularly noteworthy feature of this study is that the effect of uncertainty was considered not only on the final outcomes of the models, but also on each step of the modeling process, as follows.

1. *Testing the individual significance for a species using GLMs.* Regressions for linear, quadratic, cubic, and quartic polynomial models were fitted separately for the unperturbed data and the 10 realizations, and the level of probability determined for each realization regression and compared with that of the unperturbed data. All 424 field plots were used for this analysis.

2 *Exploring the shape of the species–environment curves using general additive models (GAMs).* These models were developed for individual predictor variables, first for comparison to the individual predictor, and then as full models to consider change due to the combination of relationships.
3 *Selecting a final model from all predictor variables by stepwise GLM procedures.* Backward stepwise selection of variables applying Akaike's information criterion (AIC) was used with a split-sample approach (in which 70% of the data were used for model development and 30% were retained for testing) for all modeled realizations to examine the effects of propagated error on this process.
4 *Assessing the accuracy of model predictions.* An initial model was constructed for each species based on GAM analysis of the unperturbed data, and this model was then applied to each of the realizations. The coefficients and their standard errors were examined and the model accuracy tested using both kappa accuracies (i.e. prediction accuracy as assessed via the kappa statistic, or Cohen's kappa) and area under the receiver operating curve (AUC; Bradley, 1997).
5 *Predicting the spatial distribution of each of the two species.* Spatial predictions for *Corymbia maculata* and *Acmena smithii* were developed with each of the stepwise regression models and used to develop three datasets based on probabilities – the minimum probability, maximum probability, and range of probability for each cell – to support displays of differences in probability estimates. The threshold that optimized kappa accuracy for each dataset was then used to reclassify the probability grids to presence/absence (1/0) grids and these were summed to give the number of realizations that predicted the species as present for each cell.

The results showed that all these steps and outcomes – including the statistical significance of environmental variables, shapes of species response curves in GAMs, stepwise model selection, coefficients and standard errors for GLMs, prediction accuracy, and the spatial extent of predictions – were greatly affected by the modeled DEM error.

The results at Step 1, for example, showed that the propagated error for all variables and both species affected both the polynomial selected for each predictor variable and whether the predictor variables were selected at all. The relationships between *C. maculata* and net solar radiation, slope and, to a lesser extent, TWI were affected by the DEM error. The relationship between *A. smithii* and TWI was also affected by the DEM error. Similarly, the examination of the shapes of the species response curves produced in Step 2 using GAMs showed not only how the shape and complexity of the response change, affecting for example the selection of a term for a GLM in Step 3, but also how the general response pattern is altered as well. This result shows how the predictor relationship might be affected by the DEM error as well.

The results of the stepwise model selection in Step 3 were also affected by DEM error. For example, seven realizations were insignificant for *C. maculata* and TWI compared with the unperturbed result (which included TWI) but the other variables (geonutrients, average air temperature, net solar radiation) were less affected due either to a stronger relationship with *C. maculata* and/or to their robustness to DEM error. Average air temperature, on the other hand, was dropped from nine realizations of the models for *A. smithii* due to a weakness in the relationship (which is consistent with the species ecology and the knowledge that average air temperature was not greatly affected by error in the DEM). The effects on the coefficients included in the final GLM models varied with species. The relationships between *C. maculata* and both geologic nutrients and elevation were strong enough to be relatively unaffected by DEM error. The results of the models of *A. smithii* with direct and indirect variables, on the the other hand, showed much greater variation because of the sensitivity of TWI to error in the DEM in the first case and the sensitivity of topographic position to errors in the DEM in the second case.

Four plots were used in Step 4 to summarize the effects of DEM error on the prediction accuracies (i.e. kappa statistics based on the *p*-optimum, a threshold value that leads to the best accuracy value as assessed by kappa accuracy) for the 10 error realizations with two species (*C. maculata*, *A. smithii*) and two sets of model variables (direct and indirect environmental variables). These results showed that the general trend was similar for the 10 realizations and the unperturbed data for the *C. maculata* and *A smithii* models that relied on direct variables and less so for the models that relied on indirect variables.

And finally, the effect of DEM error on spatial predictions of the occurence of the two species (Step 5) was captured by Van Niel and Austin (2007) in a series of maps. The results for *C. maculata* were slightly better than those for *A. smithii* but the relatively large areas of disagreement and the range of variations in the probabilities of presence showed the extent to which error in the DEM affected the spatial predictions of these two species in this study area.

The results from this study by Van Niel and Austin (2007) showed how the propagated uncertainty based on the error in the DEM may impact a number of steps and the final results from predictive vegetative mapping. The single most substantial result was the discovery of an interaction effect between the strength of the relationship (between the presence of a species and one or more environmental variables), level of error, and the sensitivity of the variable(s) to error in the DEM. This means that a strong relationship with an independent variable that has a high sensitivity can lead to the development of models that are very sensitive to error in the DEM and, as a consequence, Van Niel and Austin (2007) recommended that the sensitivity of DEM-derived environmental variables be considered before including them in any predictive vegetation modeling workflow.

### 5.2.2 Modeling Soil Erosion and Deposition

In the second example, Temme et al. (2009) examined the propagation of errors from DEMs for the slope (a local land surface parameter), the TWI (a regional land surface parameter), and the soil redistribution resulting from water erosion (a complex model output) in the Baranja Hill, Croatia watershed that was used as an example throughout the recent book on geomorphometry edited by Hengl and Reuter (2009).

The Baranja Hill DEM used for this study was a 25-m DEM derived from a 1:5000-scale contour map. The uncertainty of this DEM was determined using 5867 precise elevation measurements of which 3633 were retained (one per DEM grid cell) and used to calculate the differences between the DEM and measured elevation. The mean and standard deviation of the observed errors were 0.75 and 7.45 m, respectively and these values were used along with a second-order stationarity assumption to generate 100 simulated error maps.

The DEM errors propagated strongly to slope (the mean coefficient of variation across 100 MCSs was 42% for unfilled DEMs and 49% for filled DEMs) but only moderately for TWI (the mean coefficient of variation of TWI was 10% for unfilled and 16% for filled DEMs), although the coefficient of variation for TWI varied more spatially than that of slope. There were numerous differences in slope estimates between the filled and unfilled DEMs in flat areas as well as other settings, but the relatively uniform coefficient of variation recorded for slope meant that there was more uncertainty in slope in steep areas. The TWI results, on the other hand, show that the TWI values were less sensitive than slope to the input DEM, but this may have been influenced by the FMFD (Freeman, 1991) flow-routing algorithm, as modified by Holmgren (1994), that was used to calculate upslope contributing areas. The main reason for variations in TWI between simulations was the change in drainage pattern (caused by the rerouting of flow once the error was added to the initial elevation values) and the tendency for the standard deviation to increase with mean TWI meant we are less certain of high TWI values, which for this land surface parameter occur in flat areas and valley bottoms.

Temme et al. (2009) next used the water-erosion module of the LAPSUS (Landscape Process Modeling at Multi-dimensions and Scales) landscape evolution model (Schoorl et al. 2000, 2002) to simulate erosion and deposition in the Baranja Hill study area for 10 years. The model utilizes water flow and slope to calculate sediment transport capacity and calculates erosion and deposition by comparing this transport capacity to the predicted amount of sediment in transport. The latter was simulated with the same multiple-flow direction algorithm used to calculate TWI above and the approach of Temme et al. (2006) was used to handle the flows of water and sediment into sinks. The latter capability was important because it meant that the model could simulate erosion and deposition using both unfilled and filled DEMs. Default values were used for the slope, overland flow, and vegetation driving factors included in the LAPSUS water and erosion model (Schoorl et al., 2002) because the authors were only interested in the influence of DEM uncertainty on model performance in this particular application.

The results, at first glance, might be viewed as reassuring because the general erosion and deposition patterns were similar for unfilled and filled DEMs with erosion occurring in the upper valleys and deposition occurring in flat areas. However, the mean soil redistribution maps of the 100 simulations on unfilled DEMs showed considerably more deposition and less erosion than the filled DEMs (in part because the depressions were filled prior to the model runs in the latter case) and the results for both sets of model runs were very sensitive to errors in the DEM (the mean coefficient of variation of soil redistribution was 4600% for unfilled and 1000% for filled DEMs). Hence, the coefficients of variation were larger and more spatially variable for soil redistribution than they were for TWI and slope because the LAPSUS model results were sensitive to three forms of error in the input DEM: those associated with the error in the source DEM and those that were introduced into the slope and TWI terms by the same source error.

The results of this second case study, taken as whole, showed how the differences between the deterministic and simulated outcomes varied for all three variables in terms of pattern and value. In addition, the results showed that the influence of DEM filling was larger for TWI and soil redistribution. The upslope contributing area is an important variable in the calculation of both of these outcomes and is strongly dependent on the continuity of the drainage pattern, which is perturbed by adding error and will not be restored in the same way for all simulations by filling depressions. And finally, the authors called for better methods to simulate uncertain DEMs that better represent real-world conditions after noting various problems connected with the use of smoothed or filled DEMs and the assumptions (including the second-order stationarity assumption used for this study) that are typically used to compute the error in DEMs in a statistically valid way.

Two additional points are worth making as well: (i) the possibility that the uncertainty associated with errors from various sources may be cumulative rather than offsetting (i.e. likely to cancel each other out as is sometimes assumed); and (ii) there will be many potential sources of error in complicated modeling applications like that illustrated here in addition to those associated with the DEM error. A more recent study by Temme et al. (2011), for example, illustrates this second observation. This particular study used the same DEM and errors to evaluate the influence of DEM uncertainty on uncertainty about the relevance of the water erosion and deposition, creep, and solifluction processes (as described in Temme & Veldkamp, 2009) for describing landform evolution in the same Baranja Hills, Croatia study area. This study showed: (i) how MCS-type analyses can be used for every input to a process description for which uncertainty estimates can be made (i.e. they are not limited to DEMs); and (ii) the post-evaluation of included processes in this particular study can help to estimate the uncertainty associated with adding or removing different landscape processes.

### 5.2.3 Numerical Simulations of Landscape Development

In the third example, Shelef and Hilley (2013) examined the impact of flow routing on catchment area calculations, slope estimates, and numerical simulations of landscape development. There were three components to this study: (i) the use of five flow-direction algorithms and a series of standard geometric shapes to measure the accuracy of the drainage area calculations; (ii) the use of different flow-direction and slope calculation rules to predict the locations of geomorphologic transitions and slope failure occurrence over high-resolution airborne laser swath mapping (ALSM) topography; and (iii) the use of these same flow-direction and slope calculation rules to predict the outcomes calculated with landscape evolution models. The methods that were used and the major results of this study are described here.

The D4, D8 (O'Callaghan & Mark, 1984), and D∞ (Tarboton, 1997) single-flow direction algorithms and the FMFD (Freeman, 1991) and $D_{trig}$ (Shelef & Hilley, 2013) multiple-flow direction algorithms were used with three standard geometric shapes (an outward facing cone, an inward facing cone, and a tilted plane) to measure the accuracy of the drainage area calculations in the first component of this study. A constant flow width ($h$) was used for all cases and the RMSE of the difference between the SCA produced by each flow direction method and that expected theoretically at each point in the discretized landscape was used to compare the performance of these flow-direction algorithms. The results showed how: (i) the first two of the aforementioned single-flow direction algorithms generally did worse than the D∞ and two multiple-flow direction algorithms (as many earlier studies have shown); (ii) the differences in relative RMSE values for different surface orientations suggested that the choice of the best-performing flow-direction algorithm may depend on the orientation of the surface; and (iii) the relative performance of the flow-direction algorithms varied when the extent of the source area was increased (because the errors in the distribution of individually generated flows were combined to offset each other for the multiple-direction flow algorithms). The D∞ flow-direction algorithm is sometimes categorized as a multiple-flow direction algorithm because it calculates a single flow direction between 0 and 360° but can distribute flow to one or two downslope grid cells depending on the orientation of the grid to the calculated flow direction (see Section 3.1.3.1 for additional details).

The second part of this study visually compared flow partitioning patterns over natural surfaces by calculating flow routing with the same flow-direction algorithms used in the first part and a single grid element over a 100 × 100, 2-m ALSM DEM from the Eel River basin in the northern California Coastal Range. The authors first calculated the influence zones for one and five grid elements over a small catchment in the Eel River basin. They next prescribed equal area contributions to all grid elements and calculated the Péclet number (Pe) for each grid element (Perron, Dietrich & Kirchner, 2008; Perron, Kirchner & Dietrich, 2009; Perron et al., 2012) to explore the impact of choice of flow-direction algorithm on the spatial transition from advective (i.e. channel forming) to diffusive (i.e. soil creep) dominated portions of the landscape. The Péclet number was calculated as follows:

$$\mathrm{Pe} = \frac{K(3A)^{m+1/2}}{D} \tag{5.1}$$

where $K$ is the bedrock erodibility coefficient, $A$ is the drainage area, $m$ is the area scaling component for the detachment limited erosion rule, and $D$ is the soil diffusion coefficient.

And finally, they examined the impact of the choice of different flow-direction algorithms on process thresholds determined by both slope and drainage area using the steady-state hydrologic model of Dietrich et al. (1992) and Montgomery and Dietrich (1994). The spatial distribution of the grid elements dominated by overland flow processes (when all elements have the same area contribution) was examined with the aforementioned model using:

$$\frac{A}{h} \geq \frac{T}{q}\sin\beta \tag{5.2}$$

where $h$ is the unit length of the contour across which the catchment is draining (with $h = \Delta x$), $T$ is the soil transmissivity, $q$ is the precipitation rate, and $\beta$ is the slope angle associated with each grid element. The D8 and FMFD algorithms were used with Equation 5.2 and three values of $T/q$ (60, 240, and 540 m) for this part of the study.

The results showed how the areas with Pe >1 tracked the zones of convergence calculated with each of the five flow-direction algorithms and showed how: (i) the reduced flow convergence produced by D4 restricted Pe >1 to large valleys; (ii) D8 allowed advective processes to dominate in convergent areas farther upslope relative to D4; (iii) the diffusion associated with FMFD ensured that this algorithm also failed to predict advective processes in small upslope tributaries; (iv) the reduced dispersion of D∞ and $D_{trig}$ predicted Pe >1 in upslope, convergent parts of the landscape; and (v) the three multiple-flow direction algorithms (since they included D∞ in this group) predicted Pe >1 in isolated areas because the flow can disperse and then converge again downslope depending on the local topography with these algorithms. The extent of the SOF predicted with the D8 flow-direction algorithm and the steady-state hydrologic model (Dietrich et al., 1992, 1993) extended to higher areas of the landscape because flow was concentrated along individual flow paths, in contrast to the broader and more dispersed area predicted with the FMFD algorithm. Isolated elements of overland flow were predicted with both flow-direction methods (but usually in different locations) due to the inclusion of slope in Equation 5.2 and the tendency for elements of low slope and high drainage area to generate overland flow. Shelef and Hilley (2013) noted that the extent of landsliding within the landscape would show a similar sensitivity to the flow direction and slope rules because the local slope and the extent of subsurface saturation exerts an important control on the presence of landslides in many landscapes (e.g. Dietrich et al., 1992; Montgomery & Dietrich, 1994).

The final part of this study used several forms of correlation analysis to examine the impact of the D4, D8, FMFD, and $D_{trig}$ flow-direction algorithms on the time-invariant form of landscapes simulated using GTL-based landscape development models and a 100 × 100, 5-m DEM. The GTL model domain was uplifted at a constant rate and lowered by detachment-limited stream incision (Howard & Kerby, 1983; Seidl & Dietrich, 1992; Howard, 1994; Whipple & Tucker, 1999) and soil diffusion (Culling, 1960, 1963, 1965; Howard, 1994). The flow-direction and slope rules were incorporated in two ways, since the model included the drainage area (i.e. upslope contributing area) and the channel slope was calculated downslope in the direction of flow. The latter approach meant that channel slope was calculated using the elevation of each cell and the elevation of the cell into which the flow is routed for the D4 and D8 single-flow direction algorithms and using a weighted mean slope calculated over all downslope facing facets for the FMFD and $D_{trig}$ multiple-flow direction algorithms.

The results showed differences in the relief and channel patterns in the synthetic landscapes produced with different flow-direction and slope calculation rules but identical initial and boundary conditions. The analysis of variance results showed that: (i) the channel segment length variables were significantly influenced mostly by the flow-direction rule; and (ii) the segment orientation variables were significantly influenced by both the flow-direction and slope calculation rules, and that the impact of the slope calculation rule was greater than that of the flow-direction rule. The fact that nearly every metric Shelef and Hilley (2013) investigated was statistically distinct between simulations suggests that the artifacts arising from the flow-direction and slope calculation rules may be imprinted on the numerical simulations and that studies which seek to compare properties of simulated and real-world landscapes would benefit from a systematic exploration of the effect of flow-direction and slope calculation rules on this comparison.

The results from this study by Shelef and Hilley (2013), taken as a whole, show that the flow-direction algorithm(s), in some cases, may not only influence the results of the terrain modeling calculations, but also the numerical simulations of landscape development.

### 5.2.4 Modeling Soil–Water–Vegetation Interactions

In the fourth study, Tang et al. (2014) investigated the influence of two different flow-direction algorithms in soil–water–vegetation interactions using the LPJ-GUESS dynamic ecosystem model in a northern subarctic peatland catchment. Accurate hydrologic estimates are needed to fully capture vegetation dynamics and carbon fluxes in these kinds of environments and to predict what is likely to happen if temperatures in northern high latitudes increase quickly, as expected, over the next century.

LPJ-GUESS is a dynamic ecosystem model that simulates vegetation dynamics as well as soil biogeochemistry (Smith, Prentice & Sykes, 2001; Sitch et al., 2003). The modeled water balance in the original LPJ-GUESS model, like many other ecosystem models that focus on water movement between the atmosphere, plants and soil, did not include lateral water interactions between grid cells, which will be problematic for studies at the catchment scale. Several approaches have been developed during the past 10 years to support water communication between grid cells (Rost et al., 2008; Wolf, 2011; Tang et al., 2014). The distributed scheme proposed by Tang et al. (2014) extracts drainage area, flow direction, and slope from square-grid DEMs and describes the effective contributing area to add lateral water movement in LPJ-GUESS. Neither this new model, hereafter referred to as LPJ-distributed hydrology (LPJ-DH), nor the original LPJ-GUESS model included soil freezing or peatland processes, which may introduce additional uncertainties into the hydrology and carbon balance estimates for northern catchments like the one examined in this study. A new version of LPJ-GUESS that included mechanistic descriptions of soil temperature profiles and peatland hydrology (LPJ-GUESS-WHyMe; Wania, Ross & Prentice, 2009a,b, 2010; McGuire et al., 2012; W. Zhang et al., 2013) was developed around the same time and was used as the platform for the study reported here.

For the work at hand, the D8 (O'Callaghan & Mark, 1984) single-flow and TFM (Pilesjö & Hasan, 2014) multiple-flow direction algorithms for describing lateral flow were integrated into LPJ-GUESS-WHyMe to investigate the impact of the choice of flow-direction algorithm on the estimates of hydrologic and ecological processes within the study catchment. The two new model versions with distributed hydrology were renamed LPJ-GUESS-WHyMe-SF and LPJ-GUESS-WHyMe-TFM and used with a 50-m square-grid DEM to match the available resolution of the climate data. Four sets of observations were used for evaluation: (i) measured daily runoff at six sampling locations during the years 2007–2009 provided by Olefeldt et al. (2013) which were used to validate modeled monthly runoff; (ii) daily water table position measurements during the summer season in the Stordalen Mire from 2003 to 2012 (based on Petrescu et al., 2008); (iii) three years (2006–2008) of net ecosystem exchange (NEE) measurements which are continuously measured with an eddy covariance tower in the west part of Stordalen Mire (Christensen et al., 2012); and (iv) soil respiration data during the summer season during the years 2004, 2005, 2007, and 2008 (estimated by using darkened chamber measurements; Bäckstrand et al., 2008). These data were used with the relative RMSE (RRMSE) to evaluate the model outputs. The closer the value of RRSME to zero, the better the model performance (Stehr et al., 2008):

$$\mathrm{RRMSE}=\sqrt{\frac{\sum_{i=1}^{n}\left(S_i-O_i\right)^2}{n}}*\frac{1}{\bar{O}} \tag{5.3}$$

where $\bar{O}$ is the mean value of the observed runoff, $O_i$ is the observed runoff, $S_i$ is the simulated runoff, and $n$ is the number of observations.

The TFM extracted drainage accumulation pattern showed smoother and more continuous spatial patterns than that of D8 (as would be expected). The daily runoff observations for the years 2007–2009 were summed to get monthly totals and used to compare observed and modeled runoff from LPJ-GUESS-WHyMe-SF and LPJ-GUESS-WHyMe-TFM. The RRMSE values varied from location to location and from month to month, but generally the LPJ-GUESS-WHyMe-TFM model showed much better agreement with the observed values than LPJ-GUESS-WHyMe-SF. Those parts of the study area with semi-wet mesic and wet fen conditions were identified and used to evaluate the performance of the two models for predicting the water table position and the results showed that one or other of the two models did better than the other in predicting water table position in some years and conditions. Turning next to carbon pools and fluxes, the results from the LPJ-GUESS-WHyMe-TFM model were highly correlated with the observed NEE and showed much lower absolute RRMSE (1.571) than the LPJ-GUESS-WHyMe-SF output (RRMSE 3.353). The comparison with the dark chamber measurements was similar, with the LPJ-GUESS-WHyMe-TFM (RRMSE 0.792) producing more accurate outputs than the LPJ-GUESS-WHyMe-SF estimates (RRMSE 0.930), with more points closer to the average measured values.

The results from Tang et al. (2014), taken as a whole, showed that LPJ-GUESS-WHyMe-TFM displayed greater skill in being able to capture a variety of observations compared to LPJ-GUESS-WHyMe-SF notwithstanding the use of identical climate inputs and model settings. The TFM algorithm showed better correspondence with the measured monthly runoff and better agreement with observed $CO_2$ fluxes for most months in the 3-year observation period, demonstrating that the choice of single- or multiple-flow direction algorithm does matter for mesoscale hydrologic and ecological applications.

### 5.2.5 Modeling Global Wetlands

In the fifth and final study, Zhang et al. (2016) demonstrated the feasibility of using TOPMODEL (and the CTI embedded in this model) to capture the spatial heterogeneity of inundation at a large scale and the significance of correcting maximum wetland extent to improve the modeling of inter-annual variations in wetland area. The overarching goals of this project were to: (i) describe the improvements to the TOPMODEL implementation; (ii) estimate global wetland dynamics using the LPJ-wsl (Lund–Potsdam–Jena Wald Schnee und Landschaft version) dynamic global vegetation model; and (iii) quantify the uncertainties by comparing three DEM products at different spatial resolution and accuracy on simulated inundation dynamics.

The summary of this study offered here focuses on the role of TOPMODEL and the new sub-grid TOPMODEL parameterization scheme in this work. TOPMODEL, which includes the assumption that lateral soil water transport driven by topography follows the same exponential decline as the vertical decrease in hydraulic conductivity within soil profiles in a basin (Beven & Kirkby, 1979; Sivapalan et al., 1987), has proved a popular choice for global wetland modeling (e.g. Ducharne et al., 1999; Kleinen, Brovkin & Schuldt, 2012; Ringeval et al., 2012; Zhu et al., 2014). These TOPMODEL-based implementations have successfully captured the broad geographic distribution of wetlands and their seasonal variability (e.g. Gedney & Cox, 2003; Ringeval et al., 2012, Stocker, Spahni & Joos, 2014; Zhu et al., 2014), but have consistently overestimated both the extent of wetlands and the duration of inundation at regional and global scales compared to existing remote sensing and ground surveys (Junk et al., 2011; Prigent et al., 2007; Quiquet et al., 2015).

The CTI that was first described in Chapter 3 determines inundated areas in terrain-based hydrologic applications and plays a critical role in TOPMODEL. It measures the tendency of soils to become saturated (Beven & Cloke, 2013) and consequently it drives the accuracy of wetland area scaled to the larger grid cell (Ducharne, 2009; Mulligan & Wainwright, 2013). The first global CTI product developed at a 1-km resolution from the HYDRO1k global dataset was released by the USGS in 2000 but several studies (Sørensen & Seibert, 2007; Grabs et al., 2009; Lin, Zhang & Chen, 2010; S. Lin et al., 2013) have shown how this dataset tends to overestimate the inundation extent due to the lack of quality of the underlying DEM. That said, the primary goals of the work of Zhang et al. (2016) were to improve the modeling of dynamically varying wetland extents using: (i) a parameter constraint to match integrated satellite and inventory observations; and (ii) a better parameterization of CTI values for determining wetland seasonal cycles using new topographic data sources and grid versus catchment aggregation schemes.

The CTI, which was defined earlier in Equation 3.46, depends on the contributing area per unit contour and local surface slope, which approximates the local hydraulic gradient. Following the central equations in TOPMODEL, the relationship between local water table depth, $z_1$, and mean grid water table depth, $z_m$, can be written as:

$$\lambda_1 - \lambda_m \lambda = f\left(z_1 - z_m\right) \tag{5.4}$$

where $\lambda_m$ is the mean CTI averaged over the grid box, and $f$ is the saturated hydraulic conductivity decay factor with depth for each soil type. This equation was valuable for the work at hand because it related the local soil moisture status to the grid box mean moisture status based on sub-grid variations in topography. Higher CTI values than average indicate areas with higher than average water table depth and vice versa.

Zhang et al. (2016) used these relationships to calculate the inundated areas ($F_{wet}$) of all the sub-grid points within a grid cell that had a water table depth $z_1 \geq 0$:

$$F_{wet} = \int_{z_1}^{z_{max}} \mathrm{pdf}(\lambda) d_x \tag{5.5}$$

where instead of using the CTI values themselves, they used an exponential function that agreed well with the three-parameter gamma distribution, as applied in recent applications for modeling wetland extent, to approximate the distribution of CTI values within a grid cell in order to reduce computation costs. This approach meant that the wetland area fraction ($F_{wet}$) was represented as:

$$F_{wet} = F_{max} e^{-C_s f(\lambda_1 - \lambda_m)} \tag{5.6}$$

where $C_s$ is the coefficient representing the topographic information generated by fitting the exponential function to the discrete cumulative distribution function of the CTI, and $F_{max}$ is the maximum wetland fraction of a grid cell. The authors also explained how the value of $F_{wet}$ was modified in some instances at a subsequent step to account for the effects of permafrost on soil moisture properties.

Zhang et al. (2016) next noted how the maximum soil saturated fraction calculated from discrete cumulative distribution functions are prone to uncertainties (due to the difficulties in determining the water table depth, the hydraulic factor *f*, and the reliance on coarse-resolution DEMs, among other challenges) and proposed the following approach to calculate representative long-term maximum wetland extent, $F_{max}$, within each 0.5° grid box (*i*):

$$F_{max_i} = \max\left(A_{GLWD_i}, A_{(SWAMPS\text{-}GLWD_i)}\right) \tag{5.7}$$

where $A_{GLWD}$ represents the wetland estimate from the Global Lake and Wetland Database (GLWD; Lehner & Döll, 2004) and $A_{(SWAMPS\text{-}GLWD)}$ represents the long-term wetland estimate from a combination of the satellite-based observation Surface Water Microwave Product Series (SWAMPS; Schroeder et al., 2015) and GLWD (SWAMPS-GLWD). This calibration was used to constrain the maximum wetland fractions used in Equation 5.5.

The experimental design used by Zhang et al. (2016) and summarized in Table 5.2 incorporated two main parts. The first was the adoption and use of three DEMs with varying spatial resolutions as follows.

1 The HYDRO1k dataset at 30 arcseconds (US Geological Survey, 2000), which was developed from the USGS's 30-arcsecond DEM of the world (GTOPO30) and was the first product that supported spatially explicit hydrologic routines applied in large-scale applications. This DEM had been previously processed to remove elevation depressions that would have caused local hydrologic "sinks."

**Table 5.2** Model experiments for different parameterization schemes and corresponding DEM products used by Zhang et al. (2016).

| Model experiment | DEM | DEM source | Resolution (arcsecond) | Coverage | River basin | Aggregation type | Hydro-corrected |
|---|---|---|---|---|---|---|---|
| HYDRO1k_BASIN | Hydro1k | GTOPO30 | 30 | Global | HYDRO1k | Catchment | Yes |
| HYDRO1k_GRID | Hydro1k | GTOPO30 | 30 | Global | HYDRO1k | Grid | Yes |
| GMTED_BASIN | GMTED | SRTM + others | 15 | Global | HYDRO1k | Catchment | No |
| GMTED_GRID | GMTED | SRTM + others | 15 | Global | HYDRO1k | Grid | No |
| SHEDS_BASIN | HydroSHEDS | SRTM | 15 | <60° N | HydroSHEDS | Catchment | Yes |
| SHEDS_GRID | HydroSHEDS | SRTM | 15 | <60° N | HydroSHEDS | Grid | Yes |

*Source*: Zhang et al. (2016). http://www.biogeosciences.net/13/1387/2016/. 

2 The 2010 GMTED at 15 arcseconds (Danielson & Gesch, 2011), which was produced using seven data sources including SRTM, the global DTED, Canadian elevation data, SPOT 5 Reference3D data, and data from the Ice, Cloud, and Land Elevation Satellite (ICESat), covering nearly all global terrain. This DEM was applied without hydrologic correction.
3 The HydroSHEDS dataset at 15 arcseconds (Lehner, Verdin & Jarvis, 2008), which was developed from the SRTM dataset and offered a significant improvement in the availability of high-resolution DEMs for areas between 60° N and 60° S (i.e. the limit of SRTM). This DEM had also been previously processed to remove elevation depressions that would have caused local hydrologic "sinks."

Zhang et al. (2016) calculated their own global CTI maps using the QMFM2 (Quinn et al., 1995) algorithm from the R library *Topmodel* (Buytaert, 2011) and each of the aforementioned sources to avoid using more than one flow-direction algorithm; this meant, for example, that they did not use the new CTI data layer produced by Marthews et al. (2015) that is described in Chapter 6. They chose the QMFD2 algorithm for these calculations because multiple-flow direction algorithms (like the one chosen here) have been shown to provide better accuracy calculating CTI than single-flow direction algorithms in flat areas (Pan et al., 2004; Kopecký & Čižková, 2010). It is worth noting as well that the use of the three DEMs as described meant that the experiments provided a comparison of corrected and uncorrected DEMs.

The second part of the experimental design shown in Table 5.2 entailed the estimation of the CTI parameter using two alternative schemes:

1 a regular "grid-based" or gridded approach such that the sub-grid CTI values were averaged over 0.5° grids; and
2 an irregular "basin-based" approach, where mean CTI values were calculated over the entire catchment area in which the respective grid cell was located, and a majority algorithm was used in cases in which two or more catchments overlapped a grid cell (avoiding isolated grid cells for specific river basins) to generate a global catchment map at a 0.5° resolution.

The results, which were summarized in a series of maps and tables by Zhang et al. (2016), showed the following.

1 The parameterization of TOPMODEL based on three DEM products – HYDRO1k, GMTED, and HydroSHEDS – revealed that HydroSHEDS performed best in capturing the spatial heterogeneity and inter-annual variability of inundated areas compared to inventories, lending support to the claims by Wood et al. (2011) that higher-resolution modeling leads to better spatial representation of saturated and

non-saturated areas, even though limitations in upscaling parameterizations may potentially diminish the final gains.

2 The comparison between HydroSHEDS and GMTED also indicated that for capturing inundated areas at the same spatial resolution, the parameter maps derived from the DEM without correction (GMTED) were less accurate compared to the corrected ones (HydroSHEDS), because without these corrections, some valleys would have been interpreted as closed depressions and led to an underestimation of inundated areas (Marthews et al., 2015).
3 The basin-based parameterization schemes performed marginally but significantly better and therefore provided improved wetland area estimates.
4 The comparison of simulated mean annual minimum, maximum, and amplitude of wetland extent with observational datasets revealed that the simulated wetland area for the period 1980–2010 fell within the range of $4.37 \pm 0.99$ Mkm$^2$ (Mkm$^2 = 10^6$ km$^2$) which is very close to other published estimates.

The results of this study by Zhang et al. (2016), taken as a whole, demonstrated the feasibility of using TOPMODEL to capture the spatial heterogeneity of inundation at a large scale and highlighted the significance of correcting the maximum wetland extent using remotely sensed imagery and national surveys (which bring different strengths and weaknesses to the task at hand) to improve the modeling of inter-annual variations in wetland area which loom large in our capacity to predict the changing methane emissions on a global scale. Two future needs related to the TOPMODEL parameterizations were noted – the first concerned the choice of flow-direction algorithm [and whether or not QFMD2 (Quinn et al., 1995) provided the best estimates of upslope contributing area] and the second involved the need to assess the uncertainties introduced by aggregating sub-grid cell to grid cell areas when transforming the CTI values for 15- or 30-arcsecond grid cells to inundation estimates for 0.5° grid cells.

The five fitness-for-use studies reviewed in this section, taken as a whole, illustrate not only the kinds of creativity and preprocessing that will be needed to use the terrain modeling outputs described in this book in future work, but also the growing importance and need for multiscale analysis and cross-scale inference in the environmental applications that were highlighted by the five studies. These last two challenges, multiscale analysis and cross-scale inference, are reviewed in more detail in Section 5.3.

## 5.3 Multiscale Analysis and Cross-scale Inference

We saw in Sections 5.1 and 5.2 that the issue of scale looms large in any discussion of error and uncertainty in digital terrain modeling and how the problems associated with scale may present themselves in various guises. Many of the papers in Section 5.1 described the error associated with the discretization of space, the ways in which the elevation source data were acquired and the interpolation and/or filtering techniques that were used to construct the DEMs, building on the arguments made earlier in Sections 1.2 and 2.4 that the choice of DEM resolution should be guided by the spatial resolution of the processes and/or patterns of interest and the level of support offered by the elevation source data. In addition, several of the examples that were used in Section 5.2 to describe how the fitness-for-use concept can be implemented and used to help interpret the significance of the results of terrain modeling applications which incorporate various kinds of data spanning two or more resolutions, and show why multiscale analysis and cross-scale inference are likely to grow in importance in the future.

One factor driving the growing popularity of multiscale analysis and cross-scale inference is the increasing availability of DEMs across a broad range of scales (Drăguţ & Eisank, 2011). The growth in the number of sources and kinds of elevation source data that was documented in Chapter 2 has spawned numerous papers that have examined the impact of scale effects on the values calculated for one or more land surface parameters and how these differences were propagated to various model ouputs. However, the modeling of scale effects by itself provides few if any clues on how to select non-arbitrary scales for specific analyses and/or specific landscapes (Drăguţ & Eisank, 2011; Goodchild, 2011). This is a serious problem which calls into question the value of some of the current and many of the past terrain analysis modeling applications.

The crux of the problem is that the local terrain shape, which is usually thought of as the continuous variation of elevation values over the terrain surface from point to point, has an enormous impact on local and regional land surface parameters, but this role is influenced by the scale as well as other data and computational factors. Florinsky (1998) suggested some time ago that local parameters, such as slope, aspect and curvatures, are mathematical variables rather than real-world values. This interpretation can be extended to all local land surface parameters for two reasons. The first is that the local terrain shape may rely on different mathematical descriptions, so that calculated local land surface parameters depend on algorithm selection, as illustrated by the emergence and adoption of more than 20 single- and multiple-flow direction algorithms during the past 30 years. The second is that the terrain shape portrayed by DEMs is a function of scale, which combines the complexity of the terrain itself along with the scale or resolution of the data, and the spatial scale at which the terrain surface was observed (Deng et al., 2008). One consequence of this state of affairs is that it is possible to use the same local land surface parameters to describe terrain shape at different scales (i.e. resolutions) which may or may not reflect or inform our understanding of different processes, assuming as Goodchild (2011) does, that there is an "optimum" scale for characterizing each and every process operating at the Earth's surface.

However, it is unlikely that such knowledge would solve all of these problems because the special character of non-local primary land surface parameters is such that they rely on the terrain shape of a larger, non-neighbor area and need to be defined with reference to other non-local points. Therefore, calculating non-local attributes is more difficult because it incurs additional efforts in constructing point-to-point connections over the landscape and involves more complex algorithms and scale considerations (Desmet & Govers, 1996b; Gallant & Wilson, 2000) and it is likely that many of the driving forces that help shape the processes operating at the Earth's surface operate over a variety of scales or extents themselves. This last argument means that we should embrace multiscale analysis and focus on the development and application of methods for making cross-scale inferences when source data are gathered at different scales and/or one wants to change scales when analyzing specific characteristics or outcomes across multiple geographic areas (i.e. extents).

Indeed, some progress has been made in recent years. Deng (2007), for example, suggested that scale had become an integral part of many of the workflows adopted in terrain modeling applications in recent years. Li (2008) reviewed the approaches that have been used to examine scale dependencies in terrain modeling workflows and outlined a general strategy for their analysis, and Drăgut and Eisank (2011) described several recent studies that have proposed new strategies for scale detection and/or scale optimization. The examples here illustrate these advances, showing some of the ways researchers have: (i) looked for an appropriate scale or scales at which to conduct their work; (ii) checked whether or not the stationarity assumption (which among other things would mean that the relationships found in one study would hold true everywhere) is valid or not; (iii) worked with various kinds of source data that were gathered at different scales; and (iv) explored new ways to conduct multiscale analysis, particularly when transitioning from general geomorphometry (i.e. the land surface parameters that were discussed in Chapter 3) to specific geomorphometry (i.e. the landforms and other land surface objects that were the focus of Chapter 4).

Two examples can be used to illustrate the complexities that must be overcome in the search for an appropriate or optimum scale in environmental research. In the first of these two studies, Dingman et al. (2013) described a method for the cross-scale modeling of surface temperature and tree seedling establishment in mountain landscapes. Climate models predict temperature 2 m above ground and will often skip the processes that influence plant seedling survival and establishment, since these typically occur at heights of a few centimeters. Dingman et al. (2013) therefore combined climate models with temperature lapse rates that were derived from field temperature sensors and distributed across the landscape using a 30-m square-grid DEM. This method captured differences in solar radiation and cold air drainage and successfully estimated minimum and maximum surface temperatures on ridgelines as well as north- and south-facing slopes. Quasi-Poisson regression models were then used to predict recruitment of *Quercus kelloggii* (black oak) seedlings from temperature variables and showed that the predictions were significantly improved using these estimates of surface temperature compared to air temperature modeled at 2 m. This result brings the kinds of nested multiscale relationships envisaged in Figure 1.1 to the fore, and illustrates how biophysical processes operating across several scales help to shape the environmental regimes occurring in specific places.

The second study by Jencso et al. (2009) evaluated the validity of the common assumption in steep headwater catchments with shallow soils that the upslope contributing area serves as a first-order control on the distribution of soil water and groundwater. The hillslope–riparian water table connectivity connects the dominant catchment landscape elements (i.e. hillslopes and riparian zones) and the channel network but this connectivity is more often than not heterogeneous in space and temporally transient. Jencso et al. (2009) conducted two experiements: (i) they first tested the relationship between upslope contributing area and the existence and longevity of the hillslope–riparian–stream (HRS) shallow groundwater connectivity using quantified water table connectivity based on 84 recording wells distributed across 24 HRS transects within the Tenderfoot Creek Experimental Forest in Montana; and (ii) they applied this relationship to the entire stream network to quantify landscape-scale connectivity through time and ascertain its relationship to catchment-scale runoff dynamics. The results showed that: (i) the longevity of HRS water table connectivity and the size of each transect's upslope contributing area were correlated; and (ii) the shape of the estimated annual landscape connectivity duration curve was highly correlated to the catchment flow duration curve, suggesting that the internal catchment landscape structure (topography and topology) does act as a first-order control on runoff source area and catchment response in these types of landscapes.

Both of the aforementioned studies compared the results of terrain-based modeling workflows to field measurements and demonstrated the efficacy of the approach that was used through statistical analysis. They did not identify a single "optimum" resolution (this was beyond the scope of both studies) and the relationships they did find would likely vary with climate, geology, etc. These limitations raise a broader and more general concern that the "optimum" grid size would likely be different for different target variables (Drăguţ & Eisank, 2011), which calls into question the stationarity assumption which is often, at least implicity, accepted in many terrain modeling applications. This can be illustrated with digital soil mapping, where several recent studies have tried to optimize the grid spacing and/or neighborhood sizes used for the prediction of soil classes (e.g. Smith et al., 2006; Zhu et al., 2008; Behrens et al., 2010; Cavazzi et al., 2013). One such example is explored in more detail next.

Cavazzi et al. (2013) examined the influence of grid cell size and the interaction effect between the analysis window and grid cell sizes in three areas of Ireland that varied in terms of their geomorphology and soil type. The primary goal of this study was to examine the scale dependency of soil classification performance at the landscape scale using artificial neural networks and random forests, two machine-learning techniques that are commonly used in digital soil mapping. To do this, the authors created a series of DEMs representing different scales from the original 20-m DEM by smoothing and resampling it with different analysis window and grid cell sizes to generate a total of 143 combinations: (i) the original; (ii) 10 smoothed but not resampled; (iii) 12 resampled and not smoothed; and (iv) 120 smoothed and resampled. These datasets were then used to generate 11 terrain parameters from which four points per square kilometer were randomly extracted and used to predict soil classes. The overall prediction accuracy in the three study areas varied between 35 and 60% but it is the variability across and within the three study areas that is of interest for the story here. The grid cell spacing (i.e. size) proved significant in all areas, the interaction between analysis window and grid cell sizes proved significant in morphologically varied (i.e. rough) areas, and analysis window size was significant only in flat homogeneous areas at coarser resolutions (>140 m in this study). The general conclusions offered by Cavazzi et al. (2013) – that predictive performance was best at very fine and very coarse scales in morphologically varied areas, at coarse scales in flat homogeneous areas and relatively scale invariant in mixed areas – suggest that different scale combinations may need to be deployed in different parts of the landscape to obtain a better prediction accuracy of soil classes. The major lesson here is that the search for the "optimum" scale for a particular terrain modeling application may vary from one part of a study area (i.e. landscape) to the next.

The examples that have been introduced thus far have assumed that we start out with gridded variables at the same grid spacing. However, this will not always be the case and it is worth looking for a moment at how to proceed in these other cases. A recent study by Grohmann, Smith and Riccomini (2011), which set out to demonstrate how surface roughness varies with scale, gives a sense of how difficult working across multiple scales is at present.

Grohmann et al. (2011) used surface roughness as an expression of the variability of a topographic surface at a given scale, where the scale of analysis is determined by the size of the landforms or geomorphic features of interest after noting that no single definition exists. Six techniques for the calculation of surface roughness were selected for an assessment of the parameter's behavior across 12 different spatial scales (i.e. analysis windows) and five grid resolutions in the Midland Valley, Scotland. The area ratio, which evaluates the similarities between the surface area and a flat area of square cells or triangles defined by calculating the ratio of these values, operated independently of scale, providing consistent results across spatial resolutions. The vector dispersion (or orientation) produced results with increasing roughness and homogenization of terrain at coarser resolutions and larger window sizes. The standard deviation of residual topography highlighted local features and did not detect regional relief. The standard deviation of elevation correctly identified breaks of slope and was good at detecting regional relief. The standard deviation of slope also correctly identified breaks of slope and smooth sloping areas, providing the best results for geomorphologic analysis. The standard deviation of profile curvature also identified the breaks of slope, although not as strongly as the standard deviation of slope, and this measure was also sensitive to noise and spurious data (as has been demonstrated in many other studies as well).

These results suggest two challenges: the first is that the study generated slightly different results across the six methods and the second is that the extensibility of these results is currently unknown. In addition, there may be additional ways to conceptualize and measure surface roughness: Trevisani and Cavalli (2016), for example, recently presented an approach to illustrate the potential impact of considering directionality in the calculation of roughness indices and concluded that the use of flow-directional roughness can improve terrain modeling (e.g. sediment connectivity and surface texture modeling) and the interpretation of landscape morphology.

The less ambitious and more popular approach for working across multiple scales is to retreat to the coarsest resolution of any of the variables of interest and to use one or more generalization techniques to specify values of the variables at that particular scale. This approach builds on the long-standing interest and work on generalization in cartography and geographic information science (e.g. Buttenfield & McMaster, 1991; Weibel, 1992; Mackaness, Ruas & Sarjakoski, 2007; Palomar-Vázquez & Pardo-Pascual, 2008). Stanislawski and colleagues, for example, have published several papers to support display and delivery of the US National Map and other USGS geospatial data at multiple scales. Hence, Stanislawski (2009), Buttenfield, Stanislawski and Brewer (2011) and Stanislawski and Buttenfield (2011) developed methods to subselect, or prune, features from the multi-scale high-resolution NHD layer to automate the generation of a multiple representation database (MRDB) and to simplify, or further generalize, the remaining features for cartographic display while maintaining the connectivity of the hydrographic network and local density variations that typify physiographic or climate variations.

Some of these same ideas can be adapted and used with DEMs as well. Chen and Zhou (2013), in one noteworthy study, recently proposed a scale-adaptive DEM for multiscale terrain analysis. The goal was to develop a DEM that is self-adaptive to a given scale of an application, rather than forcing the application to fit the current scale of the DEM. The approach uses an adaptive compound point extraction (CPE) method to extract surface "significant points" from a high-resolution DEM according to their degree of importance (DOI) to the scale of the application, and a TIN model can then be generated to support the terrain analysis at the desired scale. Figures 5.2 and 5.3 show the rule set and data structure for the proposed scale-adaptive DEM algorithm, and a series of tests showed that the proposed method offered better performance than the popular bilinear raster resampling (Albani et al., 2004; Kienzle, 2004; Deng et al., 2007) and maximum $z$-tolerance (Heller, 1990; Chang, 2007) methods, by preserving basic streamlines and providing consistent values with changing scales better than these other methods.

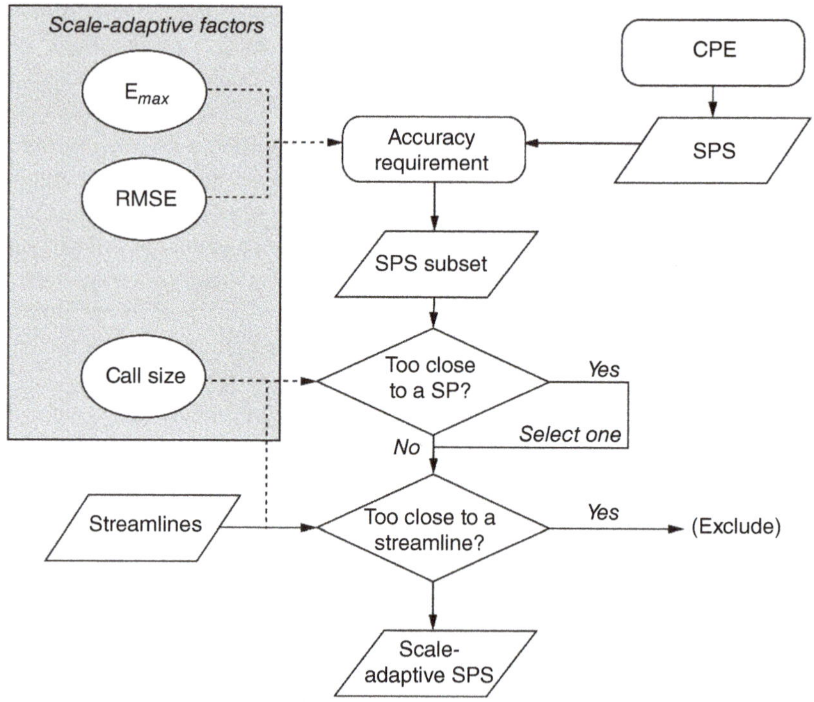


**Figure 5.2** Composition of the rule set for the scale adaptive digital elevation model (S-DEM) algorithm. *Source*: Chen & Zhou (2013, p. 1335). Reproduced with permission of the *International Journal of Geographical Information Science*.

We turn last of all to the transition from general to specific geomorphometry because most of the methods used here will use one or more forms of multiscale analysis and cross-scale inference. Drăguţ & Eisank (2011) eloquently described how three observations – (i) that most spatial modeling is performed on a raster data structure; (ii) that there is a strong and enduring interest in landforms and landform elements (e.g. Smith & Mark, 2003); and (iii) that grid cells only exist for the purpose of representation (Fisher, 1997; Cova & Goodchild, 2002; Goodchild et al., 2007) – mean that we need models to transition from the continuous land surface envisaged in Chapter 3 to the discrete entities discussed in Chapter 4. The development and application of these models is challenging because these objects vary tremendously in terms of shape, size, etc. Hence, whether we adopt classification methods to directly assign cells to landform classes (e.g. Dikau, 1989), or we follow one of the data-driven approaches by which we partition DEMs into discrete, spatially intact land surface objects that are favored by Drăguţ and Eisank (2011), this task will almost always invoke and utilize multiscale analysis. Four examples are reviewed next to show how multiscale analysis and cross-scale inference have been utilized in these works.

In the first example, Gallant and Dowling (2003) described an algorithm that used DEMs to identify valley bottoms based on their topographic

(a)

| 4566 | 4573 | 4574 | 4562 | 4543 | 4527 | 4524 | 4529 | 4539 | 4533 | 4533 | 4533 | 4533 | 4541 | 4558 | 4567 |
|---|---|---|---|---|---|---|---|---|---|---|---|---|---|---|---|
| 4573 | 4579 | 4574 | 4555 | 4523 | 4497 | 4490 | 4495 | 4519 | 4542 | 4542 | 4539 | 4547 | 4554 | 4564 | 4572 |
| 4575 | 4582 | 4568 | 4533 | 4477 | 4433 | 4420 | 4428 | 4470 | 4525 | 4549 | 4554 | 4555 | 4555 | 4563 | 4572 |
| 4561 | 4572 | 4549 | 4493 | 4418 | 4364 | 4349 | 4372 | 4426 | 4495 | 4540 | 4550 | 4552 | 4553 | 4560 | 4568 |
| 4529 | 4544 | 4513 | 4442 | 4366 | 4317 | 4304 | 4324 | 4381 | 4463 | 4525 | 4545 | 4551 | 4554 | 4558 | 4565 |
| 4490 | 4503 | 4473 | 4403 | 4330 | 4294 | 4282 | 4296 | 4348 | 4428 | 4500 | 4537 | 4548 | 4553 | 4555 | 4560 |
| 4457 | 4466 | 4441 | 4378 | 4315 | 4284 | 4271 | 4276 | 4320 | 4398 | 4481 | 4527 | 4540 | 4548 | 4550 | 4554 |
| 4433 | 4438 | 4415 | 4365 | 4309 | 4278 | 4264 | 4268 | 4315 | 4399 | 4477 | 4502 | 4530 | 4538 | 4540 | 4545 |
| 4411 | 4411 | 4391 | 4349 | 4300 | 4271 | 4259 | 4267 | 4320 | 4406 | 4433 | 4495 | 4501 | 4513 | 4524 | 4532 |
| 4389 | 4388 | 4369 | 4331 | 4291 | 4267 | 4254 | 4256 | 4299 | 4350 | 4433 | 4438 | 4429 | 4450 | 4497 | 4516 |
| 4373 | 4375 | 4358 | 4323 | 4289 | 4268 | 4251 | 4242 | 4260 | 4307 | 4340 | 4337 | 4348 | 4384 | 4416 | 4443 |
| 4365 | 4375 | 4360 | 4327 | 4292 | 4268 | 4251 | 4237 | 4235 | 4252 | 4261 | 4242 | 4231 | 4269 | 4344 | 4370 |
| 4359 | 4383 | 4373 | 4335 | 4294 | 4269 | 4252 | 4236 | 4227 | 4226 | 4222 | 4207 | 4207 | 4207 | 4249 | 4272 |
| 4342 | 4370 | 4382 | 4344 | 4298 | 4272 | 4256 | 4240 | 4228 | 4222 | 4216 | 4207 | 4207 | 4207 | 4207 | 4207 |
| 4330 | 4356 | 4359 | 4347 | 4303 | 4273 | 4258 | 4246 | 4233 | 4222 | 4215 | 4207 | 4207 | 4207 | 4207 | 4207 |
| 4330 | 4330 | 4358 | 4347 | 4306 | 4268 | 4255 | 4249 | 4238 | 4226 | 4214 | 4207 | 4207 | 4207 | 4207 | 4207 |

(b)

| 0 | 0 | 0 | 100 | 0 | 30 | 0 | 0 | 30 | 0 | 0 | 0 | 0 | 0 | 0 | 0 |
|---|---|---|---|---|---|---|---|---|---|---|---|---|---|---|---|
| 0 | 30 | 0 | 0 | 30 | 0 | 0 | 60 | 0 | 0 | 0 | 0 | 0 | 0 | 0 | 0 |
| 0 | 0 | 0 | 0 | 0 | 0 | 0 | 0 | 0 | 0 | 90 | 0 | 0 | 0 | 0 | 0 |
| 30 | 180 | 30 | 0 | 0 | 40 | 0 | 240 | 0 | 0 | 0 | 0 | 0 | 0 | 0 | 0 |
| 0 | 0 | 0 | 0 | 30 | 0 | 30 | 0 | 0 | 30 | 0 | 0 | 0 | 0 | 0 | 0 |
| 0 | 0 | 0 | 0 | 0 | 80 | 0 | 0 | 0 | 0 | 0 | 0 | 0 | 0 | 0 | 0 |
| 0 | 0 | 60 | 0 | 0 | 0 | 0 | 60 | 30 | 0 | 0 | 0 | 0 | 0 | 0 | 0 |
| 60 | 0 | 0 | 0 | 0 | 0 | 0 | 30 | 30 | 0 | 40 | 0 | 0 | 0 | 0 | 0 |
| 0 | 0 | 0 | 0 | 0 | 40 | 0 | 0 | 0 | 60 | 0 | 30 | 0 | 0 | 0 | 0 |
| 30 | 0 | 80 | 0 | 0 | 0 | 0 | 0 | 30 | 0 | 30 | 100 | 30 | 0 | 0 | 100 |
| 0 | 0 | 0 | 0 | 0 | 0 | 0 | 180 | 0 | 0 | 40 | 0 | 0 | 30 | 0 | 0 |
| 0 | 0 | 0 | 0 | 0 | 0 | 0 | 0 | 0 | 0 | 0 | 0 | 40 | 30 | 30 | 30 |
| 0 | 0 | 0 | 0 | 0 | 40 | 0 | 0 | 0 | 40 | 0 | 0 | 0 | 250 | 0 | 0 |
| 0 | 0 | 140 | 0 | 0 | 0 | 0 | 0 | 0 | 0 | 0 | 0 | 0 | 0 | 0 | 0 |
| 0 | 0 | 0 | 0 | 0 | 0 | 0 | 0 | 0 | 0 | 0 | 0 | 0 | 0 | 0 | 0 |
| 0 | 40 | 0 | 30 | 0 | 0 | 0 | 0 | 0 | 0 | 0 | 0 | 0 | 0 | 0 | 0 |

30 50 90 125 m

**Figure 5.3** Data structure for S-DEM: (a) the original DEM (the number in each cell represents elevation); (b) the index array using DOI (the number in each cell represents the largest adaptable cell size in meters). *Source*: Chen & Zhou (2013, p. 1335). Reproduced with permission of the *International Journal of Geographical Information Science*.

signature as flat low-lying areas (see Sections 3.2.1.3 and 4.3 for additional details). Valley bottoms are important because they function as hydrologic buffers that significantly affect runoff behavior, but they occur at a range of scales from a few meters to hundreds of kilometers in extent. The algorithm proposed by Gallant and Dowling (2003) operates at a range of scales and combines the results at different scales into a single multi-resolution index that classifies degrees of valley bottom flatness to accommodate this diversity.

The second and third examples focused on mountains, which Fisher et al. (2004) and Deng and Wilson (2008) described as enigmatic phenomena without precise and repeatable definition (see Section 4.3 for additional details). Both of these studies therefore modeled mountain peaks as classes of spatial fuzzy sets. Fisher et al. (2004), for example, deployed a multiscale approach to the definition of the fuzzy set membership of peaks and passes in the English Lake District and showed how the landscape elements identified in the analysis corresponded to well-known landmarks included in a place name database for the area, although many more were found in the analysis than are named in the available database. Deng and Wilson (2008), on the other hand, mapped mountain peaks as multiscale entities with modifiable boundaries and variable contents. Four semantic meanings were imported and quantified to characterize peaks across a range of spatial scales and the multi-criteria "peakness," defined as the prototypicality of identified summits and the similarity of each point (i.e. cell) to summits, was evaluated at each scale. The procedure then summarized the individual-scale peakness across considered spatial scales into a univariate membership surface, and mapped the vague peak entities as non-homogeneous peak regions whose boundaries depended on user-specified peakness thresholds.

In the fourth and final study, Lindsay et al. (2015) derived measures of terrain ruggedness and relative topographic position, which are frequently used for mapping soils, vegetation, and habitat. This pair of land surface parameters is inherently scale dependent because they are defined in the context of local neighborhoods. An integral image-based approach was used to measure the deviation from mean elevation (Dev; Equation 3.35) on a DEM that covered an extensive and heterogeneous region in eastern North America. An integral image, which is sometimes called a summed area table, is a simple transformation used to measure the sum of all values within rectangular subsets of a raster grid (Chow, 1984). These images make the calculation of elevation residuals trivial because a single image can be used to calculate the elevation residual for an entire range of scales (i.e. multiple window sizes). This property means the integral image approach is very efficient compared with traditional image filtering techniques, and that it can easily support fine-resolution multiscale analyses of elevation residuals over large areas. Lindsay et al. (2015) also described three plug-ins they developed for measuring Dev in the Whitebox Geospatial Analysis Tools Version 3.2 (see Section 6.4.9 for additional details) and a novel method for visualizing the scaling characteristics of topographic position using a multiscale topographic position color composite image by combining $Dev_{max}$ rasters from three scale ranges into the red–green–blue channels of a 24-bit color image.

These four examples show some of the ways to utilize multiscale analysis in workflows which aim to transition from general to specific geomorphometry. Drăgut and Eisank (2011) noted how this transition has not been very well conceptualized in geomorphometry so far, and proposed discrete geomorphometry as one way to move from general to specific geomorphometry.

Their proposed approach would combine multiscale pattern analysis and object delineation. Discrete geomorphometry, like specific geomorphometry, would focus on objects, defined as homogeneous areas delineated by discontinuities in land surface parameters, either on individual or combined layers, which reveal land surface patterns at specific scales or across scales (see Section 4.5 for additional details). Drăguţ and Eisank (2011) thought segmentation was a good candidate for conducting discrete geomorphometry, and superior to the clustering in property space and cell classification methods which were also introduced in Chapter 4 for example. They noted two main strategies for segmentation (one using predefined types and degrees of homogeneity and another using data-driven approaches) and the work that needs to be accomplished before these approaches would be fully operational. Given both of these approaches, the resulting objects would serve as intermediate building blocks in the transition from grid cells to landform divisions and once these objects were assigned to classes of elementary forms, or given interpretations such as "mountain," "terrace," or "valley bottom," the transition from general to specific geomorphometry would have been completed.

The concepts and examples described in this section suggest that multiscale analysis and cross-scale inference can only grow in importance over time and that they are likely to play prominent roles in terrain modeling workflows in future years.

## 5.4 The US National Water Model

This chapter closes with a brief summary of the US National Water Model (NWM; see http://water.noaa.gov/about/nwm for additional details) which was released in June 2016 to demonstrate how the land surface parameters and objects that were introduced in the earlier chapters can be incorporated and used in large science-based projects that have the potential to deliver multiple societal benefits over long timescales.

The NWM simulates observed and forecast streamflow over the entire continental USA for four time intervals (i.e. 1 hour, 18 hours, 10 days, and 30 days). The NWM relies on a complicated representation of the physical processes such as precipitation, snowmelt, infiltration, and surface runoff which vary substantially with changing elevations, soil and vegetation over short distances and times and, in turn, may cause the response of rivers and streams to change very quickly. The NWM complements the hydrologic modeling which has been performed for approximately 4000 locations across the continental USA for many years by providing information at very fine spatial and temporal scales at these locations and the vast majority of local streams that lack traditional river forecasts.

The hourly outputs are designed to simulate current conditions, and short-range forecasts that cover the next 18 hours are executed hourly as well. Medium forecasts out to 10 days and an ensemble long-range forecast out to 30 days are produced daily. All four of these model configurations provide streamflow for 2.7 million river reaches and related hydrologic information for 1-km and 250-m square grids that cover the continental USA and parts of Canada and Mexcio (see later for more precise specification).

The core of this new system is the National Center for Atmospheric Research (NCAR)-supported community Weather and Forecasting Hydrologic Model (WRF-Hydro; https://www.ral.ucar.edu/projects/wrf_hydro). This model, which was designed to link multiscale process models of the atmosphere and terrestrial hydrology, is fully parallelized and runs on clusters and high performance computing systems.

For this NWM project, the WRF-Hydro model is forced with various forms of meteorological data from several sources, depending on the configuration. The Analysis and Assimilation configuration, for example, cycles hourly and produces real-time snapshots of current streamflow and general hydrologic conditions across the nation. This configuration also produces single model restart files which are used to initialize the 18-hour, 10-day, and 30-day forecast simulations. The meteorological forcing data for this configuration are drawn from the Multi-Radar/Multi-Sensor system (MRMS; http://www.nssl.noaa.gov/projects/mrms) gauge-adjusted and radar-only observed precipitation products along with short-range Rapid Refresh (RAP; http://rapidrefresh.noaa.gov) and High Resolution Rapid Refresh (HRRR; http://ruc.noaa.gov/hrrr/) forecasts in areas where MRMS coverage does not exist. The short-range forecast is also forced with meteorological data from the RAP and HRRR models and cycles hourly to produce hourly deterministic forecasts of streamflow and hydrologic states out to 18 hours. The medium-range forecast configuration, which is executed once per day, is forced with Global Forecasting System (GFS; http://www.emc.ncep.noaa.gov/GFS/.php) output. This configuration produces 3-hourly deterministic output and extends out 10 days. The long-range forecast cylces four times per day (i.e. every 6 hours) and produces a daily 16-member, 30-day ensemble forecast of streamflow, evapotranspiration, and other hydrologic states. This configuration incorporates four ensemble members in each cycle that are forced with a different Climate Forecast System (CFS; http://cfs.ncep.noaa.gov/) member and produces 6-hourly streamflow and daily land surface output.

The WRF-Hydro model is configured in the NWM to use the Noah-MP Land Surface Model (LSM) to simulate land surface processes. Separate water routing modules perform diffusive wave surface routing and saturated subsurface flow routing on a 250-m grid, and Muskingum–Cunge channel routing down NHD (i.e. NHDPlus, Version 2) stream reaches. River analyses and forecasts are provided across a domain encompassing the continental USA and hydrologic contributing areas, while land surface output is available on a larger domain that extends beyond the continental USA into Canada and Mexico (roughly from latitude 19 to 58° N). The geographic support for this system includes the NED (see Section 2.5 for additional details), NLCD (http://www.mrlc.gov/index.php), and the NHDPlus dataset.

The NHDPlus dataset (https://www.epa.gov/waterdata/nhdplus-national-hydrography-dataset-plus) offers a suite of geospatial products that build on and extend the capabilities of the NHD, NED and Watershed Boundary Dataset (WBD). The NHDPlus integrates the vector NHD stream network and WBD hydrologic unit boundaries with the NED gridded land surface. This hydrologically conditioned surface is critical for the work at hand because it enables the delineation of catchments for each of the 2.7 million uniquely labeled NHD stream reaches. These stream reaches have an average length of 3 km and the catchments, which are used to link precipitation, temperature and runoff data with each stream reach for estimating streamflow, cover an average area of about 3 $km^2$. The USGS streamflow observations are also assimilated into the NWM analysis and assimilation configuration and all four configurations incorporate 1260 reservoirs located in the continental USA that are parameterized using a level pool scheme.

The forcings and geospatial infrastructure described above were connected using the results from three of the terrain modeling tasks introduced in previous chapters. These tasks, which are used to route the precipitation falling on the land surface to local streams, lakes and reservoirs in the NWM, comprised: (i) the removal of local depressions in the 10-m NED DEM so flow from every one of the approximately 179 billion grid cells could be routed to the nearest stream, lake, or reservoir; (ii) the preparation of a flow direction grid for the 10-m NED DEM; and (iii) the identification and specification of flows aross flat areas and the modification of the flow direction grid prepared in step (ii) to incorporate these changes.

These tasks were completed for the NWM project using TauDEM (Tarboton, 2016), which in its most recent versions, provides a suite of DEM tools for high-performance extraction and analysis of hydrologic information. The tools themselves are described in more detail in Section 6.4.8, and the focus here is on how the most recent versions of TauDEM have: (i) taken advantage of parallel processing using the message parsing interface (MPI); and (ii) leveraged high-performance computing environments through batch jobs as well as cyberinfrastructures such as the Open Topography (http://opentopo.sdsc.edu/) and CyberGIS (http://sandbox.cigi.illinois.edu/) gateways. The computational improvements reported for each of the three aforementioned terrain modeling tasks are described in turn next.

Yildirim et al. (2016) recently described how TauDEM was modified to speed up the removal of depressions (i.e. pits) from DEMs. TauDEM uses the method of Planchon and Darboux (2001) to fill the depressions in DEMs and Tarboton and colleagues had previously shown they were able to fill the depressions in the entire 10-m NED DEM in 2 days using 4096 processors on the Stampede supercomputer system hosted by the Texas Advanced Computing Center at the University of Texas, Austin. Yildirim et al. (2016) describe three recent improvements – (i) replacing the row-wise data decomposition used in traditional terrain analysis workflows and their first parallelization efforts with a block-wise data decomposition scheme; (ii) improving the communication mode by taking advantage of non-blocking MPI routines; and (iii) adopting the VRT virtual raster format for writing TauDEM output as a virtual raster dataset consisting of multiple files stored separately on disk to maximize the disk parallelism – and documented how this new version of TauDEM was able to fill the depressions in the 10-m NED DEM stored in VRT format (637,212 columns × 280,812 rows; >667 GB) in 116 minutes using 400 processors on the ROGER supercomputer system hosted by the CyberGIS Center for Advanced Digital and Spatial Studies at the University of Illinois, Urbana-Champaign.

Similarly, Survila et al. (2016) recently described a new and more efficient parallel flow direction algorithm for TauDEM that identifies flat terrain and reduces the number of sequential and parallel iterations needed to calculate the final flow direction grid. TauDEM also uses the sequential algorithm first proposed by Garbrecht and Martz (1997) to resolve the flow direction problem for DEM cells in flat regions, and this plus the D8 (O'Callaghan & Mark, 1984) and D∞ (Tarboton, 1997) flow-direction algorithms included in TauDEM are the most computationally expensive functions among the approximately 30 functions included in this software suite. Survila et al. (2016) described two recent improvements: the inclusion of a connected-component labeling approach to identify flat regions first, and the classification of flat areas to distinguish those stored on a single processor versus those spanning the partitions of several processors because communication is needed to exchange data among shared flat areas whereas local flat areas can be processed in parallel without further inter-process communication (which was a significant bottleneck with their initial parallel implementation of the D8 algorithm). The final flow direction grid is calculated with the same sequential D8 flow-direction algorithm once the problems with local and shared flat areas are resolved. The results of three numerical experiements conducted on the Roger and Stampede supercomputer systems showed the dramatic improvement in performance: the computing time required to resolve the 9,098,177 flat areas on a DEM of size 429 MB on Stampede, for example, using a single processor was reduced from 4105.27 seconds with the previous version (TauDEM, Version 5.3.4) to 8.85 seconds using the new version; in a second experiement using up to 128 processors, the computing time was reduced from 59.9 seconds to 0.23 seconds. Survila et al. (2016) noted how the new algorithm can process a 67-GB compressed DEM in less than 2 minutes – when the previous version took 2 days for D8 (O'Callaghan & Mark, 1984) to process a 36-GB DEM – and that the dramatic performance gains are owed to the improved numerical efficiency of the sequential flat resolving algorithm and the resulting reduction in communications across parallel processors.

The various NWM outputs can be accessed via an interactive map viewer on the National Water Center website (http://water.noaa.gov/map) and the full set of NWM output and forcing files can be retrieved from the NOAA Operational Model Archive and Distribution System (NOMADS; http://nomads.ncep.noaa.gov/pub/data/nccf/com/nwm) and the National Centers for Environmental Protection (NCEP) FTP server at ftp://ftpprd.ncep.noaa.gov/pub/data/nccf/com/nwm.

This project not only gives a sense of the kinds of terrain modeling applications that are likely to dominate in the years ahead, but also some of the opportunities afforded by high-performance computing and sensor systems to reimagine how land surface parameters and objects are conceptualized and calculated.

Ten of the most robust and popular terrain modeling software suites for generating terrain modeling products and services, including TauDEM, are described next in Chapter 6.

# 6 Terrain Modeling Software and Services

Wood (2009a) distinguished two fundamental approaches for calculating land surface parameters and landform objects in his review of geomorphometry software eight years ago. The first relied on domain-specific application software that was specially tailored to the analysis of surface models and the second on generic geographic information systems.

The specialty software range from free and fully open-source software products such as ILWIS and SAGA to free but closed source software products such as LandSerf and MicroDEM and commercial packages such as RiverTools (Wood, 2009a). These software suites can be used to calculate large numbers of land surface parameters with a variety of methods and the occasional workaround. Wood (2009a) distinguished the software aimed at hydrologic analysis from that aimed toward general geomorphometry and provided a ternary diagram to show the variability of the individual software solutions using these two criteria plus the number and variety of GIS tools provided with specific software solutions. This approach is not replicated here, but to note that one could also just as easily make the case for specialty software aimed at ecological analysis nowadays.

A quick review of the work highlighted throughout this book shows how most of the early work on the analysis of DEMs was conducted with specialist stand-alone software programs. Some of these programs have been continuously improved and updated and are still in use today (e.g. ILWIS, RiverTools, TauDEM), whereas some have stalled (e.g. LandSerf, MicroDEM) and others have come and gone (e.g. TAPES-C, TAPES-G, TAS), and some new programs have emerged as well (e.g. SAGA, Whitebox GAT).

The full GIS, on the other hand, offers the advantages of a wider range of functionality, larger and more active user communities, and more substantial support in terms of supporting documentation, help, etc. but at the cost of higher license fees and a steeper learning curve (in terms of finding appropriate tools, crafting solutions, etc.) for those new to the field

*Environmental Applications of Digital Terrain Modeling*, First Edition. John P. Wilson.

(Wood, 2009a). The best-known examples of full GIS include ArcGIS (a commercial software ecosystem produced by Esri in Redlands, California), GRASS, QGIS, and TerrSet.

Indeed, the popularity of long-standing GIS like ArcGIS, GRASS, QGIS, TerrSet, and related software like MATLAB and R has soared during the past 20 years and this, coupled with the inclusion of greater terrain modeling functionality in these systems, has seen more and more terrain modeling work performed with these systems. Some specialized software programs have been built on top of the ArcGIS (ArcGIS Geomorphometry Toolbox; ArcGIS Toolbox for Surface Gradient; Geomorphometric Analysis; and Arc-Geomorphometry) and other GIS platforms as well.

The computational power that is available across a variety of computing platforms (i.e. server, desktop, tablet, and mobile devices) has also grown enormously over this period and in those cases where software development and application have stalled, this has occurred more often than not because there were not sufficient resources to reinvent the enabling software as the computer hardware and relevant computer operating system paradigms and languages evolved throughout this period. Today's leading commercial GIS platform, ArcGIS, for example, has evolved tremendously during the past 25 years. The capabilities and "look-and-feel" of the original command-line ArcInfo workstation flagship product was first combined with the Windows-based ArcView product and then merged and rebranded in the Windows-based ArcGIS platform launched in 1999. Various programming languages and related tools have been supported over the years, ranging from the application-specific Arc Macro Language (AML) and Avenue languages for ArcInfo and ArcView, respectively, in the early years to C++ and Visual Basic and most recently Python (see Zandbergen, 2012 for additional details). Bridges have also been provided to support collaborative work using a variety of complementary products such as Land Modeler and R. The ArcInfo coverage that served as the dominant file format 10–15 years ago was first replaced by the shapefile, which quickly became the de facto standard for storing and supporting geographic information, and then the geodatabase, which uses an object-relational database approach for managing and storing spatial data. Finally, and perhaps the most important message thus far, is that more changes can be expected in the years ahead since the speed and magnitude of the changes in information technology show no signs of slowing down.

With this in mind, the remainder of this chapter is divided into five sections. The first describes the pervasive changes to data capture and computing systems that are underway now and how these are likely to affect the ways in which scientists and practitioners will perform digital terrain modeling work in the future. The second describes the kinds of terrain modeling tools that are currently included in the ArcGIS ecosystem and the third describes the roles and characteristics of a series of third-party extensions that have been developed to extend the native capabilities of the ArcGIS ecosystem itself. The fourth reviews the roles and characteristics of some of the other proprietary and free and open-source terrain analysis and modeling products, some of which are continuously evolving, and all of which still attract sizable user communities. The fifth and final section offers some concluding thoughts for how most of us are likely to conduct terrain analysis and modeling work in the next 10–20 years.

## 6.1 Changes in Data Capture and Computing Systems

The computer hardware and software that can be used to support terrain analysis and modeling has grown and evolved tremendously since Evans (1972) first described the general and specific land surface parameters in his landmark paper published nearly 50 year ago. These changes can be observed from a variety of perspectives, such as the tremendous increases in the speed with which calculations can be performed and the two- and three-dimensional color visualizations that can be rendered, the size of the datasets that can be used, the kinds of human interactions that are now commonplace [i.e. the switch, for example, from command-line and script-based languages which were used to modify or extend the functionality and/or alter the user interaction to graphical user interfaces (GUIs) and application program interfaces (APIs) using languages like C++, Java, and Python], and high-performance computing infrastructures on which the software can be deployed, and the operating systems and file formats that are now supported.

To understand the need for, and significance of, nimble and faster computational platforms and resources we need to go back to some of the foundational concepts and milestones noted in Chapters 1 and 2. This need for change starts with the opportunities afforded by active remote sensing platforms for the capture of repetitive and fine-resolution elevation data across most and, in some instances, the entire surface of the Earth. The SRTM, ASTER GDEM and WorldDEM products, for example, offer relatively fine-resolution (3–30 m) gridded elevation data for the majority of the Earth's land surface at the same time that scientists have begun to grapple with the enormous, and in some cases irreversible, changes that human population growth, development and modernization have brought to the physical systems and ecosystem services that support life as we know it.

Land surface models, for example, have been used in a variety of settings to help explain and predict global climate change and considerable effort has been expended to better characterize the physical processes that control the water cycle in some of this work. See Coe (1998), Gedney and Cox (2003), Coe, Costa and Soares-Filho (2009), Dadson and Bell (2010), Dadson et al. (2010), Dadson, Bell and Jones (2011), and Zulkafli et al. (2013) for recent examples. Many of these applications have used the popular hydrology model TOPMODEL (Beven & Kirkby, 1979) in which the steady-state TWI plays a pivotal role.

In one such study, Marthews et al. (2015) recently seized the opportunity provided by these new global elevation datasets to produce a new high-resolution, spatially consistent data layer of steady-state TWI values for all ice-free land pixels worldwide. This new data layer is based on the hydrologically conditioned HydroSHEDS data layers (Lehner et al., 2008) which were derived from raw SRTM data at a 3-arcsecond pixel resolution (approximately 90 m at the Equator) through the application of multiple hydrologic conditioning steps (Lehner, 2013). The authors acquired and used the HydroSHEDS DEM and a layer of precalculated upstream contributing areas for each 15-arcsecond pixel and then upscaled these data to a resolution of 15 arcseconds (approximately 450 m at the Equator) to calculate the new TWI layer. The tiles for these two layers were next mosaicked into a global layer using ArcGIS 10.1 and then converted to NetCDF format using GDAL (Geospatial Data Abstraction Library; OSGF-Open Source Geospatial Foundation, 2011).

The TWI values were calculated by Marthews et al. (2015) for each pixel using the GA2 algorithm, which is a slightly modified version of the GRIDATB algorithm written by Beven in 1983 and described in Quinn et al. (1991, 1995), that was written specifically for the work at hand based on the basic loop structure implemented in R by Buytaert (2011) and with four modifications to allow for the use of HydroSHEDS data. The modifications implemented by Marthews et al. (2015) were as follows: (i) they applied the correction for DEM resolution suggested by Ducharne (2009) to allow calculations to be carried out at continental scales; (ii) they used the direction-of-steepest-descent (D8; O'Callaghan & Mark, 1984) flow routing algorithm in place of the multiple-flow direction algorithms of Quinn et al. (1991, 1995) because the D8 algorithm was used to derive the UPLAND layer in the HydroSHEDS database; (iii) they used a routine written by Oki in 1996 (Dadson & Bell, 2010) to estimate average cell sizes because the HydroSHEDS DEM used a native geographic projection where pixel dimensions varied with latitude (i.e. the real width of the pixels gets steadily shorter relative to the height towards the poles); and (4) they followed the US Geological Survey (2000) and Evans (2003) and added a small slope gradient increment in flat areas with no inflow or outflow. Florinsky (2017) recently described a series of equations that can be used to calculate TWI with the native geographic projection and thereby avoided Step (iii) and the various problems this approximation would have generated.

Sticking with Marthews et al. (2015) for the moment, the GA2 algorithm was run on the ARCUS server, which is a 1344-core cluster at the Oxford Research Center in the UK, because of the large layer size and the need for calculations to be carried out over domains composed of complete watersheds. Masks were used to identify and delete values for those parts of the Earth's surface covered with water in its various forms. Zonal histograms were plotted using ArcGIS 10.1 and subsequent statistics were calculated using R. This example, along with the US NWM introduced in the previous chapter, and the discussion that follows will show how this approach, combining multiple software tools on top of high-performance computing, is rapidly becoming more and more common. The key discovery to note with the example application described here is that the 15-arcsecond resolution is four times the resolution of the previous best HYDRO1k TWI layers (US Geological Survey, 2000) and that the quality of the source elevation data and the preprocessing steps used to prepare the data before calculating the TWI values means that this new layer represents the most accurate global-scale estimates of TWI values prepared thus far (notwithstanding the potential shortcomings raised by Florinsky, 2017).

The other trend related to elevation data sources that follows from the review provided in Chapter 2 is the rapid emergence of LiDAR as the source of choice for regional and national datasets. This is illustrated by the 3DEP initiative that aims to use LiDAR to develop and share a 1-m digital elevation model for the conterminous USA during the next decade. This and similar initatives in other countries have important consequences for the kinds of computing platforms that will be needed to work with these data. The following comparisons highlight some of the consequences that will follow from the acquisition and use of fine-resolution LiDAR-based datasets.

1. Keeping the same geographic extent but reducing the grid spacing from 1 km to 30 m would increase the size of a raster 1111 times.
2. Keeping the same geographic extent but reducing the grid spacing from 1 km to 1 m would increase the size of a raster 1,000,000 times.
3. Keeping the same geographic extent but reducing the grid spacing from 100 to 1 m would increase the size of a raster 10,000 times.
4. Keeping the same geographic extent but reducing the grid spacing from 30 to 1 m would increase the size of a raster 900 times.
5. Keeping the same geographic extent but reducing the grid spacing from 10 to 1 m would increase the size of a raster 100 times.

These consequences speak to just a part of the challenge because the fine-resolution LiDAR-based data products means, perhaps for the first time, that we may have more resolution than we need to accurately describe one or more of the land surface processes of interest (e.g. Leempoel et al., 2015).

This last observation elevates both the rationale and need for multiscale analysis as well. The fact that many of the land surface processes we are interested in operate at different scales and that we often cannot know these scales in advance was noted in Chapter 1. This, coupled with the fact that DEMs come with vastly different resolutions nowadays, means that we will often need to modify the data or the analytical approaches or both. The values of the slope and other land surface parameters calculated by passing a $3 \times 3$ moving window over an elevation layer will likely mean something completely different using a 1-m square grid versus a 10- or 30-m square grid for example. The good news is that these new elevation data sources afford many new opportunities to find the operational scales but we cannot do this unless the terrain modeling software and services support cross-scale inference and multiscale analysis and visualization.

Several of the predictions about the future of geomorphometry offered by Gessler et al. (2009) have already been realized. For example, LiDAR has already emerged as the preferred source of mass-produced DEMs at regional and national scales (extents) and scientists have quickly taken up increasingly complicated and diverse applications that use more types as well as volumes of data along with advanced spatial and geostatistical analysis techniques (e.g. Marthews et al., 2015; National Water Center, 2016). Gessler et al. (2009) also predicted continuous improvements in the storage capacity and processing speeds of our computing platforms but thought that these improvements might not keep pace with the demands created by new challenges and goals. And finally, Gessler et al. (2009) imagined a geomorphometric atlas of the world, for example, that would distribute land surface parameters and objects for any area, online and with global coverage, similar to the USGS's global 1-km HYDRO1k Elevation Derivatives database (https://edna.usgs.gov/) but with more raster and vector layers across a range of scales. The construction of such an atlas would support the scaling up or down of mechanistic processes as well as linkages with broader physical and biological patterns in space and time, as illustrated by the US NWM application described in Chapter 5.

More recently, Guth (2013) has argued that the availability of new datasets coupled with 64-bit operating systems and large amounts of RAM on desktop computers and large bandwidth for data downloads will change the practice of geomorphometry. He showed, for example, how a desktop PC with a 64-bit operating system, 64 GB of RAM and clever indexing could load the entire SRTM DEM into memory for rapid, random access and used this dataset with an automated algorithm to find dune spacing (crest separation), dune height, and crest orientation. Many other scholars, however, do not share Guth's optimism. Ortega and Rueda (2010), Tesfa et al. (2011), Qin and Zhan (2012), Wang et al. (2012), Qin, Zhan and Zhu (2014a), Qin et al. (2014b) and Yildirim et al. (2015) have all proposed parallel methods to circumvent bottlenecks in computational efficiency and memory capacity. Yildirim et al. (2015), for example, has proposed a new approach for the management of memory in parallelized hydrologic terrain analysis workflows that used a user-level virtual memory system for shared memory, multi-threaded systems. Their method included tailored memory management of raster-based calculations for datasets that are larger than available memory and a novel order-of-calculations approach to parallelize hydrologic terrain modeling workflows. They used a modified version of the DEM pit filling approach of Planchon and Darboux (2001), as used in TauDEM (Tarboton et al., 2015a), to demonstrate how their approach could support raster-based computations using very large DEM datasets on a single machine with limited memory.

A much more promising and ambitious approach is one that uses web-based cyberinfrastructure to take work off the desktop or laptop computer and onto cloud-based data and processing servers (Wang, 2010; Wright & Wang, 2011; Wang et al., 2013). As precursors to what will almost certainly become routine in the next few years, Tarboton et al. (2015a) have demonstrated web- and client-based use of data services that support the TauDEM tools for watershed delineation and generation of hydrology-based terrain information such as topographic wetness and stream networks. In a similar vein, Tarboton et al. (2015b) also recently described the HydroShare collaborative environment and web-based services (Horsburgh et al., 2016) that are being developed to support the sharing and processing of hydrologic data and models.

That said, the next three sections review the terrain analysis and modeling software and services that have found widespread use during the past decade or more, starting with Esri's ArcGIS ecosystem (Section 6.2) before moving to third-party add-ons (Section 6.3) and finishing up with other terrain analysis and modeling software and services (Section 6.4).

## 6.2 Esri's ArcGIS Ecosystem

The contrast between the terrain modeling tools currently provided by the Esri product line and those documented by Reuter and Nelson (2009) shows just how quickly the computational platforms and tools have changed over the past decade.

Their chapter described the geomorphometry tools provided by ArcView 3 and the ArcGIS suite just eight years ago. ArcView was painted as an easy-to-use system for importing, creating and analyzing DEMs. The Spatial Analyst and 3D Analyst extensions provided special tools for working with rasters and TINs and the Avenue scripting language was supported from 1996 and 2002 and used by many to build and share customized terrain modeling tools. The ArcGIS suite, which was initially released in December 1999, combined the ease of use of ArcView with the functionality and flexibility of ArcInfo. This last product was launched in 1982 and supported on several proprietary UNIX workstation platforms. This command line software included the GRID raster modeling environment, with commands for the analysis of grids and TINs, and the AML scripting language.

The ArcGIS Desktop product which was launched in 1999 runs on top of Windows and includes a GUI based on COM objects. This product includes Spatial and 3D Analyst extensions to support grid and TIN analysis and numerous opportunities for users to author and/or adapt and execute programs or scripts to add new functionality using several languages (C++, Visual Basic, Java, Python, etc.). One only needs to take a quick tour of the functionality, organization, and user experience of the ArcInfo, ArcView, ArcGIS and now ArcGIS Professional (Pro) software products to appreciate what this means for terrain analysis and modeling.

Reuter and Nelson (2009) described examples of geomorphometry workflows using each of the first three aforementioned platforms. These workflows provided the functionality for importing DEMs as well as other elevation data, creating DEMs using one or more approaches and options (IDW, TOPOGRID, filling sinks, etc.), calculating land surface parameters using standard tools as well as predefined scripts downloaded from a variety of websites (e.g. http://codesharing.arcgis.com/, http://geomorphometry.org), and exporting data into other formats and/or software. They showed how the standard raster (i.e. grid) tools could be used to calculate slope gradient, aspect, plan and profile curvature, flow direction, flow accumulation, and how two or more of these parameters could be combined to calculate secondary land surface parameters such as the SPI and TWI.

These authors also showed how a series of scripts could be executed by clicking on a special terrain analysis toolbox and used to calculate 28 different land surface parameters. They noted how the generic zonal and focal raster tools in ArcGIS could be used to query raster elevation datasets and report the highest elevation in a watershed and the difference between the highest and lowest elevations in a watershed or some other user-defined window. Finally, they noted that the TIN analysis tools were limited to the calculation of slope and aspect and they suggested converting the TIN to a raster and performing analysis there for those starting with TIN surface models.

The release of the ArcGIS 10.5 and ArcGIS Pro 1.4 ecosystems in the first quarter of 2017 shows how much has changed during the past decade. The latest release is organized around ArcGIS Pro, a 64-bit desktop application with a new graphics engine that supports multiple views, map layouts, multi-threaded processing, and the publication and sharing of web maps through ArcGIS Online or Portal for ArcGIS. The key features of the current ecosystem for the terrain modeling work at hand include the number and variety of tools and how they are organized, the elevation datasets that are supported, the ways in which tools and data can be combined and used to calculate land surface parameters and land surface objects, and the roles of ArcGIS Pro and/or ArcGIS Online in supporting terrain analysis and modeling workflows.

The ArcGIS Pro tool reference is organized into tool and environment sections. These sections, which can be browsed and searched, complement one another. The tools are grouped into toolsets and then collected into toolboxes whereas the environment settings are applied to the tools and will affect the results of using one or more tools as part of some terrain analysis and/or modeling workflow.

The tools themselves can be executed via the geoprocessing pane, ModelBuilder, or Python. To run a tool in the geoprocessing pane, the user must select the tool, set the input and output as well as any required parameters and then click run. Environment settings are optional and if set they will only be applied to that specific execution of the tool. Optional parameters may be set or left blank to use the default values. ModelBuilder provides a visual programming environment that allows users to graphically link geoprocessing tools into new tools and workflows that are referred to as models. These tools can be validated and executed directly in ModelBuilder (after setting the parameters similar to what happens in the geoprocessing pane) or exported to scripting languages that can then execute the tool(s) in batch mode. Finally, the user can also open the Python window, select and add the tool they wish to use to the code, enter valid parameter values, and then run the tool from this user interface.

The ArcGIS ecosystem is built around a geodatabase. The geodatabase is used to store a variety of different kinds of datasets linking spatial features and attributes. These may include coverages which store topology information about the spatial data but suffer limitations in how they handle some types of features, and shapefiles within which each file can only handle one type of feature. The geodatabase, on the other hand, can store multiple feature classes (i.e. types of features) within one file and can model the behaviors of features with rules on how features relate to one another. The terrain analysis and modeling tools described later rely on a simple raster (i.e. an array of equal-sized cells arranged in rows and columns where each cell contains an attribute value and location coordinates) and several custom datasets (terrain, TIN, and LAS for example) for storing elevation data as part of a geodatabase.

The terrain dataset is the first and most complicated of the three custom datasets. The terrain dataset is a multi-resolution, TIN-based surface built from measurements stored as features in a geodatabase. A terrain dataset does not actually store a surface as a raster or TIN but instead organizes the data for fast retrieval and derives a TIN surface on the fly. This organization involves the creation of terrain pyramids that are used to quickly retrieve only the data needed to construct a surface at the required LoD for a given area of interest (AOI) from the database. The terrain dataset, along with its collection of supporting tools, provide the means to explore and use terrain surfaces in a variety of workflows.

The second custom dataset is a TIN. This dataset stores a single-resolution TIN consisting of elevation points and the topologic relationships connecting nearby points and neighboring triangles. These datasets can be loaded into terrain datasets and both raster and TIN datasets can be extracted from terrain datasets based on some user-defined AOI and LoD.

The third and final custom LAS dataset uses a public file format to support the sharing of three-dimensional point cloud data among data users. This file format was developed for the storage and exchange of airborne LiDAR point cloud data. An Esri LAS dataset stores references to one or more LAS files on disk, as well as additional surface features, and allows users to examine LAS files in their native format using a variety of statistical analysis and display tools. The LAS dataset simply stores references to the source LAS files and surface constraints and can be constructed very quickly using these references plus the header information for each LAS file that needs to be accessed. This means that a LAS dataset can be used to conduct an initial quality assurance/quality control check on the LiDAR data and to determine whether or not the LiDAR data meet the desired quality standards before importing the data into a terrain dataset.

Many examples of these kinds of datasets may be able to be accessed and used directly with the tools from ArcGIS Online or Portal for ArcGIS. ArcGIS Online is a web application for storing and sharing geographic information, including content published by Esri, ArcGIS users, and other authoritative data providers.

And finally, it is important to note that many of the terrain analysis and modeling tools described here can only be accessed via one or other of the Spatial Analyst and 3D modeling extensions (i.e. toolboxes in ArcGIS Desktop 10.5) which are sold and licensed separately on top of the core ArcGIS Pro 1.4 platform. With this background, we can now explore the tools that are available and explain how the various elevation datasets can be acquired and used with these tools in the ArcGIS ecosystem.

The Spatial Analyst extension toolbox provides more than 170 tools in 23 toolsets for performing spatial analysis and modeling (Table 6.1). The tools included in the Interpolation, Surface, Hydrology, and Solar Radiation toolsets, in particular, provide numerous opportunities for supporting the calculation of land surface parameters.

The Interpolation toolset, for example, includes four deterministic methods (IDW, Natural Neighbor, Spline, and Trend), the Kriging geostatistical method, and the Topo-to-Raster and Topo-to-Raster-by-File tools for creating continuous surfaces from contour lines and/or point data that support hydrologic analysis and modeling (see Section 2.3 for additional details) (Table 6.2).

The Surface toolset provides 12 tools for quantifying and visualizing the land surface represented by a DEM. Starting with a raster (i.e. gridded) elevation surface as input, the user can calculate contours, slope gradients and aspects, shaded relief and visibility surfaces with these tools (Table 6.3). These new surfaces may serve as an endpoint (e.g. Figure 2.1) or as input for additional terrain analysis and modeling workflows.

The Hydrology toolset provides 11 tools for modeling the flow of water across the land surface (Table 6.4). These tools can be applied individually to identify and/or fill sinks, to calculate flow directions, upstream and downstream flow lengths, and flow accumulation or used in sequence to create stream networks and/or delineate watersheds.

The Solar Radiation toolset includes three tools that can be used to calculate the insolation across a landscape extent specified by the user (Table 6.5). The tools calculate the visible sky, track the changes in the sun's position in the sky over time, and modify the insolation estimates to account for topographic shading and reflection. The documentation accompanying this toolset explains not only how to use each of the tools but also how to interpret the results (the current versions of the tools, for example, assume clear sky conditions and therefore cannot simulate the effects of clouds on solar radiation).

Table 6.1 List of Spatial Analyst toolsets and tools.

| Toolset | Tools |
|---|---|
| Conditional | Con ◦ Pick ◦ Set Null |
| Density | Kernel Density ◦ Line Density ◦ Point Density |
| Distance | Corridor ◦ Cost Allocation ◦ Cost Back Link ◦ Cost Distance ◦ Cost Path ◦ Euclidean Allocation ◦ Euclidean Direction ◦ Euclidean Distance ◦ Path Distance ◦ Path Distance Allocation ◦ Path Distance Back Link |
| Extraction | Extract by Attributes ◦ Extract by Circle ◦ Extract by Mask ◦ Extract by Points ◦ Extract by Polygon ◦ Extract by Rectangle ◦ Extract Multi Values to Points ◦ Extract Values to Points ◦ Sample |
| Generalization | Aggregate ◦ Boundary Clean ◦ Expand ◦ Majority Filter ◦ Nibble ◦ Region Group ◦ Shrink ◦ Thin |
| Groundwater | Darcy Flow ◦ Darcy Velocity ◦ Particle Track ◦ Porous Puff |
| Hydrology | Basin ◦ Fill ◦ Flow Accumulation ◦ Flow Direction ◦ Flow Length ◦ Sink ◦ Snap Pour Point ◦ Stream Link ◦ Stream Order ◦ Stream to Feature ◦ Watershed |
| Interpolation | IDW ◦ Kriging ◦ Natural Neighbor ◦ Spline ◦ Topo to Raster ◦ Topo to Raster by File ◦ Trend |
| Local | Cell Statistics ◦ Combine ◦ Equal to Frequency ◦ Greater Than Frequency ◦ Highest Position ◦ Less Than Frequency ◦ Lowest Position ◦ Popularity ◦ Rank |
| Map Algebra | Raster Calculator |
| Math (General) | Abs ◦ Divide ◦ Exp ◦ Exp10 ◦ Exp2 ◦ Float ◦ Int ◦ Ln ◦ Log10 ◦ Log2 ◦ Minus ◦ Mod ◦ Negate ◦ Plus ◦ Power ◦ Round Down ◦ Round Up ◦ Square ◦ Square Root ◦ Times |
| Math Bitwise | Bitwise And ◦ Bitwise Left Shift ◦ Bitwise Not ◦ Bitwise Or ◦ Bitwise Right Shift ◦ Bitwise XOr |
| Math Logical | Boolean And ◦ Boolean Not ◦ Boolean Or ◦ Boolean XOr ◦ Combinatorial And ◦ Combinatorial Or ◦ Combinatorial XOr ◦ Diff ◦ Equal To ◦ Greater Than ◦ Greater Than Equal ◦ In List ◦ Is Null ◦ Less Than ◦ Less Than Equal ◦ Not Equal ◦ Over ◦ Test |
| Math Trigonometric | ACos ◦ ACosH ◦ ASin ◦ ASinH ◦ ATan ◦ ATan2 ◦ ATanH ◦ Cos ◦ CosH ◦ Sin ◦ SinH ◦ Tan ◦ TanH |
| Multivariate | Band Collection Statistics ◦ Class Probability ◦ Create Signatures ◦ Dendrogram ◦ Edit Signatures ◦ Iso Cluster ◦ Iso Cluster Unsupervised Classification ◦ Maximum Likelihood Classification ◦ Principal Components |
| Neighborhood | Block Statistics ◦ Filter ◦ Focal Flow ◦ Focal Statistics ◦ Line Statistics ◦ Point Statistics |
| Overlay | Fuzzy Membership ◦ Fuzzy Overlay ◦ Weighted Overlay ◦ Weighted Sum |
| Raster Creation | Create Constant Raster ◦ Create Normal Raster ◦ Create Random Raster |
| Reclass | Lookup ◦ Reclass by ASCII File ◦ Reclass by Table ◦ Reclassify ◦ Rescale by Function ◦ Slice |
| Segmentation and Classification | Classify Raster ◦ Compute Confusion Matrix ◦ Compute Segment Attributes ◦ Create Accuracy Assessment Points ◦ Segment Mean Shift ◦ Train Iso Cluster Classifier ◦ Train Maximum Likelihood ◦ Train Support Vector Machine Classifier ◦ Update Accuracy Assessment Points |
| Solar Radiation | Area Solar Radiation ◦ Points Solar Radiation ◦ Solar Radiation Graphics |
| Surface | Aspect ◦ Contour ◦ Contour List ◦ Contour with Barriers ◦ Curvature ◦ Cut Fill ◦ Hillshade ◦ Observer Points ◦ Slope ◦ Viewshed ◦ Viewshed2 ◦ Visibility |
| Zonal | Tabulate Area ◦ Zonal Fill ◦ Zonal Geometry ◦ Zonal Geometry as Table ◦ Zonal Histogram ◦ Zonal Statistics ◦ Zonal Statistics as Table |

*Source*: modified from http://pro.arcgis.com/en/pro-app/tool-reference/spatial-analyst/an-overview-of-the-spatial-analyst-toolbox.htm

The 3D Analyst extension or toolbox provides more than 100 tools in 11 toolsets to support the creation and analysis of surface data represented in raster, terrain, TIN, and LAS dataset formats (Table 6.6). These include tools for creating and managing different types of datasets, the conversion from one dataset format to another, interpolation tools that can produce

**Table 6.2** List of Interpolation tools.

| Tools | Descriptions |
|---|---|
| IDW | Interpolates a raster surface from points using an inverse distance weighted (IDW) technique |
| Kriging | Interpolates a raster surface from points using kriging |
| Natural Neighbor | Interpolates a raster surface from points using a natural neighbor technique |
| Spline | Interpolates a raster surface from points using a two-dimensional minimum curvature spline technique. The resulting surface passes exactly through the input points |
| Topo-to-Raster | Interpolates a hydrologically "correct" raster surface from point, line, and polygon data |
| Topo-to-Raster- by-File | Interpolates a hydrologically "correct" raster surface from point, line, and polygon data using parameters specified in a file |
| Trend | Interpolates a raster surface from points using a trend technique |

*Source*: modified from http://pro.arcgis.com/en/pro-app/tool-reference/spatial-analyst/an-overview-of-the-interpolation-tools.htm

**Table 6.3** List of Surface tools.

| Tools | Descriptions |
|---|---|
| Aspect | Derives aspect from a raster surface. The aspect identifies the downslope direction of the maximum rate of change in value from each cell to its neighbors |
| Contour | Creates a line feature class of contours (isolines) from a raster surface |
| Contour List | Creates a feature class of selected contour values from a raster surface |
| Contour with Barriers | Creates contours from a raster surface. The inclusion of barrier features allows you to independently generate contours on either side of a barrier |
| Curvature | Calculates the curvature of a raster surface, optionally including profile and plan curvature |
| Cut Fill | Calculates the volume change between two surfaces, typically for cut and fill operations |
| Hillshade | Creates a shaded relief from a surface raster by considering the illumination source angle and shadows |
| Observer Points | Identifies which observer points are visible from each raster surface location |
| Slope | Identifies the slope (gradient, or rate of maximum change in *z*-value) from each cell of a raster surface |
| Viewshed | Determines the raster surface locations visible to a set of observer features |
| Viewshed2 | Determines the raster surface locations visible to a set of observer points, using geodesic methods |
| Visibility | Determines the raster surface locations visible to a set of observer features, or identifies which observer points are visible from each raster surface location |

*Source*: modified from http://pro.arcgis.com/en/pro-app/tool-reference/spatial-analyst/an-overview-of-the-surface-tools.htm

**Table 6.4** List of Hydrology tools.

| Tools | Descriptions |
|---|---|
| Basin | Creates a raster delineating all drainage basins |
| Fill | Fills sinks in a surface raster to remove small imperfections in the data |
| Flow Accumulation | Creates a raster of accumulated flow into each cell. A weight factor can optionally be applied |
| Flow Direction | Creates a raster of flow direction from each cell to its steepest downslope neighbor |
| Flow Length | Calculates the upslope or downslope distance, or weighted distance, along the flow path for each cell |
| Sink | Creates a raster identifying all sinks or areas of internal drainage |
| Snap Pour Point | Snaps pour points to the cell of highest flow accumulation within a specified distance |
| Stream Link | Assigns unique values to sections of a raster linear network between intersections |
| Stream Order | Assigns a numeric order to segments of a raster representing branches of a linear network |
| Stream to Feature | Converts a raster representing a linear network to features representing the linear network |
| Watershed | Determines the contributing area above a set of cells in a raster |

*Source*: modified from http://pro.arcgis.com/en/pro-app/tool-reference/spatial-analyst/an-overview-of-the-hydrology-tools.htm

**Table 6.5** List of Solar Radiation tools.

| Tools | Descriptions |
|---|---|
| Area Solar Radiation | Derives incoming solar radiation from a raster surface |
| Points Solar Radiation | Derives incoming solar radiation for specific locations in a point feature class or location table |
| Solar Radiation Graphics | Derives raster representations of a hemispherical viewshed, sun map, and sky map, which are used in the calculation of direct, diffuse, and global radiation |

*Source*: modified from http://pro.arcgis.com/en/pro-app/tool-reference/spatial-analyst/an-overview-of-the-solar-radiation-tools.htm

**Table 6.6** List of 3D Analyst toolsets and tools.

| Toolsets | Descriptions |
|---|---|
| 3D Features | Provides tools for evaluating geometric properties and relationships between three-dimensional features |
| CityEngine | Contains tools that expose some capabilities of CityEngine without requiring Esri CityEngine to be installed |
| Conversion | Contains tools that convert feature classes, files, LAS datasets, rasters, TINs, and terrains to other data formats. The tools are organized into toolsets based on the data type being converted |
| Data Management – LAS Dataset | Provides tools for creating and managing LAS datasets |
| Data Management – Terrain Dataset | Provides tools for creating and managing terrain datasets |
| Data Management – TIN Dataset | Provides tools for creating and managing TIN datasets |
| Functional Surface | Provides analysis tools that evaluate elevation information from raster, terrain, and TIN surfaces |
| Raster Interpolation | Provides numerous interpolation tools that can produce continuous raster surfaces from a given set of sample points, including hydrologically "correct" surface models |
| Raster Math | Features tools that perform mathematical operations on raster datasets |
| Raster Reclass | Contains tools that enable reclassification of raster data |
| Raster Surface | Provides analysis tools that enable the determination of raster surface properties, such as contours, slope, aspect, hillshade, and difference calculation |
| Triangulated Surface | Provides analysis tools that enable the determination of surface properties of TIN, terrain, and LAS datasets, such as contours, slope, aspect, hillshade, difference calculation, volumetric computations, and outlier detection |
| Visibility | Feature tools that enable visibility analysis to be performed using various types of observer features and obstruction sources that include surfaces, multipatches, which are used for representing buildings and three-dimensional features |

*Source*: modified from http://pro.arcgis.com/en/pro-app/tool-reference/3d-analyst/an-overview-of-the-3d-analyst-toolbox.htm

continuous raster surfaces from a given set of sample points, including hydrologically "correct" surface models, and tools to determine slope, aspect and other surface properties of TIN, terrain, and LAS datasets.

The Data Management toolset provides three collections of geoprocessing tools that operate on terrain, TIN, and LAS datasets. The Terrain dataset collection provides tools for creating and analyzing a terrain dataset, which is a multi-resolution triangulated surface consisting of feature-based measurements stored in a feature dataset (Table 6.7). The TIN dataset collection provides tools for creating and editing a TIN dataset, which is a single resolution surface consisting of feature-based measurements (Table 6.8). The LAS dataset collection provides tools for analyzing and managing the classification of LiDAR data referenced by a LAS dataset (Table 6.9).

The Raster Interpolation toolset can be used to create a continuous surface from sampled point values and provides exactly the same functionality as the Interpolation toolset provided in the Spatial Analyst toolbox (which is provided under a separate Spatial Analyst license). This means that the four deterministic (IDW, Natural Neighbor, Spline, Trend), one geostatistical (Kriging), and two specifically designed hydrologic (Topo-to-Raster, Topo-to-Raster-by-File) interpolation tools can be accessed from either the Spatial Analyst and/or the 3D Analyst toolboxes.

**Table 6.7** List of the Data Management – Terrain Dataset tools.

| Tools | Descriptions |
|---|---|
| Add Feature Class To Terrain | Adds one or more feature classes to a terrain dataset |
| Add Terrain Pyramid Level | Adds one or more pyramid levels to an existing terrain dataset |
| Append Terrain Points | Appends points to a point feature referenced by a terrain dataset |
| Build Terrain | Performs tasks required for analyzing and displaying a terrain dataset |
| Change Terrain Reference Scale | Changes the reference scale associated with a terrain pyramid level |
| Change Terrain Resolution Bounds | Changes the pyramid levels at which a feature class will be enforced for a given dataset |
| Create Terrain | Creates a new terrain dataset |
| Delete Terrain Points | Deletes points within a specified area of interest from one or more features that participate in a terrain dataset |
| Remove Feature Class from Terrain | Removes reference to a feature class participating in a terrain dataset |
| Remove Terrain Pyramid Level | Removes a pyramid level from a terrain dataset |
| Replace Terrain Points | Replaces points referenced by a terrain dataset with points from a specified feature class |

*Source*: modified from http://pro.arcgis.com/en/pro-app/tool-reference/3d-analyst/an-overview-of-the-data-management-toolset.htm

**Table 6.8** List of the Data Management – TIN Dataset tools.

| Tools | Descriptions |
|---|---|
| Copy TIN | Creates a copy of a TIN dataset |
| Create TIN | Creates a TIN dataset |
| Delineate TIN Data Area | Redefines the data area, or interpolation zone, of a TIN based on its triangle edge length |
| Edit TIN | Loads data from one or more input features to modify the surface of an existing TIN |

*Source*: modified from http://pro.arcgis.com/en/pro-app/tool-reference/3d-analyst/an-overview-of-the-data-management-toolset.htm

Table 6.9 List of the Data Management – LAS Dataset tools.

| Tools | Descriptions |
|---|---|
| Change LAS Class Codes | Modifies the class code values of the LAS files referenced by a LAS dataset |
| Classify LAS By Height | Reclassifies LiDAR points based on their height from the ground surface |
| LAS Point Statistics By Area | Evaluates the statistics of LAS points that overlay the area defined by polygon features |
| Locate LAS Points By Proximity | Identifies LiDAR points within a three-dimensional proximity of *z*-enabled features while also providing the option to reclassify the points and export them to an output feature class |
| Set LAS Class Codes Using Features | Classifies data points in LAS files referenced by a LAS dataset using point, line, and polygon features |

*Source*: modified from http://pro.arcgis.com/en/pro-app/tool-reference/3d-analyst/an-overview-of-the-data-management-toolset.htm

Table 6.10 List of Triangulated Surface tools.

| Tools | Descriptions |
|---|---|
| Decimate TIN Nodes | Creates a TIN dataset using a subset of nodes from a source TIN |
| Extrude Between | Creates three-dimensional features by extruding each input feature between two TIN datasets |
| Interpolate Polygon To Multipatch | Creates surface-conforming multipatch features by draping polygon feature class over a surface |
| Locate Outliers | Identifies anomalous elevation measurements from terrain, TIN, or LAS datasets that exceed a defined range of elevation values or have slope characteristics that are inconsistent with the surrounding surface |
| Polygon Volume | Calculates the volume and surface area between a polygon and terrain or TIN surface |
| Surface Aspect | Creates polygon features that represent aspect measurements derived from a terrain, TIN, or LAS dataset surface |
| Surface Contour | Creates contour lines derived from a terrain, TIN, or LAS dataset surface |
| Surface Difference | Calculates the volumetric difference between two surface models stored as either a TIN or terrain dataset |
| Surface Slope | Creates polygon features that represent ranges of slope values for triangulated surfaces |

*Source*: modified from http://pro.arcgis.com/en/pro-app/tool-reference/3d-analyst/an-overview-of-the-triangulated-surface-toolset.htm

The Raster Surface toolset provides eight of the 11 tools provided by the Surface toolset in Spatial Analyst. The three visibility tools (Observer Points, Viewshed2, Visibility) are not included but the remainder of the tools support the calculation of contours, slope gradients and aspects, and shaded relief.

The Triangulated Surface toolset provides nine tools that operate on terrain, TIN, and LAS datasets (Table 6.10). These tools can be used to extract surface properties, such as slope, aspect and contours, identify outliers in data points, calculate volumes and areas, and create three-dimensional feature classes that model the surface. Some of these tools replicate those found in the Spatial Analyst's Surface toolset but the tools included in this particular toolset work with all three of the custom datasets described earlier.

And finally, the documentation for the Triangulated Surface toolset lists four tools for TINs that were available prior to ArcGIS Desktop Version 10.0 and recommends the tools among those currently available that can be used to perform the same tasks.

The value of the aforementioned capabilities (i.e. tools) has been greatly enhanced by the new elevation and hydrologic analysis geoprocessing services that use ArcGIS Online's data and analysis capabilities. The user specifies a few input features and all of the other data needed for the analysis, as well as the computations, are hosted by ArcGIS Online. Most analyses need relatively small amounts of data to be passed to and from the services so they are fast; for example, one can calculate a viewshed anywhere in the world in less than 5 seconds.

To access these ready-to-use services, you would open your ArcGIS Pro 1.4 or ArcGIS Desktop 10.5 and log into your ArcGIS Online organizational account. You would then double-click one of the services listed near the bottom of the Catalog window (Figure 6.1) and run it like any other geoprocessing tool. The elevation tools available as of December 2016 allow a user to draw a viewshed or elevation profile for a location, calculate the slope of a road or a channel, and the predominant aspect of a forest plot (or some other area of interest). The hydrology tools include services to delineate watersheds and to trace where water flowed to or came from. There are other tools that might be leveraged as well; for example, the US Soils Available Water Storage

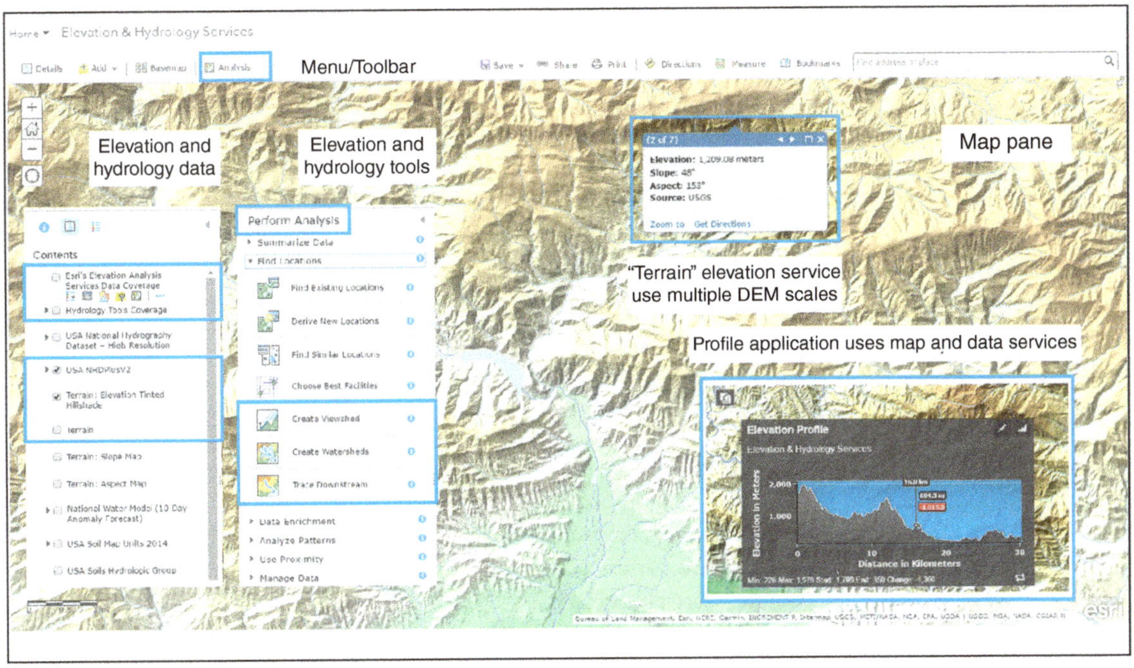


**Figure 6.1** Schematic showing some of the capabilities and how the elevation and hydrology tools are accessed in Esri's ArcGIS Online platform (as of February 2017).

layer could be used to calculate the mean soil water storage per watershed or subwatershed once these units have been delineated. An Esri user can also use them directly in their ModelBuilder models just like any other tool and developers can also use these services in their applications via the representational state transfer (REST) web services architecture.

The elevation services work alongside global data at resolutions of approximately 1 km and 90 m, and with data from Canada, Mexico and the USA at a resolution of 30 m plus data for the continental USA at a resolution of 10 m. The hydrology services in the USA are based on NHDPlus Version 2.1 (Moore & Dewald, 2016) which has been optimized for fast web service performance, and globally at 90-m resolution based on the HydroSHEDS project (Lehner, 2013). The elevation and hydrology content for ArcGIS Online grows continuously through Esri's Community Maps program and as new areas and resolutions are covered, they will be included in these analytic services. There are many other tools and data that might be useful for specific projects as well, such as the planning and assessment tools and accompanying datasets that were recently published as part of Esri's National Green Infrastructure Initiative (see http://www.esri.com/about-esri/greeninfrastructure for additional details).

The Esri user can therefore expect both the number and sophistication of these ready-to-use web services and the accompanying online data to grow steadily over time.

## 6.3 Third-party Esri Add-ons

Several groups have endeavored to build additional terrain analysis and modeling functionality and/or a different user experience on top of the current ArcGIS platform or one of its predecessors. However, the rapid evolution of the Esri platform means continuous development will be required for these add-ons to achieve their goals over the long term. Three such add-ons are described in this section.

### *6.3.1 ArcGIS Geomorphometry Toolbox*

The enormity of the aforementioned development challenge is illustrated by the ArcGIS Geomorphometry toolbox, which was created with the goal of simplifying terrain analysis and modeling for people who need to calculate land surface parameters quickly (Reuter, 2009). The tools included in this toolbox were authored using Esri's AML to assist with data preparation, the calculation of numerous land surface parameters, and the classification of landforms. Most of the land surface parameters are calculated by moving a 3 × 3 cell window across a DEM in ArcGIS but some of the secondary land surface parameters, such as TWI, topographic position index (TPI), mass balance index, and elevation residuals, can be calculated using a range of window extents (i.e. using ArcGIS focal functions) and the openness land surface parameter can be calculated using varying window extents up to 9 × 9 cells in size (Reuter, 2009). The cost of this add-on is minimal (at least when it is acquired and used for research) but the AML scripts can only be used with an installation of ArcInfo workstation that is distributed with a full ArcGIS license but seldom used nowadays. The geomorphometric functions are organized under two menus labeled Terrain Parameters and Landforms. The former includes both primary and secondary land surface parameters but the most unique component, and therefore the most compelling reason to acquire and use this add-on today, is the 11 algorithms for land surface classification proposed by Pennock, Anderson and de Jong (1994), MacMillan and Pettapiece (1997), MacMillan et al. (2000), Meybeck, Green and Vorosmarty (2001), Park, McSweeney and Lowery (2001), Weiss (2001), Reuter (2004), Dobos et al. (2005), and Iwahashi and Pike (2007) that were coded and included under the Landforms menu. This add-on requires a Spatial Analyst license as well as the basic ArcGIS license (and the ArcInfo workstation software product distributed for free with the standard ArcGIS license).

### *6.3.2 ArcGIS Geomorphometry and Gradient Metrics Toolbox*

The second toolbox was authored by Evans and colleagues and is part of a larger suite of quantitative methods for spatial ecology (Evans et al., 2014). The latest version (2.0) was released in 2014 and the authors plan to release the analog R package that utilizes GRASS and QGIS to move this effort into the open-source community in the near future. The current version of the ArcGIS Geomorphometry and Gradient Metrics toolbox is compatible with ArcGIS Desktop 10.2 and higher.

The functions for calculating land surface parameters are grouped under menus labelled Directionality, Statistics, Texture and Configuration, and Temperature and Moisture. The first two menus include general purpose utilities and statistical functions. The Texture and Configuration menu includes functions for the calculation of topographic attributes such as dissection (Evans, 1972), hierarchical slope position (Murphy, Evans & Storfer, 2010), surface curvature (Bolstad & Lillesand, 1992), roughness, slope position (Gallant & Wilson, 2000), and the surface relief ratio (Pike & Wilson, 1971). The Temperature and Moisture menu includes functions for the calculation of secondary topographic attributes such as TWI (I.D. Moore et al., 1991), heat load index (McCune & Keon, 2002), IMI (Iverson et al., 1997), and a site exposure index (Balice et al., 2000). The scripts for calculating slope gradients, aspects and curvatures rely on the basic ArcGIS functions and a 3 × 3 moving cell window and others use ArcGIS focal functions operating at a range of window extents. The add-on requires a

Spatial Analyst license in addition to the standard ArcGIS Desktop license and the metrics have found widespread use in a number of landscape ecology, species distribution, predictive vegetation, and remote sensing applications.

### *6.3.3 ArcGeomorphometry Toolbox*

The third and newest of the three toolboxes was authored by Rigol-Sanchez et al. (2015) and provides a series of Python/NumPy processing functions presented through an easy-to-use graphical menu for ArcGIS. This toolbox allows GIS users to flexibly apply several multiscale analysis methods to parameterize and classify a DEM into discrete land surface units by varying the extent, the resolution of the source DEM, and/or the size of the analysis window. This add-on requires a Spatial Analyst as well as the standard ArcGIS Desktop license, similar to the first two add-ons.

The authors programmed the geomorphometry functions listed in Table 6.11 in Python using the Numerical Python (NumPy) library for scientific computing, which includes support for powerful N-dimensional array objects. This new functionality is made accessible via a Python toolbox that can be installed, shared, and modified. Once loaded, the ArcGeomorphometry menus and help pages appear seamlessly within ArcGIS and this means, among other things, that these new functions can be run in a stand-alone mode by calling them from a geoprocessing pane, ModelBuilder, or a Python window.

The ArcGeomorphometry toolbox runs on any desktop capable of running ArcGIS and Spatial Analyst. The tools use linear map units, such as feet or meters, which assumes that the input DEM has a projected coordinate system. The toolbox currently provides functions for true multiscale land surface analysis and classification based on the methods proposed by Evans (1972, 1979, 1980), Shary (1995), Wood (1996a,b), Blaszczynski (1997), Shary et al. (2002), and Yokoyama et al. (2002). These functions are grouped by method under four menus labelled Average Relief, Openness, Evans–Wood Method, and Shary Method (Table 6.11). The calculations under the Average Relief and Openness function menus make use of NumPy array indexing functions, whereas the functions listed under the Evans–Wood Method and Shary Method function menus make use of NumPy functions to fit a bivariate quadratic polynomial for each DEM cell to elevation values contained within the user-specified kernel by least squares. Polynomial parameters are then used to obtain geomorphometric variables.

The three multiscale components are handled in unique ways as well. The initial geographic extents and grid resolutions are chosen when the elevation source dataset is selected but the user, of course, can use standard ArcGIS functions to clip this dataset and/or to modify the grid resolution used. The grid resolution is often relaxed when working with large geographic extents (see Marthews et al., 2015 for a recent project that executed this strategy). The third element that can be varied, the size of the analysis

**Table 6.11** Terrain analysis and modeling functions included in ArcGeomorphometry.

| Functions | Descriptions |
|---|---|
| **Average relief functions** | |
| Average slope | Calculate average slope percent |
| Classified average relief | Reclassify signed average local relief grid using user-defined slope and signed average local relief cut-offs |
| Signed average relief | Calculate signed average local relief |
| **Openness functions** | |
| Negative openness | Calculate eight-direction average minimum elevation angle below surface |
| Positive openness | Calculate eight-direction average maximum elevation angle above surface |
| **Evans–Wood Method functions** | |
| Aspect | Compute slope orientation or aspect |
| ElevationSmoothed | Return elevation smoothed by quadratic function |
| CrossCurvature | Computer cross-sectional curvature |
| Feature | Classify DEM into surface-specific elements (pits, peaks, ridges, channels, passes, plains) using user-defined slope and curvature thresholds |
| LongCurvature | Compute longitudinal curvature |
| MaxProfCurvature | Compute maximum profile curvature |
| MinProfCurvature | Compute minimum profile curvature |
| Modified Evans-Young | Modified Evans–Young (pre-filtering) algorithm |
| PlanCurvature | Compute plan curvature |
| ProfileCurvature | Compute profile curvature |
| Slope | Compute slope steepness |
| **Shary Method functions** | |
| Aspect | Compute slope orientation or aspect |
| CrossCurvature | Compute cross-profile curvature |
| LongCurvature | Compute longitudinal curvature |
| MaxProfCurvature | Compute maximum profile curvature |
| MeanCurvature | Compute mean curvature |
| MinProfCurvature | Compute minimum profile curvature |
| PlanCurvature | Compute plan curvature |
| ProfileCurvature | Compute profile curvature |
| Rotor | Compute rotor |
| TangentialCurvature | Compute tangential curvature |
| TotalCurvature | Compute total curvature |
| TotalGaussianCurvature | Compute total Gaussian curvature |
| TotalRingCurvature | Compute total ring curvature |
| Slope | Compute slope steepness |
| Unsphericity | Compute unsphericity |

*Source*: Rigol-Sanchez et al. (2015, p. 159). Reproduced with permission of Elsevier.

window, is set when implementing one of the ArcGeomorphometry functions. The user next selects the input DEM and must then specify the size of the processing kernel (i.e. analysis window) for land surface analysis by typing in the desired square dimension (or a circle diameter in the case of openness). Any positive odd kernel size is allowed, which means that the maximum size of the analysis window is limited only by the geographic extent of the input DEM or the available system resources.

This toolbox takes advantage of the power of ArcGIS Desktop to handle large DEM datasets while retaining ease of navigation through its custom interface to a more sophisticated set of terrain analysis and modeling functions that includes support for multiscale analysis of DEMs. The results of running each function can be graphically displayed in Esri's ArcMap 2D or ArcScene 3D display environments and combined with other raster or vector datasets to promote further assessment of the efficacy of the methods and the significance of the results.

## 6.4 Other Software Choices

The comprehensive review of geomorphometry software provided by Wood (2009a) and the software utilized to produce the related works listed in the References at the end of this book point to the large number and variety of terrain analysis and modeling software choices that are available nowadays. This section describes the current status of the seven packages reviewed in Hengl and Reuter (2009) plus two others that offer novel and unique functionality and/or have attracted large numbers of users during the past decade. The details offer further proof that the distinction often drawn by commentators between proprietary and free and open-source solutions is becoming more and more blurred with time. Several of the software platforms reviewed in this section, similar to the Esri ecosystem and add-ons described in the previous sections, combine both proprietary and free and open-source software in a variety of ways.

### *6.4.1 GRASS*

GRASS (Geographic Resources Analysis Support System) is a general-purpose GIS for the management, analysis, modeling, and visualization of many types of geographic information. This is also an open-source product and, as such, it provides complete access to its source code written in the ANSI C programming language (Hofierka, Mitaášová & Neteler, 2009). GRASS has more than 350 modules (i.e. tools and workflows) and Neteler and Mitášová (2008) provide detailed descriptions of the organization, functionality, and uses of GRASS, including the functions that support land surface analysis and modeling. GRASS is a command-line software which uses a simple and intuitive series of naming conventions (see Hofierka et al., 2009, p. 388, Table 1 for additional details).

GRASS provides several functions for importing, displaying, and computing DEMs from a variety of elevation data sources. Grid-based DEMs can be displayed as two-dimensional raster maps and three-dimensional views and GRASS data can be viewed using several external programs. For example, QGIS, which is discussed in more detail in Section 6.4.5, can be used to view two-dimensional maps. QGIS can also read the GRASS raster and vector data directly and its GRASS plug-in offers a toolbox, providing GUI access to important GRASS commands for data analysis (Hofierka et al., 2009).

The GRASS software also includes several interpolation functions for computing a DEM from contours or point data, including the Regularized Spline with Tension (RST; Mitas & Mitášová, 1999; Neteler & Mitášová, 2008) function that uses a series of parameters related to tension, smoothing, anisotropy, and the minimum and maximum distances between points to customize the behavior of the resulting surface. The five local land surface parameters described in the next paragraph can be computed simultaneously with the RST interpolation tools because this interpolation function is differentiable to all orders (Mitášová et al., 1995).

GRASS provides functions to calculate many of the land surface parameters described in Chapter 3. The local parameters include slope and aspect as well as profile, tangential and mean curvatures. Plan curvature can also be derived from the tangential curvature and the sine of the slope angle (Mitášová & Hofierka, 1993). GRASS also provides modules for computing several of the regional land surface parameters that are frequently used for the analysis of mass flows over the land surface. The following basic flow parameters – flow accumulation, upslope contributing area, stream network, watershed boundaries, and flow-path length – can be calculated with several single- and multiple-flow routing algorithms and a variety of functions are provided to handle sinks and/or flow over flat areas. There are also modules for simulating overland flow and sediment transport and erosion/deposition based on the Water Erosion Prediction Project (WEPP) theory (Mitas & Mitášová, 1998). Hofierka et al. (2009) showed how shell scripting, map algebra, and various land surface parameters could be used inside GRASS to calculate a topographic index of soil erosion and deposition using a modified LS factor for the USLE. Finally, GRASS provides several modules related to solar radiation, including one to calculate all three components of solar irradiance/radiation (direct beam, diffuse, and reflected) for clear skies as well as overcast conditions (Šúri & Hofierka, 2004). The r.sun function requires a few mandatory input parameters, such as elevation, slope and aspect of the terrain, day number and, optimally, a local solar time. The remainder of the input parameters are either internally computed (solar declination) or the values can be overridden by explicitly defined settings to fit specific user needs (Linke atmospheric turbidity, ground albedo, the direct beam and diffuse components of a clear-sky index, and the time step used for the calculation of daily radiation from sunrise to sunset). The clear-sky index is used to parameterize the attenuation of cloud cover and calculate overcast irradiance/radiation from clear-sky raster maps.

GRASS is a mature, general-purpose, open-source GIS that provides many functions for the analysis, modeling, and visualization of land surface parameters and landform objects. The software has many users and a dedicated team led by Hofierka, Mitas and Mitášová, among others, has a long history of helping to lead the development of new functions and capabilities that advance the underlying science and application of terrain analysis and modeling tools.

### 6.4.2 ILWIS

The Integrated Land and Water Information System (ILWIS) is a Windows-based GIS and remote sensing software platform. The software was developed by ITC up to version 3.3 in 2005 but starting in July 2007 it has been distributed as a free open-source software (ILWIS Open, including binaries and source code) under the 52°N Initiative. Successive versions have incorporated new functionality and bug fixes and the most recent versions (3.8 and up) have incorporated a new web server system so ILWIS can now function as a web processing server.

Hengl, Maathuis and Wang (2009) provided a detailed description of the functions for working with DEMs, hydrologic modeling, and surface energy balances. The software includes functions to import DEMs, derive DEMs from sample elevations, and to identify and clean artifacts from DEMs using methods proposed by Gorte and Koolhoven (1990), Pilouk and Tempfli (1992), Hengl et al. (2004), and Reuter et al. (2009).

The software also includes a built-in hydro-processing module that can be used to prepare DEMs for hydrologic applications. The functions included in this module can be used to: (i) remove or fill sinks; (ii) burn drainage features into the DEM; and (iii) specify flow paths in lakes, reservoirs, and other flat areas with undefined flow directions (based on the method proposed by Garbrecht & Martz, 1997). This module also includes functions to specify the drainage networks, the accompanying catchment boundaries, and a series of hydrologically relevant land surface parameters using methods documented by Maathuis and Wang (2006). The D8 (O'Callaghan & Mark, 1984) single-flow direction algorithm is used to: (i) calculate flow directions; (ii) calculate flow accumulation and extract the basic drainage network; (iii) calculate various kinds of flow lengths; and (iv) calculate three secondary topographic attributes (TWI, SPI, STI) and a series of plots used to characterize hydrologic systems.

The basic functionality can also be extended using ILWIS scripts. Hengl et al. (2009, pp. 323–329), for example, shared three scripts to derive local land surface parameters (slope, aspect, profile, plan and mean curvature, slope-adjusted northness, and solar radiation for given angles), secondary land surface parameters (TPI, SPI, STI, and shape complexity index) based on the multiple-flow direction algorithm described by Quinn et al. (1991), and a series of generic landform shapes (channels, ridges, terraces, slopes, and pits).

Hengl et al. (2009) concluded their extensive review of this software by noting it offered some novel computational capabilities as well as some challenges. The latter included the small numbers of supported data formats, the lack of a website for users to exchange scripts and user-built modules, its continued reliance on a relatively unfriendly command-line interface and a relatively poor three-dimensional viewer, and a small user community that is dominated by former ITC students and collaborators.

### 6.4.3 LandSerf

The first version of the LandSerf platform was authored by Joseph Wood in 1996 to perform scale-based analysis of DEMs so that land surface parameters such as slope gradients, aspects and curvatures could be measured and visualized over a range of scales (Wood, 1996a,b, 2009b). The LandSerf platform is focused on visualization and, as a result, it has provided a rich interactive interface to explore the effects of scale on the parameterization of land surface variables from the outset.

The software is written in Java and includes a text-based editor, its own scripting language and both two- and three-dimensional viewers. The supporting documentation includes a user manual and an API is provided for Java programmers. The software supports contour, grid and TIN surface models and all the operations performed on these data are carried out in memory in order to improve the speed of visual interaction. This approach means that the software works best with gridded DEMs with no more than $6000 \times 6000$ pixels. The calculation of slope, aspect and curvature in LandSerf uses the frequently adopted method of taking first derivatives and partial second derivatives of a bi-quadratic polynomial representing a local patch of the surface (Evans, 1980) and then provides a novel method by which the six coefficients in the polynomial expression can be estimated in a user-specified window of any arbitrary size using methods documented in Wood (1996b, pp. 92–97). The software supports interactive three-dimensional viewing and "fly-throughs" of surfaces on platforms that support OpenGL. The interactive visualization techniques include various lighting and shade models, multiple image blending, LoD rendering, and dynamic graphical query.

This software works well where visual interaction is considered important and where the effects of scale are to be considered (Wood, 2009b). The software is free but not open and the choice of Java means that it can be implemented on Windows, Linux, UNIX, and MacOSX platforms. There have been 11 new releases of the software since the initial release in 1996 and several have added new functionality (e.g. vector handling in 1998, attribute tables in 2003, and raster and vector overlay in 2004). However, the latest version (2.3) was released back in September 2007 and the LandSerfing Forum on the LandSerf website (http://www.landserf.org) was suspended in April, 2010.

And last but not least, the multiscale surface characterization that was championed with this platform can now be performed in two generic GIS: ArcGIS, using the third-party ArcGeomorphometry extension authored by Rigol-Sanchez et al. (2015), and GRASS using the r.param.scale module.

### 6.4.4 MicroDEM

The first version of the MicroDEM software was written by Peter Guth and released in 1985. The software has focused on the DTED maintained by the US National Imagery and Mapping Agency (NIMA) and now its successor the NGA, and has sought to support the analysis of DEMs in their native format (i.e. horizontal spacing reported in arcseconds) since the outset. This meant that many of the terrain analysis and modeling algorithms described in this book for planar grids have been modified to use the native format. The functionality of the software has increased over time and the software now reads both geographic and square grids (Guth, 2009). The current version is written in Delphi (Object Pascal) using a 64-bit architecture and distributed as freeware without source code. The software has been downloaded more than 1000 times and the website (http://www.usna.edu/Users/oceano/pguth/website/microdem/microdem.htm) shows an active discussion forum. Users can create their own scripts to provide additional functionality.

The software provides various methods to calculate and visualize more than 30 primary and secondary land surface parameters (slope, aspect, curvatures, openness, etc.) and a novel eigenvector technique to quantify terrain organization (the degree to which ridges and valleys align, their preferred orientation, etc.). The early versions of MicroDEM focused on slope maps, three-dimensional oblique views, line-of-sight profiles and viewsheds. It has since become a general-purpose GIS, integrating imagery and shapefiles with DEMs, but it retains its strong focus on geology and geomorphology. Guth (2009), for example, described how MicroDEM was used to produce an atlas of geomorphic parameters for the SRTM dataset and, more recently, he showed how the new version that runs on a 64-bit Windows operating system can be used to conduct global-scale terrain analysis work in main memory (Guth, 2013).

The main benefit of this software is its continued reliance on geographic units. The conversion of global elevation datasets from decimal degrees to meters or feet will create errors that will grow in magnitude with the geographic area of interest and the reliance on geographic units will be essential for work conducted across large regions and, of course, the Earth as a whole (Guth, 2009; Florinsky, 2017).

### 6.4.5 QGIS

This software (QGIS, previously known as Quantum GIS) is another general-purpose, free and open-source GIS for the management, analysis and modeling of geographic information. Gary Sherman began the development of Quantum GIS in 2002 and it became an incubator project of the Open Source Geospatial Foundation in 2007. The first version of the software was released in Janaury 2009 and the latest version (2.18) was released in October 2016. The software is written in C++ and incorporates a variety of other open-source software, including GDAL, the Geometry Engine – Open Source software suite (GEOS; https://trac.osgeo.org/geos/), and the MySQL, PostgreSQL/PostGIS, and SQLite database engines. QGIS has been translated into more than 50 languages to date and it runs on multiple operating systems, including MacOSX, Linux, UNIX and Microsoft Windows. The core QGIS software is maintained by volunteer developers who regularly release updates and bug fixes and the core capabilities have been extended with a variety of plug-ins written in Python or C++, as illustrated by the GRASS and SAGA GIS plug-ins described here. The various web resources (user guides, tutorials and videos) have been supplemented with the publication of a series of print resources in the past few years (e.g. Bruy, 2015; Lawhead, 2015; Menke et al., 2015; Graser, 2016).

The terrain modeling tools available in QGIS take two forms. The first type uses the Raster Based Terrain Analysis plug-in provided with QGIS which can be used to calculate slope, aspect, a ruggedness index, and total curvature using the first- and second-order derivatives calculated on a $3 \times 3$ moving window. Slope is calculated in degrees of inclination. Aspect is calculated in degrees from North (0°) in a counter-clockwise direction to indicate the direction a hillslope is facing, so that west and east are represented by 90 and 270° instead of the usual convention (270 and 90°, respectively). The ruggedness index proposed by Riley et al. (1999) is used to summarize the terrain heterogeneity by calculating the relief within each $3 \times 3$ moving window and using the class limits summarized in Table 6.12 to represent the terrain ruggedness (see Section 3.1.9 for additional details). The calculation of total curvature relies on the second derivatives and uses positive and negative values to report surfaces that are upwardly convex and concave, respectively, and values of 0 to indicate areas where the surface is flat (and no curvature exists).

The second type of tool uses third-party plug-ins that must be selected, added and installed prior to use in QGIS. The Shaded Relief plug-in, for example, is a relatively simple tool that can be used with a fixed sun location and the horizon calculated from three-dimensional DEM data in QGIS to accurately display areas of bright sun exposure as well as low shaded (dark) areas. This third-party plug-in functionality is very important because this same approach can be used to add, install, and run most of the terrain modeling tools provided in the GRASS, SAGA, and TauDEM software platforms as well.

**Table 6.12** Class limits used in QGIS to classify ruggedness index values into categories that describe different types of terrain.

| Ruggedness classes | Ruggedness index class limits |
|---|---|
| Level | 0–80 m |
| Nearly level | 81–116 m |
| Slightly rugged | 117–161 m |
| Intermediately rugged | 162–239 m |
| Moderately rugged | 240–497 m |
| Highly rugged | 498–958 m |
| Extremely rugged | ≥959 m |

*Source*: modified from http://www.gistutor.com/quantum-gis/20-intermediate-quantum-gis-tutorials/48-quantum-gis-qgis-raster-based-terrain-analysis-techniques.html

The GRASS plug-in provides access to more than 400 GRASS modules (i.e. tools or workflows) for managing, analyzing and visualizing various forms of geographic information. The QGIS user must create a GRASS data directory which is often called "grassdata" and within which data are organized by projects in subdirectories called "locations" for which the coordinate system, map projection, and geographic boundaries must be defined. These locations, in turn, can have several mapsets which can be used to subdivide the project and the accompanying "locations" into different topics or subregions and/or to support workspaces for individual team members (Neteler & Mitášová, 2008). The complete list of GRASS modules available in QGIS Version 2.18 can be found at https://grasswiki.osgeo.org/wiki/GRASS-QGIS_relevant_module_list and includes many of the terrain modeling tools noted in Section 6.4.1 for example.

The SAGA plug-in similarly provides access to more than 200 tools organized in 34 module libraries. This particular plug-in provides more than 40 tools spread across six to eight module libraries for calculating many of the land surface parameters and objects described in Chapters 3 and 4 (see Table 6.13 and Section 6.4.7 for additional details).

And finally, there are the TauDEM tools which focus on the derivation of hydrologic information from DEMs. These can also be installed and used in QGIS, as described in Section 6.4.8.

## 6.4.6 *RiverTools*

RiverTools is a proprietary, user-friendly GIS application for the analysis and visualization of digital terrain, watersheds and river networks distributed by Rivix, LLC (Peckham, 2009). The latest 64-bit version (4.0) runs on top of the MS Windows, MacOSX, and Linux operating systems.

Many of the grids computed by RiverTools are required as inputs for distributed hydrologic models and a relatively new hydrologic model, TopoFlow, can be used as a plug-in to RiverTools to create a one-stop hydrologic modeling and visualization environment. Both RiverTools and TopoFlow are written in the Interactive Data Language (IDL; Exelis Visual Information Systems, Inc., Broomfield, CO). The mosaicking, visualization, and analysis routines included in RiverTools can be accessed using an intuitive GUI and IDL provides a fast and robust scripting language for extending RiverTools.

RiverTools provides the ability to import many DEM file formats and the calculations can use one of a number of Earth ellipsoid models. The user can use geographic or planar coordinates and the area, length and slope calculations performed on DEMs using geographic coordinates account for latitude dependence. Flow directions, total and specific contributing areas, river networks, and channel slopes can be derived with the D8 (O'Callaghan & Mark, 1984), D∞ (Tarboton, 1997), and Mass Flux (Gruber & Peckham, 2009) flow-routing algorithms and multiple functions are provided to work with or around sinks and flow directions across flat areas. Grids can also be derived for flow distances; longest channel lengths; plan, profile, tangential, streamline, mean and Gaussian curvature; the TWI, basin relief, Horton–Strahler stream orders, and Pfafstetter basin codes. Finally, specific features, such as lakes and reservoirs, can be masked and watersheds can be delineated and characterized using a variety of tools.

The novel features of the latest version of RiverTools include the addition of TopoFlow, an open-source, D8-based, spatially distributed hydrologic model with a user-friendly point-and-click interface. Its main purpose is to model a variety of physical processes in a watershed with the goal of accurately predicting how various hydrologic variables will evolve in time in response to climatic forcings. The outputs include time evolutions for single pixels (such as hydrographs), collections of user-selected pixels, or entire grids (as animations). The new graphics systems included in RiverTools 4.0 give users the ability to toggle between faster Direct Graphics and publication-quality Object Graphics.

### 6.4.7 SAGA

The System for Automated Geoscientific Analyses (SAGA) software was initially developed at the University of Göttingen, Germany and most of the source code was published under a series of open-source software licenses in 2004 (Olaya & Conrad, 2009). The software itself is organized around four components: (i) an API that provides all the basic functions for conducting geographic analysis; (ii) a set of 234 modules that are organized in 42 module libraries and represent the geoscientific functions; (iii) a GUI that provides the software's front-end and through which the user manages data and executes functions (i.e. modules); and (iv) a command-line interpreter that

**Table 6.13** List of SAGA module libraries and modules focused on calculation of terrain parameters and objects.

| Module libraries | Modules (i.e. tools and workflows) |
|---|---|
| Grid Gridding | Inverse Distance Weighted ∘ Kernal Density Estimation ∘ Modified Quadratic Sherpard ∘ Natural Neighbor ∘ Shapes to Grid ∘ Triangulation |
| Grid Spline | B-spline Approximation ∘ Cubic Spline Approximation ∘ Multilevel B-spline Approximation ∘ Multilevel B-spline Interpolation ∘ Thin Plate Spline (Global) ∘ Thin Plate Spline (Local) ∘ Thin Plate Spline (TIN) |
| Simulation Hydrology | Overland Flow – Kinematic Wave D8 ∘ Water Retention Capacity |
| Terrain Channels | Channel Network & Drainage Basins ∘ Channel Network ∘ Overland Flow Distance to Channel Network ∘ Strahler Order ∘ Vertical Distance to Channel Network ∘ Watershed Basins |
| Terrain Hydrology | Burn Stream Network into DEM ∘ Catchment Area (Flow Tracing) ∘ Catchment Area (Recursive) ∘ Catchment Area ∘ Cell Balance ∘ Edge Contamination ∘ Fill Sinks ∘ Fill Sinks (Wang & Liu) ∘ Fill Sinks XXL (Wang & Liu) ∘ Flat Direction ∘ Flow Path Length ∘ Flow Width & Specific Catchment Area ∘ Lake Flood ∘ LS Factor ∘ SAGA Wetness Index ∘ Sink Drainage Route Detection ∘ Sink Removal ∘ Slope Length ∘ Stream Power Index (SPI) ∘ Topographic Wetness Index (TWI) ∘ Upslope Area |
| Terrain Lighting | Analytical Hillshading ∘ Sky View Factor ∘ Topographic Correction |
| Terrain Morphometry | Convergence Index ∘ Convergence Index (Search Radius) ∘ Curvature Classification ∘ Diurnal Anisotropic Heating ∘ Downslope Distance Gradient ∘ Effective Air Flow Heights ∘ Hypsometry ∘ Land Surface Temperature ∘ Mass Balance Index ∘ Morphometric Protection Index ∘ Multiresolution Index of Valley Bottom Flatness (MRVBF) ∘ Real Area Calculation ∘ Relative Heights & Slope Positions ∘ Slope, Aspect, Curvature ∘ Surface Specific Points ∘ Terrain Ruggedness Index (TRI) ∘ Topographic Position Index (TPI) ∘ TPI-based Landform Classification ∘ Vector Ruggedness Measure (VRM) ∘ Wind Effect |
| Terrain Profiles | Cross Profiles ∘ Profile from Points Table ∘ Profiles from Lines |

*Source*: modified from https://docs.qgis.org/2.8/en/docs/user_manual/processing_algs/saga/index.html

can be used to not only execute existing modules but also to write scripts to automate complicated workflows (Böhner, McCloy & Strobl, 2006; Olaya & Conrad, 2009; Cimmery, 2010a,b; Conrad et al., 2015).

The current SAGA software suite is a fully fledged GIS that focuses on spatial analysis and visualization. It provides powerful tools for the analysis of DEMs and can be used to calculate many of the land surface parameters and objects described in Chapters 3 and 4 (Table 6.13). The software provides functions to import DEMs in multiple formats (including all those supported by GDAL), five functions to create gridded DEMs from point samples (Nearest Neighbor, Triangulation, Inverse Distance, Modified Quadratic, and Ordinary Kriging), and additional functions to smooth or sharpen the elevation surface. The latter includes several filter algorithms and two modules for the removal of sinks, one of which implements the method proposed by Planchon and Darboux (2001).

The slope, aspect and various curvatures can be calculated using multiple approaches along with the convergence index (Köthe & Lehmeier, 1993), which uses the aspect values of neighboring cells to characterize flow convergence and divergence. The latter index is similar to plan curvature, but does not rely on the absolute height difference (Olaya & Conrad, 2009). There are also modules to calculate the real-world surface area (as opposed to the projected area) and the landform classification based on plan and profile curvature proposed by Dikau (1989). The solar radiation functions sum incoming energy for user-defined time periods by taking into account topographic shading, reflection from surrounding slopes, and atmospheric effects (following the methods used in the SRAD program that was part of the now defunct TAPES-G suite; Wilson & Gallant, 2000b).

A large number and variety of modules are also available for calculating land surface parameters specific to hydrology. The D8 (O'Callaghan & Mark, 1984), D∞ (Tarboton, 1997), and FMFD (Freeman, 1991) algorithms are provided to calculate catchment areas and these, plus the ADK (Lea, 1992) and DEMON (Costa-Cabral & Burges, 1994) algorithms, are provided to calculate flow directions. Most of the other hydrology-related modules use either the D8 or FMFD flow-routing algorithms to calculate other land surface parameters. The list covers a variety of primary land surface parameters (i.e. upslope and downslope areas, flow-path lengths and flow depths), elevation residuals (similar to Wilson & Gallant, 2000b), and secondary land surface parameters, namely two forms of the TWI, the SPI, and the LS factor for the USLE (Wischmeier & Smith, 1978; Renard et al., 1991). The SAGA Wetness Index is based on a modified upslope contributing area calculation, which does not consider the flow as a very thin film (Böhner et al., 2002). Olaya and Conrad (2009, p. 304, Figure 9) show how this particular index predicts more realistic higher potential soil moisture for cells situated on valley floors with a small vertical distance to a channel compared to the standard TWI calculation.

### 6.4.8 *TauDEM*

The TauDEM (Terrain Analysis Using Digital Elevations Models) is a suite of tools for the extraction and analysis of hydrologic information from topography as represented by a DEM (Tarboton, 2016). The software runs on MS Windows personal computers and UNIX clusters and performs calculations in geographic as well as planar (i.e. projected) coordinates. The latest release (Version 5.3.1) includes pre-compiled versions in both 32- and 64-bit and the software can partition the domain into stripes to take advantage of parallel processing using the MPI.

This software provides a full suite of terrain analysis tools for hydrologic applications, including (i) the development of hydrologically conditioned DEMs using the flooding approach; (ii) the calculation of flow directions, slopes, and contributing areas using single- and multiple flow-direction algorithms; (iii) the delineation of stream networks using multiple methods, including topographic form-based methods sensitive to spatially variable drainage density; (iv) the delineation of watersheds and subwatersheds draining to each stream segment and the linkages between watershed and segment attributes for setting up hydrologic models; and (v) a series of specialized functions such as the ability to calculate the slope–area ratio and the TWI, the ability to calculate the distance up to ridges and down to streams in a variety of ways, and to evaluate upslope contributions that may be subject to decay or attenuation.

The software also includes a set of stand-alone command-line executable programs and an ArcGIS toolbox GUI written in Python. This GUI makes system calls to the command-line executables to allow ArcGIS users to access the tools as standard ArcGIS tools. The TauDEM command-line tools can also be accessed and used in QGIS. However, the user will need to install the SEXTANTE plug-in first and then download the TauDEM plug-in from the QGIS official repository. MS Windows users can use the pre-compiled binaries and installation instructions from the TauDEM website (http://hydrology.usu.edu/taudem/taudem5/index.html) and some additional instructions and workarounds are provided for Linux users in QGIS itself.

The novelty and value of this software is linked to its hydrologic focus on the one hand and its use of both the D8 (O'Callaghan & Mark, 1984) and D∞ (Tarboton, 1997) flow-routing algorithms on the other hand. Flow directions and contributing areas can be calculated with both of these algorithms and the software includes functions to calculate avalanche run-out areas; concentration-limited, decaying, reverse, and transport-limited accumulation; distance up and down slope; and upslope dependence using the D∞ flow-routing algorithm. Some of these TauDEM outputs were used to create the gesospatial infrastructure which helps to support the US NWM that was launched in June 2016 (see Section 5.4 for additional details).

### *6.4.9 Whitebox GAT*

The Whitebox Geospatial Analysis Tools (Whitebox GAT) project began in 2009 and was conceived as a replacement for the Terrain Analysis System (TAS) (Lindsay, 2005, 2009). TAS ran on top of Windows and was developed over a number of years to support the calculation of 42 primary and secondary land surface parameters (see Lindsay, 2009, pp. 384–386 for a list of the land surface parameters derived by TAS) and several stream network and basin parameters calculated from DEM-extracted streams and watersheds (Lindsay, 2009).

The Whitebox GAT has a broader focus than its predecessor because it has been positioned as an open-source desktop GIS and remote sensing software package for general applications of geospatial data analysis and visualization (Lindsay, 2014, 2016c). There have been three major releases thus far. The 2.0 release was developed using a combination of programming languages targeting the Java runtime environment, including Java, Groovy, and Python. The switch to Java means that the software now runs on top of MS Windows, MacOSX and Linux, and all other operating systems with a Java runtime. The current version of Whitebox GAT contains more than 360 plug-in tools for the analysis of geospatial data, including functions for conducting terrain analysis, spatial hydrologic processing (e.g. watershed and stream network extraction), and working with and interpolating LAS files.

The software website (http://www.guelph.ca/~hydrogeo/Whitebox/) stresses the unique open-access development philosophy adopted for this relatively new project. For example, one key aspect of the design allows users to view the source code associated with each tool directly from the tool's dialog simply by clicking the View Code button. The software encourages rapid development because new functionality can be created simply by modifying the existing code. For scripting-based plug-in tools, users can open the tool's source code, modify it, and save the modified code as a new plug-in tool. Users may experiment without destroying the existing functionality because each script-based tool can be automatically restored simply by updating from the central code repository.

The current version of Whitebox GAT runs on a Java virtual machine and relies heavily on its native raster data format. The latter makes it difficult to export Whitebox rasters to other users, software, and formats. Lindsay (2014) recently described the software design and philosophy and reported on the results of surveys of users and publications that have used the software. The survey of publications showed that "… the Whitebox GAT is frequently used in these studies to perform DEM-based analysis, which likely reflects the fact that this functionality is particularly well developed, owing to the project's origins in the Terrain Analysis System" (Lindsay, 2014, p. 6).

In addition, the survey numbers show that this new software has attracted a large and diverse number of users in just six years and recent crowd-sourced efforts to translate much of the text that appears on the user interface into 11 languages will likely lead to further adoptions in the future.

The potential of using this software for performing terrain analysis and modeling is exemplified by Lindsay and colleagues who have used the software tools to (i) measure the significance of a divide for local drainage patterns (Lindsay & Seibert, 2013); (ii) model surface drainage patterns in altered landscapes using LiDAR (Lindsay & Dhun, 2015); (iii) propose an integral image approach for performing multiscale topographic position analysis (Lindsay et al. 2015); (iv) propose efficient hybrid breaching-filling sink removal methods for flow-path enforcement in digital elevation models (Lindsay, 2016a); and (v) revisit the practice of DEM stream burning (Lindsay, 2016b).

## 6.5 Future Trends

The review of current terrain modeling software platforms provided in this chapter highlights three major trends. The first is that the advent and rising popularity of cloud-based computing platforms, coupled with cyberinfrastructure protocols and tools as well as the LiDAR, SAR, and other new digital data sources described in Chapter 2, means that most scientists and practitioners will conduct their terrain modeling work with a series of ready-to-use software services and online datasets in the years ahead. These developments will provide new opportunities to work at finer scales using geographic extents ranging from small watersheds to the Earth as a whole, and they will almost certainly mean that this work will draw tools and services from a variety of software providers and platforms in the coming years.

The second trend supports the first because the reviews of the 10 software platforms provided in this chapter shows how (i) many of the land surface parameters and landform objects described in Chapters 3 and 4 can be calculated with a variety of software products (i.e. ArcGIS, QGIS, SAGA, TauDEM, and Whitebox GAT); and (ii) some of the land surface parameters and landform classifications can be calculated with just one or possibly two of these software platforms. Some of these platforms offered unique solutions when they were first released that have since been incorporated in other platforms and some have focused on the calculation of novel and unique land surface parameters that have not found widespread use thus far. This means that the software platform chosen for a specific project should be chosen carefully with an eye to the kinds of capabilities provided by the software and the goals or needs of the project at hand.

The third and final trend is that there will likely be winners and losers among the 10 software platforms reviewed in this chapter in the coming years. ArcGIS and QGIS look like winners given their development trajectories and their track record of building bridges to other terrain modeling platforms (i.e. third-party extensions and TauDEM in the case of ArcGIS, and the GRASS, SAGA and TauDEM extensions in the case of QGIS) and related software. ILWIS, LandSerf, MicroDEM and perhaps RiverTools, on the other hand, are likely to wither given their relatively small user communities and relative lack of development in the case of the first three products and the proprietary nature of the RiverTools product and the lack of scientific testing and validation of the unique aspects of this software platform (i.e. the Mass Flux flow-direction algorithm that was described in Section 3.1.3.2). However, GRASS, SAGA, TauDEM, and Whitebox GAT will likely continue to find users for the following reasons: (i) GRASS and SAGA because they both provide some unique terrain modeling capabilities and they are accessible from QGIS; (ii) TauDEM because its development path is nicely aligned with the cloud-based computing platforms and the cyberinfrastructure protocols and tools that have emerged during the past decade and this product is accessible from both ArcGIS and QGIS; and (iii) the Whitebox GAT platform because of the special opportunities it provides for practitioners and students to learn more about the ways in which various land surface parameters and landform objects are calculated.

# 7 Conclusions

Digital terrain analysis and modeling have blossomed over the past 40–50 years spurred on by new sources of digital elevation data, the increasing adoption and use of theory to guide digital terrain analysis and modeling workflows, the specification and calculation of many new land surface parameters, the identification and extraction of landforms and other land surface objects, the improving characterization of error and uncertainty, and the development and distribution (i.e. sharing) of computer code to facilitate and support digital terrain analysis and modeling workflows. The first six chapters and references which follow this chapter show just how much progress has been made and how the rate of development has accelerated during the past 10–15 years, motivated by the large numbers of environmental problems and challenges that are expressed at the meso- and topo-scales (Figure 1.1) and the rapidly evolving computational resources that are now available to support digital terrain analysis and modeling applications. Table 7.1 lists 25 influential digital terrain analysis and modeling papers that helped to shape the preparation of this book and one cannot help but notice that 21 of the papers included in this list were published during the past 15 years. The only exceptions were the early papers by Evans (1972), Hutchinson (1989), I.D. Moore et al. (1991), and Tarboton (1997) that have helped to sow the seeds and shape the field of geomorphometry as we have come to know it today.

That said, the remainder of this chapter is divided into three parts. The first summarizes the major accomplishments of the past 10–15 years and the current state of the art (Section 7.1) and the second describes the kinds of innovations that are close at hand and will be required to push digital terrain analysis and modeling to even greater heights in the years ahead (Section 7.2). The third and final part (Section 7.3) serves as a call to action and briefly describes the kinds of collaborations that will be needed to make progress in the immediate future.

*Environmental Applications of Digital Terrain Modeling*, First Edition. John P. Wilson.

**Table 7.1** List of 25 influential digital terrain analysis and modeling papers.

Buchanan, B.P., Fleming, M., Schneider, R.L., Richards, B.K., Archibald, J., Qiu, Z. & Walter, M.T. (2014) Evaluating topographic wetness indices across central New York agricultural landscapes. *Hydrology and Earth Systems Science*, 18, 3279–3299

Clarke, K.C. & Romero, B.E. (2016) On the topology of topography: A review. *Cartography and Geographic Information Science*, 44, 271–282

Drăguţ, L. & Eisank, C. (2011) Object representations at multiple scales from digital elevation models. *Geomorphology*, 129, 183–189

Evans, I.S. (1972) General geomorphometry, derivatives of altitude, and descriptive statistics. In: R.J. Chorley (ed.) *Spatial Analysis in Geomorphology*, pp. 17–90. London, UK: Harper & Row

Fisher, P.F., Wood, J. & Cheng, T. (2004) Where is Helvellyn? Fuzziness of multi-scale landscape morphometry. *Transactions of the Institute of British Geographers NS*, 29, 106–128

Florinsky, I. (2017) Spheroidal equal angular DEMs: The specificity of morphometric treatment. *Transactions in GIS*, 21 (in press).

Gallant, J.C. & Hutchinson, M.F. (2011) A differential equation for specific catchment area. *Water Resources Research*, 47, W05535.

Hebeler, F. & Purves, R.S. (2009) The influence of elevation uncertainty on derivation of topographic indices. *Geomorphology*, 111, 4–16

Hutchinson, M.F. (1989) A new procedure for gridding elevation and stream line data with automatic removal of spurious pits. *Journal of Hydrology*, 106, 211–232

Lane, S.N., Brookes, C.J., Kirkby, M.J. & Holden, J. (2004) A network-index-based version of TOPMODEL for use with high-resolution digital topographic data. *Hydrological Processes*, 18, 191–201

Leempoel, K., Parisod, C., Geiser, C., Daprà, L., Vittoz, P. & Joost, S. (2015) Very high-resolution digital elevation models: Are multi-scale derived variables ecologically relevant? *Methods in Ecology and Evolution*, 6, 1373–1383

Lindsay, J.B. (2016) Whitebox GAT: A case study in geomorphometric analysis. *Computers and Geosciences*, 95, 75–84

Marthews, T.R., Dadson, S.J., Lehner, B., Abele, S. & Gedney, N. (2015) High-resolution global topographic index values for use in large-scale hydrological modeling. *Hydrology and Earth System Sciences*, 19, 91–104

Minár, J. & Evans, I.S. (2008) Elementary forms for land surface segmentation: The theoretical basis of terrain analysis and geomorphological mapping. *Geomorphology*, 95, 236–259

Moore, I.D., Grayson, R.B. & Ladson, A.R. (1991) Digital terrain modeling: A review of hydrological, geomorphological, and biological applications. *Hydrological Processes*, 5, 3–30

Qin, C.-Z., Ai, B.-B., Zhu, A.-Z. & Liu, J.-Z. (2017) An efficient method for applying a differential equation to deriving the spatial distribution of specific catchment area from gridded digital elevation models. *Computers and Geosciences*, 100, 94–102

Reuter, H.I., Nelson, A. & Jarvis, A. (2007) An evaluation of void-filling interpolation methods for SRTM data. *International Journal of Geographical Information Science*, 21, 983–1008

Rexer, M. & Hirt, C. (2014) Comparison of free high resolution digital elevation data sets (ASTER GDEM2, SRTM v2.1/v4.1) and the validation against accurate heights from the Australian National Gravity Database. *Australian Journal of Earth Sciences*, 61, 213–222.

Shelef, E. & Hilley, G.E. (2013) Impact of flow routing on catchment area calculations, slope estimates, and numerical simulations of landscape development. *Journal of Geophysical Research: Earth Science*, 118, 2105–2123

Tang, J., Miller, L.A., Crill, P.M., Olin, S. & Pilesjö, P. (2015) Investigating the influence of two different flow routing algorithms on soil–water–vegetation interactions using the dynamic ecosystem model LPJ-GUESS. *Ecohydrology*, 8, 570–583

Tarboton, D. (1997) A new method for the determination of flow directions and upslope areas in grid digital elevation models. *Water Resources Research*, 33, 309–319

Temme, A.J.A.M., Heuvelink, G.B.M., Schoorl, J.M. & Claessens, L. (2009) Geostatistical simulation and error propagation in geomorphometry. In: T. Hengl & H.I. Reuter (eds) *Geomorphometry: Concepts, Software, Applications*, pp. 121–140. Amsterdam, Netherlands: Elsevier.

Van Niel, K.P. & Austin, M.P. (2007) Predictive vegetation modeling for conservation: Impact of error propagation from digital elevation data. *Ecological Applications*, 17, 266–280

Winchell, M.F., Jackson, S.H., Wadley, A.M. & Srinivasan, R. (2008) Extension and validation of a geographic information system-based method for calculating the Revised Universal Soil Loss Equation length–slope factor for erosion risk assessments in large watersheds. *Journal of Soil and Water Conservation*, 63, 105–111

Zhang, Z., Zimmermann, N.E., Kaplan, J.O. & Poulter, B. (2016) Modeling spatiotemporal dynamics of global wetlands: Comprehensive evaluation of a new sub-grid TOPMODEL parameterization and uncertainties. *Biogeosciences*, 13, 1387–1408

## 7.1 Current State of the Art

A comparison of the contents of this and earlier books by Wilson and Gallant (2000a), Li et al. (2005), Zhou et al. (2008), and Hengl and Reuter (2009) shows just how much the field of geomorphometry has changed during the past decade. This first six chapters of this book document the major innovations and accomplishments that have helped to spur the growth and evolution of geomorphometry, and this work is used here to describe the current state of the art. This story is laid out in a linear flow, starting with Chapter 1 and working forward in an iterative fashion from one chapter to the next.

The first chapter set the stage for the reminder of the book by surveying the ways in which terrain analysis and modeling products have been used thus far and describing the role of DEMs and scale in geomorphometry. The survey of applications served two purposes. The first was to document how terrain analysis and modeling have been adopted and used in several environmental domains, including biogeography, climatology, ecology, geology, geomorphology, hydrology, pedology, and natural hazards, during the past 70 years. The second was to remind the reader to think about how much the data sources and computational methods have changed over this same period and why one might want to proceed carefully when using the results from some of the early terrain analysis and modeling publications to guide and/or inform their current or future work.

DEMs were introduced in Chapter 1 because nearly every project incorporating digital terrain analysis and modeling will start with the acquisition and/or construction of DEMs and the challenge is to be able to construct DEMs that account for the known and/or inferred properties of the terrain surface. Most believe that the surface shape and drainage structure constitute the most important properties and that these properties will have important implications for how one constructs, evaluates, and uses various elevation data sources (e.g. Hutchinson, 1989; Hutchinson et al., 2013; Liu et al., 2015). The need for more sophisticated assessments of the strengths and weaknesses of the new airborne and space-borne remote sensing elevation data sources than was provided for the early cartographic datasets, given the focus on horizontal resolution and vertical accuracy in those early assessments, was also articulated here.

The importance of the scale, or degree of detail, was also introduced in Chapter 1 since nearly all work involving digital terrain analysis and modeling is scale-specific. Goodchild (2011), for example, has noted how larger cells usually yield smaller estimates of slope and Fisher et al. (2004) have shown how the morphometric class at a point will vary with the scale over which it is measured (see Figure 4.7 for additional details). These problems can be attributed, at least in part, to the lack of theory about scale and therefore the difficulty of formalizing it. This state of affairs ensures that the scientist will be left with the dilemma that they cannot determine whether the failure of a model to fit perfectly is due to the model itself, or to the relatively "coarse" resolution of the elevation data, or both. The early papers sought to address this problem by documenting the sensitivity of selected land surface parameters to the choice of data source, data structure, and/or grid resolution, but a smaller group of recent papers has taken a better approach and tried to identify the level of detail needed to use land surface parameters to support the characterization of the key processes operating in specific landscapes. In one of these papers, Pain (2005) showed how a grid resolution of 5 m was required to capture the scale of the fluvial processes, and therefore the likely variation of related phenomena such as soil attributes, operating in a small catchment near Picton in New South Wales, Australia.

The role of scale and the importance of the known and/or inferred properties of the terrain surface were also considered by Clarke and Romero (2016) in their recent review on the topology of topography. These authors provided a chronological review of the development of surface network and critical point theory, the study of topography, and the progression of terrain analysis with particular consideration given to the application of surface network theory to represent the topology of topography. They also noted the importance of surface shape and drainage structure, and the difficulty of jumping from such logical constructs and the supporting theories to the development of tools and mapped products representing the measured topology of the topography. Finally, they observed how relatively little attention has been given to the topologic relationships among topographic features and that this state of affairs helps to explain why all computed surface networks are scale-dependent, fuzzy, and vague and why their undisputed calculation remains elusive.

The second chapter on the construction of DEMs reviewed the growing numbers of elevation data sources and the kinds of preprocessing tasks that will need to be performed with these sources to produce DEMs that account for the known and/or inferred properties of the terrain surface. Much was also made of the transition from cartographic to remote sensing data sources and how the accompanying consequences – the large coverage and fine resolutions afforded by many of these new DEMs – has brought numerous changes to the typical digital terrain modeling workflow (see Figure 2.1). The potential benefits are enormous: the SRTM and ASTER DEMs, for example, provide coverage for much of the globe at the 30–100 m resolutions that could only be used for small and moderate-sized catchments until relatively recently and LiDAR provides large numbers of high-density mass points which, if processed appropriately, may provide 1–3 m bare-earth DEMs with high vertical accuracy and the preservation of the terrain structure (i.e. shape). The observation was also offered that these new sources will frequently require more work, since the preprocessing tasks will typically be larger in number and more onerous because one will need to filter the noise from these remotely sensed sources and the early challenges of identifying and removing unwanted depressions, handling unresolved flow directions on flat terrain, and reconciliation of the drainage lines acquired from some other source(s) with those calculated with the DEMs will still remain.

The USGS NED was highlighted toward the end of Chapter 2 to illustrate both the role of the web for the storage and distribution of these DEMs and the magnitude and sophistication of the missions needed to gather these remotely sensed data in the first place. This last observation means that most terrain scientists and practitioners will not collect their own source data and prepare their own DEMs going forward. This will place an increased emphasis on the provenance of both the original data and the methods applied to them and the expertise of the people contributing the DEMs. This focus on provenance can not only improve the accountability (reproducibility in this instance) but also allow scientists and practitioners to consider questions surrounding the credibility (i.e. fitness-for-use) of the content for the problem or opportunity at hand (McKenzie et al., 2016). These kinds of considerations with likely elevate the importance of certain data sources like the NED in future years.

The third and largest of the chapter described more than 100 of the most popular primary and secondary land surface parameters. The latter distinction is important because the primary land surface parameters are derived directly from DEMs without additional inputs, whereas the secondary land surface parameters are derived from DEMs using two or more of the primary land surface parameters and additional inputs in some instances. However, both groups of parameters are typically calculated from square-grid DEMs and record site-specific, local and regional (i.e. at-a-distance) interactions on or near the Earth's surface: Table 3.1, for example, lists two site-specific, 46 local and 20 regional primary land surface parameters; Table 3.2 lists nine single-flow direction and 15 multiple-flow direction algorithms; and Table 3.4 lists 15 regional secondary land surface parameters focused on water flow and soil redistribution (11) or energy and heat regimes (4) at or near the Earth's surface.

The fact that 24 flow-direction algorithms have been proposed during the past 35 years points to one area that has attracted considerable attention. Several of the newer flow-direction algorithms combine square grids and TINs to avoid some of the shortcomings associated with square-grid DEMs and to take advantage of the additional discretization provided by TINs. However, evaluation of the efficacy of one or more of these flow-direction algorithms, and the impact on the flow accumulation and secondary land surface parameters which incorporate these flow directions, is difficult to execute well because (i) most real-world terrain surfaces contain a complicated mix of terrain forms that may span several scales; (ii) the multiple-flow direction algorithms tend to work better in upland areas near peaks and saddles where flow dispersion occurs and the single-flow direction algorithms tend to work better in lowland valleys and depressions where flow convergence occurs; and (iii) the errors that result from the choice of flow-direction algorithm may be difficult to separate from and/or sometimes completely obscured by the errors in the DEMs themselves. These observations suggest the final choice of flow-direction algorithm will often be a compromise, notwithstanding the progress that has been made in developing and testing new flow-direction algorithms.

The secondary land surface parameters are more complicated than the primary land surface parameters and were described in considerable detail toward the end of Chapter 3. The TWI featured prominently in this discussion given its popularity and the realization that six of the 15 secondary land surface parameters listed in Table 3.4 are connected to this secondary land surface parameter in one way or another. There are, in fact, many more options since the calculation of the TWI requires slope and upslope contributing area estimates and both of these parameters can be calculated with one or more of several different algorithms. The various options bring the need to understand what these choices mean in terms of the fitness-for-use of the TWI values. This will often be an onerous task. Buchanan et al. (2014), for example, calculated TWIs using more than 400 unique approaches that considered different DEM resolutions, the vertical precision of the DEM, flow direction and slope algorithms, smoothing versus low-pass filtering and the inclusion of relevant soil properties, and compared the resulting topographic wetness maps with observed soil moisture across numerous agricultural fields in New York State. Some commentators might argue that a similar approach should be used to demonstrate the efficacy of the preferred method for calculating each of the primary and secondary land surface parameters described in this chapter before using it.

The importance of Chapter 3 cannot be underestimated, notwithstanding this last challenge, given that one or more of the land surface parameters described in this chapter were adopted and used in the large numbers of domain-specific applications that were introduced in Chapter 1 and there can be no doubt that these land surface parameters constitute the heart of geomorphometry as we know it today.

The fourth chapter focused on the extraction and classification of landforms and land surface objects. Some of the applications can be used to delineate specific landform elements and others rely on flow variables to delineate drainage networks and basin boundaries. Fuzzy classification methods were given special prominence because many of the land surface objects and landforms that are of interest have fuzzy boundaries. Fisher et al. (2004) and Deng and Wilson (2008), for example, used these fuzzy classification methods to delineate mountains as an example of specific (fuzzy) landforms. And last, but not least, several applications have attempted to automate and extend Hammond's (1964) map of repeating landform patterns for the conterminous USA to the globe (e.g. Sayre et al., 2014; Karagulle et al., 2017).

Three aspects of the applications reviewed in this chapter were particularly noteworthy. The first was their use of varying neighborhood analysis windows that were matched to either the land surface parameter and/or the surface texture of the landscape. The second followed from the realization that the new elevation data sources provide 1–3 m DEMs and these afford the opportunity to start over in some instances. The delineation of flow networks, for example, requires the locations of channel heads and several recent studies have proposed new data-driven methods based on theory to mark the locations where the landscape changes from hillslope to channel topography as one moves down a profile (e.g. Clubb et al., 2014). The third and perhaps most substantial case concerns the recent applications that borrow and use concepts from remote sensing and data science to first segment the DEM and then classify the objects (e.g. Drăguţ & Eisank, 2011), and thereby avoid the problems of working directly with the DEM grid cells (pixels) when extracting and classifying landforms and other land surface objects.

The fifth chapter tackled sources of error, the various ways uncertainty can be estimated and handled in terrain analysis and modeling workflows, how this knowledge can be used to help assess fitness-for-use in specific applications, and the new opportunities for multiscale analysis and cross-scale inference afforded by the increasing availability of DEMs across a broad range of scales. The US NWM was also introduced in this chapter to illustrate how the land surface parameters and objects that were described in Chapters 3 and 4 can be incorporated and used in large science-based projects that have the potential to deliver substantial societal benefit over long timescales.

The sources of error were divided into five classes: (i) those related to the choice of data model; (ii) the discretization of space, choice of elevation data source, and the interpolation and/or filtering approach that was used; (iii) the preprocessing of the source data to remove unwanted depressions, resolve flow directions in flat terrain, and merge the DEM with streamlines acquired from some vector hydrography source; and (iv) the methods used to calculate primary and secondary land surface parameters, including (v) the flow-based parameters which must be processed in the correct order. These provide a roadmap of sorts for those interested in best practices and opportunities to avoid or minimize specific errors.

The attention then shifted to the ways in which the uncertainty embedded in terrain analysis and modeling outcomes can be described. A series of early studies was used to illustrate how the level of knowledge of the spatial structure of the error helps to inform the choice of methods and a case study by Hebeler and Purves (2009) was reviewed in some detail to illustrate how statistical modeling can be used to propagate error with MCS. However, this particular study stopped short of clarifying how the MCS results could be used with more complex process models to guide knowledge discovery in other domains, such as ecology, geomorphology or hydrology.

The concept of fitness-for-use was revisited in Chapter 5 and several stellar case studies were used to show how the measurement of error and uncertainty accompanying terrain analysis and modeling workflows might be used to improve our understanding of predictive vegetation modeling (Van Niel & Austin, 2007) and soil redistribution resulting from water erosion (Temme et al., 2009); how catchment area calculations, slope estimates and numerical simulations of landscape development (Shelef & Hilley, 2013) and the soil–water–vegetation interactions in the LPJ-GUESS dynamic ecosystem model (Tang et al., 2014) are influenced by the choice of flow-direction algorithm; and how a new sub-grid TOPMODEL parameterization and the associated uncertainties influence the modeling of the spatiotemporal dynamics of global wetlands (Zhang et al., 2016). Each of these examples, in their own way, represents the cutting edge of digital terrain analysis and modeling work today.

The next part of Chapter 5 used a series of case studies to demonstrate why we should embrace multiscale analysis and focus on the development and application of methods for making cross-scale inferences when source data are gathered at different scales and/or one wants to change scales when analyzing specific characteristics or outcomes across multiple geographic extents. The case studies show some of the ways researchers have looked for an appropriate scale or scales at which to conduct their work, checked whether or not the stationarity assumption (which, among other things, would mean that the relationships found in one study would hold true everywhere) is valid or not, worked with various kinds of source data that were gathered at different scales, and explored new ways to conduct multiscale analysis, particularly when transitioning from the land surface parameters that were discussed in Chapter 3 to the landforms and other land surface objects that were the focus of Chapter 4.

A strong case can be made that the reader interested in the topics covered in Chapters 3 and/or 4 should read Chapters 3, 4 and 5 as a bundle before moving forward with their own digital terrain modeling projects.

The sixth chapter describes 10 software platforms that have been frequently used to calculate the land surface parameters and/or extract and classify the landforms and other land surface objects that were featured in Chapters 3, 4 and 5. The functionality varies from one package to the next and those interested in using one or more of these software platforms therefore incurs the burden of choosing the best platform for the work at hand. Although all 10 of these platforms have attracted relatively large numbers of users, six – ArcGIS (especially when coupled with one or more of three third-party extensions), GRASS, QGIS, SAGA, TauDEM, and the Whitebox GAT – stand out because of the large numbers of terrain tools included and/or other noteworthy features (i.e. availability of GIS functions, numbers of data formats supported, and the level of interoperability). The ArcGIS platform, for example, brings numerous advantages including three third-party geomorphometry extensions, a burgeoning online archive of ready-to-use geospatial datasets, and a TauDEM plug-in. QGIS offers a similar user experience to ArcGIS and provides access to the terrain analysis tools in GRASS, SAGA, and TauDEM via plug-ins as well. SAGA is a full service GIS and the terrain analysis tools includes several that were first shared using the TAPES-G and SRAD software platforms. TauDEM is noteworthy because many of the tools have been rewritten to capitalize on the computational resources offered by cloud computing, cyberinfrastructure, and the software-as-a-service model and the Whitebox GAT offers its users the opportunity to write new and/or modify the existing code to try out new terrain analysis and modeling tools. These platforms generally work with small to moderately sized catchments and planar coordinates, but some of the individual tools have been rewritten to work with geographic coordinates over much larger geographic extents during the past decade and as Florinsky (2017) has noted, those working in these spaces should find and use those tools.

This review shows that much has been accomplished during the past few decades and we turn our attention next to future needs and opportunities in Section 7.2.

## 7.2 Future Needs and Opportunities

The future digital terrain modeling needs and opportunities are intrinsically linked to one another and, as a result, they can be combined and briefly described under the eight headings that follow.

### *7.2.1 Finding Ways to Use Provenance, Credibility, and Digital Terrain Modeling Application-context Knowledge*

The rapid emergence of the web and all this entails (e.g. web portals for sharing geospatial datasets, the provision of software as a service) coupled with advances in our knowledge and understanding of error and uncertainty and how these concepts can be used to clarify the fitness-for-use of digital terrain modeling tools and data for specific applications speaks to the importance of both provenance and credibility. These elements have traditionally been handled by metadata in the spatial science domain (e.g. FGDC, 2017) and it will be important for the geomorphometry community to adopt and use metadata to describe terrain software tools and digital elevation datasets going forward. However, there are other ways to gather this information as well, what Qin et al. (2016) called "application-context" knowledge, using case-based formalization and reasoning methods.

The work of Qin et al. (2016) can be used to illustrate these methods. These authors started with the observation that the typical digital terrain modeling workflow (see Figure 2.1) relies heavily on knowledge of the match between the algorithm (and its particular settings) and the application context, which includes the DEM resolution, target task, and the terrain in the study area. However, most of the digital terrain modeling tools that were introduced in earlier chapters cannot use application-context knowledge because this type of knowledge has not been formalized and made available for inference when using the tools. Qin et al. (2016) then noted that this situation makes the digital terrain analysis workflow building process difficult for users and especially new or non-expert users. Of course, the inclusion of routines for managing error and uncertainty, as advocated in Chapter 5, will add to the magnitude of this challenge for nearly all involved in most instances.

To respond to these challenges, Qin et al. (2016) proposed a case-based formalization for digital terrain modeling application-context knowledge and a corresponding case-based reasoning method. A case in this method consists of a series of indices that formalize the digital terrain modeling application-context knowledge and a suite of similarity calculation methods to support the case-based reasoning. The method was illustrated by selecting 124 cases of drainage network extraction (50 for evaluation and 74 for reasoning) from peer-reviewed journal articles and using these cases to determine the catchment area threshold for extracting drainage networks. The results of this experiment showed that the proposed case-based method provided an effective way to capture and use the relevant digital terrain modeling application-context knowledge while simultaneously reducing the modeling burden for users.

### *7.2.2 Rediscovering and Using What We Already Know!*

The development of the SRTM and ASTER elevation datasets represents an important milestone because these two projects have provided 30-m DEMs with near-global coverage which, in turn, has spawned new interest in calculating land surface parameters and landform classifications over continental and global extents (e.g. Sayre et al., 2008, 2009, 2013, 2014; Karagulle et al., 2017). These developments have brought into focus what geographers have long known, namely the need to choose your map projections and coordinate systems carefully to suit the geographic extent of the study area of interest.

The general strategy should be one in which spherical equal angular DEMs are chosen for large study areas (i.e. the globe, continents, and catchments for many of the globe's large and iconic rivers) and planar square-grid DEMs are chosen for small and moderate-sized catchments. Most of the methods presented in this book were developed for planar square-grid DEMs, notwithstanding the fact that Guth (2009, 2010) and others have proposed algorithms for calculating some land surface parameters for spherical equal angular DEMs. This distinction is important for a variety of reasons and those working at continental and global extents should search for and implement the appropriate tools. Two recent papers are discussed to illustrate why.

The first paper by Florinsky (2017) demonstrates why the methods and algorithms intended for planar square-gridd DEMs should not be directly applied to spheroidal equal angular DEMs. This study used two DEMs – the first a medium-resolution DEM of a relatively small, high-mountainous area (Mount Elbrus) that was extracted from the SRTM1 DEM and the second a low-resolution DEM of a vast region with diverse topography (the central and western regions of Kenya) that was extracted from the SRTM30_PLUS DEM – to demonstrate the possibilities of direct calculation of local (slope gradient), non-local (catchment area), and combined land surface parameters (topographic wetness) from spheroidal equal angular DEMs, and to show the computational errors that result when algorithms for planar square-grid DEMs are unreasonably applied to spheroidal equal angular DEMs. There is no need to apply square-grid algorithms to spheroidal equal angular DEMs and only unnecessary errors can follow from continuing this practice.

The second paper by Netzel, Jasiewicz and Stepinski (2016) shows that the choice of the appropriate map projection and coordinate system may occur in other terrain-related settings as well. These authors recently proposed TerraEx, a geoweb application to support the worldwide content-based search and distribution of elevation and landform data. This system uses a global 3-arcsecond DEM for elevation data (a compilation of the SRTM DEM with DEMs from other sources for areas not covered by the SRTM; de Ferranti, 2014) and the search is performed on the DEM classified into 10 landform elements (flat, footslope, hollow, peak, pit, ridge, shoulder, slope, spur, and valley) using the geomorphon method (Jasiewicz & Stepinski, 2013). The system identifies places in the world (local blocks, referred to by the authors as motifels) with landscapes similar to a user-defined query based on the Ruzicka similarity measure (Deza & Deza, 2014) for two histograms which summarize the local pattern of the 10 landform elements. Each query returns a similarity map in about 10 seconds, and the queries themselves are executed using the GeoPAT 2.0 computational engine for pattern-based geoprocessing (Jasiewicz, Netzel & Stepinski, 2015), which is written in C and uses parallel processing based on the Open-Multi-Processing (OpenMP) library.

The TerraEx system works as follows. The searches start with the 3-arcsecond global grid of elevation but the topography is severely distorted at high latitudes and the search requires a comparison of undistorted landscapes. The TerraEx system therefore combines grids with geographic and planar coordinates using five steps to address this problem as follows.

1. The 3-arcsecond geomorphons dataset is divided into motifels, which contain 160 cells (~14 km), and are organized into a global grid with dimensions of 10,800 × 5400 and significant overlap.
2. An appropriate UTM coordinate system zone is determined for each motifel and a region of the geomorphon map large enough to cover a motifel after transformation to this UTM zone is selected.
3. The region is next transformed to the local UTM.
4. The co-occurrence histogram is calculated from the transformed map.
5. The histogram is saved in the appropriate position on the 3-arcsecond grid.

These tasks were performed off-line and the resultant grid of histograms has been saved in server memory to support the relatively fast and efficient query noted above. This is a clever approach that could easily be replicated in other digital terrain modeling applications that incorporate both local (small area) and global (large area) geographic extents.

And for those who may be interested, the TerraEx system also doubles as a convenient source of global, geographically referenced elevation data since the user can navigate to a location of interest (with or without a prior exploration using the search) and download a 3-arcsecond DEM, a map of geomorphons, and a map of terrain relief in GeoTIFF format.

### 7.2.3 Developing New Digital Terrain Methods

The more or less continuous development of new digital terrain methods constitutes one of the overarching themes of this book and there is every reason to expect that this trend will continue to yield important benefits. Three recent examples are highlighted here to show what is possible and why this kind of innovation should be encouraged.

The first study by Krebs et al. (2015) presented a new method to assess the vertical transverse and profile curvature that combines real-scale visualization and the possibility to measure these two land surface parameters over a large range of scales. This new method can be contrasted with the traditional methods that were introduced in Chapter 3 because the traditional methods calculate plan and profile curvature for every single cell of a DEM and thereby incorporate a high level of simplification in that they more than likely ignore the complex and multiscalar character of the surface curvature itself. This new method provides the opportunity to: (i) measure and visualize the shape and size of the curvature; (ii) obtain a realistic representation of the average curvature for different subsets of spatial points; and (iii) control the conditions in which the curvature is calculated.

The second study by Byun and Seong (2015) proposed a new algorithm, which they called the maximum depth tracing algorithm (MDTA), to extract more accurate stream longitudinal profiles (SLPs) using depression-unfilled DEMs. The traditional DEM preprocessing approach described in Chapter 2 fills the depressions to ensure the continuity of flow across the DEM before calculating flow-related land surface parameters. Byun and Seong (2015) noted that this inevitably introduces distortions to the SLP, such as stair steps, bias of elevation values, and inaccurate stream paths. Their new MDTA method therefore starts with the assumption that depressions in DEMs are not necessarily artifacts to be removed and that elevation values within them may more accurately represent the real-world landscape, and proposes a new two-step procedure to calculate SLPs and to ensure the continuity of flow across unfilled DEMs. The performance of this new MDTA method was compared with that of the hydrologic functions implemented in ArcGIS and the results showed that the MDTA method provided more accurate stream paths on depression areas and thereby reduced the distortions of the SLPs derived from these stream paths.

The third and final study by Buttenfield et al. (2016) compared planar distance with eight measures of surface-adjusted distance, moving from distance "as the crow flies" to distance "as the horse runs" using five transects ranging in length from 40 to 108 km in length and six square-grid DEMs with resolutions of 3, 10, 30, 100, 1000, and 5000 m in part of Pisgah National Forest, North Carolina. All eight of the methods tested incorporated elevation but differed in their inclusion of contextual information about the surrounding pixels to gain a progression of surface adjustment as follows.

1. *Pixel-to-Pixel distance* traversed sampled points in sequence along each transect and a three-dimensional Euclidean calculation summed the distances between sampled points along each transect.
2. *Closest centroid distance* assigned the elevation at the pixel centroid to any point along the selected path that fell within the pixel and was used to measure lengths on the 3-m LiDAR benchmark to validate transect distances at coarser resolutions.
3. *TIN distance* partitioned the DEM and interpolated elevations for points within each triangular facet from the three local vertices.
4. *Natural neighbor* partitioned Thiessen neighbors. The sampled transect points were next used to seed a second layer of Thiessen polygons and the proportion of overlap between the two layers was used to weight the interpolation of the elevations of all the Thiessen polygons.
5. *Weighted average distance* computed the average elevation of eight surrounding pixels weighted by the length between corresponding pixel centroids.
6. *Bilinear interpolation distance* fitted a first-order polynomial to the pixel containing the sampled point and the four neighboring pixels.
7. *Biquadratic interpolation distance* fitted a second-order polynomial to nine pixels in a $3 \times 3$ moving window.
8. *Bicubic interpolation distance* fitted a third-order polynomial to 17 surrounding cells.

The results showed how all eight methods deviated from the benchmark, with the weighted average and bicubic interpolation methods performing best (lowest RMSEs) and the pixel-to-pixel and closest centroid methods producing the highest magnitude errors and consistent overestimation for all transects. Distance metrics are included in the calculation of several of the land surface parameters described in Chapter 3 and the results from this exploratory study, taken as a whole, showed how the error magnitudes varied with pixel size and the method of surface adjustment. This suggests that some additional work is required to check whether or not the common assumption – that the improvements in distance estimation are so small that surface adjustment is not warranted – is true (or not) for specific terrain analysis and modeling applications.

### 7.2.4 Clarifying and Strengthening the Role of Theory

The most exciting result of the recent introduction of fine-resolution elevation data, such as LiDAR, is the opportunity it affords to reassess fundamental questions about landscape form and evolution in geomorphology (Clubb et al., 2014) as well as other domains. There are at least two ways to proceed here. The first uses theory to guide the workflow that was chosen, whereas the second conducts experiments to test one or more of our existing theories.

The first approach seeks to take advantage of our existing knowledge of how processes work. Austin et al. (2013), for example, modeled the mean monthly radiation surfaces for Australia at a 1-arcsecond resolution and made small modifications to the SRAD model that was used to accept fractional vegetation cover, cloud fraction and albedo as surfaces each month rather than single-value parameters, and morning and afternoon sunshine fractions in place of a single value for the whole day. Clubb et al. (2014), on the other hand, proposed a new algorithm for predicting channel head locations that was partly informed by the stream power equation (i.e. theory), which is a detachment-limited model that proposes that the fluvial incision rate is proportional to stream power, which represents the energy expenditure of the flow (Howard, 1994; Sklar & Dietrich, 1998).

The second approach seeks to test existing theory. This approach can be illustrated using the work of Jencso et al. (2009) which tested the relationship between upslope contributing area and the existence and longevity of the HRS shallow groundwater connectivity for a series of transects and the stream network in the Tenderfoot National Forest, Montana. The results showed how the internal catchment landscape structure does act as a first-order control on runoff source area and catchment response in these types of landscapes (see Section 5.3 for additional details).

The work needed to clarify and strengthen the role of theory in geomorphometry will often be difficult, notwithstanding the increased computing power at our disposal and the increasing availability of fine-resolution elevation data. Gallant, Hutchinson and Wilson (2000) also noted how we need to develop a better understanding of the links between terrain and surface processes that vary in space and time. They noted how many of the relationships that are derived statistically may appear useful but are difficult to interpret in terms of processes because many of the environmental parameters have co-evolved or, at least, vary together. For example, the vegetation, soil properties and water availability typically vary down a hillslope in a systematic way that may be well described by land surface parameters such as slope and contributing area but the existence of such a relationship may not help to clarify the causal links between the different components of this landscape system.

The development and use of land surface parameters and landform classifications that are based on theory offer the best chance of overcoming these kinds of obstacles.

### *7.2.5 Developing High-fidelity, Multi-resolution Digital Elevation Models*

The importance of multiscale analysis and cross-scale inference (described in Section 5.3) will continue to grow over time. These kinds of analyses increasingly rely on the availability of high-fidelity, multi-resolution DEMs and methods to build such DEMs. Some progress has been made but most of this work is motivated by the need or desire for accurate topographic representation across a relatively narrow range of geographic scales (as exemplified by the hydrographic generalization work conducted by Stanislawski, 2009; Buttenfield et al., 2011; Stanislawski & Buttenfield, 2011; Stanislawski, Buttenfield & Doumbouya, 2015a; Stanislawski, Falgout & Buttenfield, 2015b).

There is also a need for high-fidelity, multi-resolution DEMs that work with global environmental simulations which typically adopt sub-grid schemes to express topographic heterogeneity (Wilby & Wigley, 1997; Kumar et al., 2012; Fiddes & Gruber, 2014). These sub-grid schemes are typically designed for empirical parameterization rather than accurate topographic representation and the more or less singular focus on the latter outcome may lead to greater uncertainties and bias. Duan et al. (2017) recently proposed a high-fidelity, multi-resolution DEM with guaranteed grid quality for Earth systems. This new method starts with a TIN which is optimized using an energy-minimized centroidal Voronoi tessellation (CVT) and refined further using the mean curvature as a density function (cCVT) due to the curvature's flexibility in capturing shape characteristics and in supporting shape evolution (Kennelly, 2008). Duan et al. (2017) compared the performance of the cCVT model with the classical heuristic model and showed how the new model outperforms the classical heuristic DEM generalization in terms of both surface approximation precision and surface morphology retention. More work is needed because the current model uses a planar coordinate base (and, of course, global applications will require a geographic coordinate base) and there is a need to test the efficacy of this approach over a wider range of domains and scales.

### *7.2.6 Developing and Embracing New Visualization Opportunities*

The tools and methods to visualize the digital terrain modeling results have not kept pace with the rapidly evolving computational resources and the availability and use of fine-resolution DEMs which cover large areas. The map generalization projects funded as part of the NED research program in the USA (see Stanislawski, 2009; Buttenfield et al., 2011; Stanislawski & Buttenfield, 2011; Stanislawski et al., 2015a,b for additional details) and the new tangible geospatial modeling system proposed by Petrasova et al. (2015)

may help to fill this gap. This new tangible geospatial modeling system is powered by the GRASS GIS and produces physical as well as virtual models of landscapes which can be used for landscape management, trail planning, wildfire management, and other applications.

The key challenge is to be able to share and communicate the digital terrain analysis and modeling results with diverse audiences to motivate and shape economic development and environmental sustainability.

### 7.2.7 Adopting and Using New Information Technologies and Workflows

The rapid advances in computational power and the changing models of computing in all their various guises (i.e. cloud computing, cyberinfrastracture, interoperability, software as a service) offer tremendous opportunities to not only develop new analytical tools but also to expand the geographic extent and significance of digital terrain modeling projects. The opportunities are illustrated using three recent projects – those by Netzel et al. (2016), Survila et al. (2016), and Qin et al. (2017) – among the many that have been published during the past decade. All three examples, in various ways, show how the adoption and use of these computing platforms brings the need to modify the digital terrain modeling workflows that many of us have used with stand-alone personal computers during the past few decades.

The first example is the TerraEx application developed by Netzel et al. (2016) that was described in Section 7.2.2. This freely available, full service geoweb app provides the first global system to locate landscapes that are similar to a user-selected query and doubles as a conveneient portal to support the distribution of 3-arcsecond DEMs and global maps of geomorphons and terrain relief.

The second example is the scalable high-performance topographic flow-direction algorithm developed by Survila et al. (2016) that was described in Section 5.4. Flow direction is one of the most computationally intensive functions in the current implementation of TauDEM (Tarboton, 2016) because this essentially local operation is transformed into a global operation to route flow across flat regions. The new algorithm proposed by Survila et al. (2016) eliminated this bottleneck by first identifying the flat areas and then using this information to reduce the number of sequential and parallel iterations needed to calculate flow direction. This new algorithm outperformed the existing parallel D8 (O'Callaghan & Mark, 1984) single-flow direction algorithm in TauDEM by two orders of magnitude.

The third and final example by Qin et al. (2017) provides an efficient solution to calculate the differential equation for calculating SCA proposed by Gallant and Hutchinson (2011) from gridded DEMs. Gallant and Hutchinson (2011) advocated using their new equation to measure the performance of one or more of the 24 approximate methods for estimating SCA (assuming that each of the single- or multiple-flow direction algorithms described in

Section 3.1.3 leads to different estimates of upslope contributing areas and/or flow widths and therefore SCA) because solving this new equation numerically was too computationally onerous to derive the overall spatial distribution of SCA for small and moderate-sized catchments. Qin et al. (2017) developed a parallel algorithm using the OpenMP programming model that is widely used in the symmetric multiprocessor parallel computing devices commonly found in standard personal computers and showed how their method:

1. achieved a sufficient speedup to use the numerical SCA solution provided by Gallant and Hutchinson (2011) to derive the spatial distribution of SCA from gridded DEMs for large areas; and
2. produced much lower errors than the MFD-md multiple-flow direction algorithm (Qin et al., 2007), which was treated as representative of conventional grid-based flow-direction algorithms, on two artificial surfaces with theoretically derived SCA values and a more complex real-world terrain surface in northeast China.

Qin et al. (2017) reviewed the numerical solution for SCA proposed by Gallant and Hutchinson (2011) and found that the method used to calculate the SCA for a cell may be independent of that of any other cell, thereby supporting high parallelizability. The parallel algorithm was implemented using the OpenMP version of the parallel raster-based geocomputation operators (PaRGO) (Qin et al., 2014b) and a series of experiments was conducted to document the effects of DEM grid size and integration step length on the numerical SCA solution. Six DEM resolutions (measuring 0.5, 1, 2.5, 5, 10, and 20 m on a side) and six step length ratios (0.1, 0.3, 0.5, 0.7, 0.9, and 1.0) were used with 1, 2, 4, 8, 12, 16, and 32 threads for these experiments. The maximum speedup ratio in these experiments reached 8.46 (with 32 threads) and the proposed algorithm performed best using a 5-m DEM and an integration step length ratio of 0.5 (i.e. half of the DEM's grid spacing). The experimental results based on the two artificial surfaces also showed that the proposed parallel algorithm generally yielded far fewer errors than a representative grid-based multiple-flow direction algorithm (MFD-md; Qin et al., 2007) in tests at different resolutions and that the proposed algorithm was not very sensitive to the choice of integration step.

### 7.2.8 Solving "Wicked" Problems of Varying Magnitudes

The final pitch is a simple one: to encourage all involved to adopt and apply the best digital terrain modeling practices to help solve one or more of the many "wicked" problems that threaten human well-being and environmental sustainability across the Earth. Many of these problems are multiscalar and expressed at local as well as regional and sometimes global scales. The solutions can be conceived in similar ways and resources gathered to solve them at a variety of scales. Two examples can be used to illustrate these possibiliites and motivate others to pursue digital terrain analysis and modeling projects that will produce actionable knowledge and outcomes.

In the first example, Woodrow, Lindsay and Berg (2016) examined the impacts of DEM grid resolution, elevation source data, and conditioning techniques on the spatial and statistical distribution of field-scale hydrologic attributes for a 12,000-ha watershed of an agricultural area in southwestern Ontario, Canada. The results showed how the decision to use one DEM conditioning technique over another, and the constraints of available DEM data resolution and source, can greatly impact the modeled surface drainage patterns at the scale of individual fields. These kinds of results are important because they can help to guide efforts to optimize best management practices for reducing soil erosion and runoff contamination within agricultural watersheds and thereby help to manage non-point source pollution at the source. The hope is that this or some other group will be able to replicate and use these results to drive reductions in non-point source pollution loads across these and similar landscapes.

The second and final example is the US NWM project. This large, interdisciplinary and collaborative project simulates and forecasts streamflow over the whole of the continental USA at intervals of 1 hour, 18 hours, 10 days and 30 days for 2.7 million stream reaches. The NWM uses the WFS-Hydro and Noah-MP land surface models to simulate meteorological conditions and terrestrial hydrology. Several of the terrain modeling functions in TauDEM (Tarboton, 2016) provide the essential "glue" and are used as part of this modeling framework to route water across the land surface to the nearest stream channel. The NWM outputs can be accessed via a series of interactive maps on the National Water Center website (http://water.noaa.gov/map) and once incorporated in the daily workflows of the relevant agencies (public safety, water resources, etc.) the NWM outputs will fundamentally change the ways in which the local, state, and federal governments prepare for and anticipate floods and related water challenges.

## 7.3 Call To Action

The progress achieved during the past decade resulted, in part, because of the increased collaboration among the geomorphometry and geographic information science communities. Their guiding principles and core concepts can be found scattered throughout this book and modern digital terrain analysis and modeling workflows are better for it. Future advances will require continued collaboration among these groups and computer scientists and various kinds of domain specialists in order to produce the fast and nimble solutions and actionable science outcomes that are desperately needed to solve the problems that threaten human well-being and environmental sustainability.

# References

Abd Aziz, S. (2008) *Development of digital elevation models (DEMs) for agricultural applications*. Unpublished PhD dissertation, Iowa State University, Iowa City, IA.

Abd Aziz, S., Steward, B.L., Tang, L. & Karkee, M. (2009) Utilizing repeated GPS surveys from field operation for development of agricultural field DEMs. *Transactions of the American Society of Agricultural and Biological Engineers*, 52, 1057–1067.

Abdulla, F.A. & Lettenmier, D.P. (1997) Development of regional parameter estimation equations for a macroscale hydrologic model. *Journal of Hydrology*, 197, 230–257.

Abrams, M., Bailey, B., Tsu, H. & Hato, M. (2010) The ASTER global DEM. *Photogrammetric Engineering and Remote Sensing*, 76, 344–348.

Agnew, L.J., Lyon, S., Gérard-Marchant, P., Collins, V.B., Lembo, A.J., Steenhuis, T.S. & Walter, M.T. (2006) Identifying hydrologically sensitive areas: Bridging the gaps between science and application. *Journal of Environmental Management*, 78, 63–76.

Aguilar, F.J. & Mills, J.P. (2008) Accuracy assessment of LiDAR-derived digital elevation models. *Photogrammetric Record*, 23, 148–169.

Aguilar, F.J., Agüera, F., Aguilar, M.A. & Carvajal, F. (2005) Effects of terrain morphology, sampling density, and interpolation methods on grid DEM accuracy. *Photogrammetric Engineering and Remote Sensing*, 71, 805–816.

Aguilar, F.J., Mills, J.P., Delgado, J., Aguilar, M.A., Negreiros, J.G. & Pérez, J.L. (2010) Modelling vertical error in LiDAR-derived digital elevation models. *ISPRS Journal of Photogrammetry and Remote Sensing*, 65, 103–110.

Ahn, C.W., Baumgartner, M.F. & Biehl, I.I. (1999) Delineation of soil variability using geostatistics and fuzzy clustering analysis of hyperspectral data. *Soil Science Society of America Journal*, 63, 142–150.

Ahokas, E., Kaartinen, H. & Hyyppä, J. (2003) A quality assessment of airborne laser scanner data. *International Archives of Photogrammetry, Remote Sensing, and Spatial Information Sciences*, 34(3/W/W13), 6.

Akpa, S.I.C., Ugbaje, S.U., Bishop, T.F.A. & Odeh, I.O.A. (2016) Enhancing pedotransfer functions with environmental data for estimating bulk density and effective cation exchange capacity in a data-sparse situation. *Soil Use and Management*, 32, 644–658.

Albani, M., Klinkenberg, B., Andison, D.W. & Kimmins, J.P. (2004) The choice of window size in approximating topographic surfaces from digital elevation models. *International Journal of Geographical Information Science*, 18, 557–593.

*Environmental Applications of Digital Terrain Modeling*, First Edition. John P. Wilson.

Allord, G.J., Walter, J.L., Fishburn, K.A. & Shea, G.A. (2014) *Specification for the U.S. Geological Survey Historical Topographic Map Collection*. Reston, VA: US Geological Survey.

Anders, N., Seijmonsbergen, A. & Bouten, W. (2009) Multi-scale and object-oriented image analysis of high-res LiDAR data for geomorphological mapping in Alpine mountains. In: R. Purves, S. Gruber, R. Straumann & T. Hengl (eds) *Proceedings of Geomorphometry 2009*, pp. 61–65. Zurich, Switzerland: University of Zurich.

Antonić, O., Pernar, N. & Jelaska, S.D. (2003) Spatial distribution of main forest soil groups in Croatia as a function of basic pedogenetic factors. *Ecological Modeling*, 170, 363–371.

Ardizzone, F., Cardinali, M., Galli, M., Guzzetti, F. & Reichenbach, P. (2007) Identification and mapping of recent rainfall-induced landslides using elevation data collected by airborne LiDAR. *Natural Hazards and Earth System Science*, 7, 637–650.

Ariza-Villaverde, A.B., Jiménez-Hornero, F.J. & Gutiérrez de Ravé, E. (2013) Influence of DEM resolution on drainage network extraction: A multifractal analysis. *Geomorphology*, 241, 243–254.

Ariza-Villaverde, A.B., Jiménez-Hornero, F.J. & Gutiérrez de Ravé, E. (2015) Multifractal analysis applied to the study of the accuracy of DEM-based stream derivation. *Geomorphology*, 243, 85–95.

Arnold, N. (2010) A new approach for dealing with depressions in digital elevation models when calculating flow accumulation values. *Progress in Physical Geography*, 34, 781–809.

Arrell, K.E., Fisher, P.F., Tate, N.J. & Bastin, L. (2007) A fuzzy c-means classification of elevation derivatives to extract the morphometric classification of landforms in Snowdonia, Wales. *Computers and Geosciences*, 33, 1366–1381.

Arun, P.V. (2013) A comparative analysis of different DEM interpolation methods. *Egyptian Journal of Remote Sensing and Space Science*, 16, 133–139.

Aryal, S.K. & Bates, B.C. (2008) Effects of catchment discretization on topographic index distributions. *Journal of Hydrology*, 359, 150–163.

Aspie, J. (1989) *Influence of groundwater on streambank soil moisture content, stream runoff production, and sediment transport in a semi-arid watershed*. Unpublished MS thesis, Montana State University, Bozeman, MT.

Austin, J.M., Gallant, J.C. & Van Niel, T. (2013) Mean monthly radiation surfaces for Australia at 1 arc-second resolution. In: *Proceedings of the 20th International Congress on Modeling and Simulation*, pp. 1589–1595. Adelaide, Australia: MSSANZ.

Austin, M.P. & Meyers, J.A. (1996) Current approaches to modelling the environmental niche of eucalypts: Implications for management of forest biodiversity. *Forest Ecology and Management*, 85, 95–106.

Austin, M.P. & Smith, T.M. (1989) A new model for the continuum concept. *Vegetation*, 83, 35–47.

Austin, M.P., Cunningham, R.B. & Fleming, P.M. (1984) New approaches to direct gradient analysis using environmental scalars and statistical curve-fitting procedures. *Vegetatio*, 55, 11–27.

Baatz, M. & Schäpe, A. (2000) Multiresolution segmentation: An optimization approach for high quality multi-scale image segmentation. In: J. Strobl, T. Blaschke & G. Griesebner (eds), *Angewandte Geographische Informationsverarbeitung*, pp. 12–23. Heidelberg, Germany: Wichmann-Verlag.

Bäckstrand, K., Crill, P.M., Mastepanov, M., Christensen, T.R. & Bastviken, D. (2008) Non-methane volatile organic compound flux from a subarctic mire in northern Sweden. *Biogeosciences*, 5, 111–121.

Bader, M.Y. & Ruijten, J.J.A. (2008) A topography-based model of forest cover at the alpine tree line in the tropical Andes. *Journal of Biogeography*, 35, 711–723.

Bailly, J.S., Lagacherie, P., Millier, C., Puech, C. & Kosuth, P. (2008) Agrarian landscapes linear features detection from LiDAR: Application to artificial drainage networks. *International Journal of Remote Sensing*, 29, 3489–3508.

Baker, M.E., Weller, D.E. & Jordan, T.E. (2006a) Improved methods for quantifying potential nutrient interception by riparian buffers. *Landscape Ecology*, 21, 1327–1345.

Baker, M.E., Weller, D.E. & Jordan, T.E. (2006b) Comparison of automated watershed delineations: Effects on land cover areas, percentages, and relationships to nutrient discharge. *Photogrammetric Engineering and Remote Sensing*, 72, 159–168.

Baker, M.E., Weller, D.E. & Jordan, T.E. (2007) Effects of stream map resolution and measures of riparian buffer distribution and nutrient retention potential. *Landscape Ecology*, 22, 973–992.

Baldi, P., Bonvalot, S., Biole, P. & Marsella, M. (2000) Digital photogrammetry and kinematic GPS applied to the monitoring of Vulcano Island. *Geophysical Journal International*, 142, 801–811.

Baldi, P., Bonvalot, S., Biole, P., Coltelli, M., Gwinner, K., Marsella, M., Puglisi, G. & Remy, D. (2002) Validation and comparison of different techniques for the derivation of digital elevation models and volcanic monitoring (Vulcano Island, Italy). *International Journal of Remote Sensing*, 23, 4783–4800.

Balice, R.G., Miller, J.D., Oswald, B.P., Edminister, C. & Yool, S.R. (2000) *Forest Surveys and Wildfire Assessment in the Los Alamos, 1998–1999*. Los Alamos, NM: Los Alamos National Laboratory.

Band, L.E. (1986) Topographic partition of watersheds with digital elevation models. *Water Resources Research*, 22, 15–24.

Band, L.E. (1989) A terrain based, watershed information system. *Hydrological Processes*, 3, 151–162.

Band, L.E. (1993) Effect of land surface representation on forest water and carbon budgets. *Journal of Hydrology*, 150, 749–772.

Band, L.E., Vertessey, R. & Lammers, R.B. (1995) The effect of different terrain representation schemes and resolution on simulated watershed processes. *Zeitschrift fur Geomorphologie, Suppl-Bd*, 101, 187–199.

Band, L.E., Hwang, T., Hales, T.C., Vose, J. & Ford, C. (2012) Ecosystem processes at the watershed scale: Mapping and modeling ecohydrological controls of landslides. *Geomorphology*, 137, 159–167.

Barling, R.D. (1992) *Saturation zones and ephemeral gullies on arable land in southeastern Australia*. Unpublished PhD dissertation, University of Melbourne, Melbourne, Australia.

Barling, R.D., Moore, I.D. & Grayson, R.B. (1994) A quasi-dynamic wetness index for characterizing the spatial distribution of zones of surface saturation and soil water content. *Water Resources Research*, 30, 1029–1044.

Bater, C.W. & Coops, N.C. (2009) Evaluating error associated with LiDAR-derived DEM interpolation. *Computers and Geosciences*, 35, 289–300.

Bates, P.D., Anderson, M.G. & Horrit, M. (1998) Terrain information in geomorphological models: Stability, resolution, and sensitivity. In: S.N. Lane, K.S. Richards & J.H. Chandler (eds) *Landform Monitoring, Modeling, and Analysis*, pp. 279–310. New York, NY: John Wiley & Sons, Inc.

Bauer, J., Rohdenburg, H. & Bork, H.R. (1985) Ein Digitales ReliefModell als Voraussetzung fuer ein deterministisches Modell der Wasser- und Stoff-Fluesse. In: H.R. Bork & H. Rohdenburg (eds) *Parameteraufbereitung fuer Deterministische*

*Gebiets-Wassermodelle, Grundlagenarbeiten zu Analyse von Agrar-Oekosystemen*, pp. 1–15. Braunschweig, Germany: Technische Universität Braunschweig.

Bayramin (2000) Using geographic information system and remote sensing techniques in making pre-soil surveys. In: *Proceedings of the International Symposium on Desertification*, pp. 27–33. Ankara, Turkey.

Beattie, C. (2014) *3D visualization models as a tool for reconstructing the historical landscape of the Ballona watershed*. Unpublished MS thesis, University of Southern California, Los Angeles, CA.

Begg, J.G. & Mouslopoulou, V. (2010) Analysis of late Holocene faulting within an active rift using LiDAR, Taupo Rift, New Zealand. *Journal of Volcanology and Geothermal Research*, 190, 152–167.

Behrens, T., Zhu, A.X., Schmidt, K. & Scholten, K. (2010) Multi-scale digital terrain analysis and feature selection for digital soil mapping. *Geoderma*, 155, 175–185.

Bell, J.C., Grigal, D.F. & Bates, P.C. (2000) A soil-terrain model for estimating spatial patterns of soil organic carbon. In: J.P. Wilson & J.C. Gallant (eds), *Terrain Analysis: Principles and Applications*, pp. 295–310. New York, NY: John Wiley & Sons, Inc.

Bergeström, S. (1976) *Development and Application of a Conceptual Runoff Model for Scandinavian Catchments*. Lund, Sweden: University of Lund, Department of Water Resources Engineering, Lund Institute of Technology Bulletin Series A, No. 52.

Bergeström, S. (1992) *The HBV Model: Its Structure and Applications*. Norrköping, Sweden: Swedish Meteorological and Hydrological Institute.

Bergeström, S. (1995) The HBV model. In: V.P. Singh (ed.) *Computer Models of Watershed Hydrology*, pp. 443–476. Highlands Ranch, CO: Water Resources.

Berry, P.A.M., Garlick, J.D. & Smith, R.G. (2007) Near-global validation of the SRTM DEM using satellite radar altimetry. *Remote Sensing of Environment*, 106, 17–27.

Beven, K.J. & Cloke, H.L. (2013) Comment on "Hyperresolution global land surface modeling: Meeting a grand challenge for monitoring Earth's terrestrial water" by Eric F. Wood et al. *Water Resources Research*, 48, WR010982.

Beven, K.J. & Kirkby, M.J. (1979) A physically-based, variable contributing area model of basin hydrology. *Hydrological Sciences Bulletin*, 24, 43–69.

Bezdek, J.C., Ehrlich, R. & Full, W. (1984) FCM: The fuzzy c-means clustering algorithm. *Computers and Geosciences*, 10, 191–203.

Bhang, K.J., Schwartz F.W. & Braun, A. (2007) Verification of the vertical error in C-Band SRTM DEM using ICESat and Landsat-7, Otter Tail County, MN. *IEEE Transactions on Geoscience and Remote Sensing*, 45, 36–44.

Bishop, M.P. & Shroder, J.F. (2000) Remote sensing and geomorphometric assessment of topographic complexity and erosion dynamics in the Nanga Parbat massif. In: M.A. Khan, P.J. Treloar, M.P. Searle & M.Q. Jan (eds) *Tectonics of the Nanga Parbat Syntaxis and the Western Himalaya*, pp. 181–200. London, UK: Geological Society.

Bishop, M.P. & Shroder, J.F. (eds) (2004) *Geographic Information Science and Mountain Geomorphology*. Chichester, UK: Springer-Praxis.

Bishop, M.P., Shroder, J.F. & Colby, J.D. (2003) Remote sensing and geomorphometry for studying relief production in high mountains. *Geomorphology*, 55, 345–361.

Bishop, M.P., Bush, A.B.G., Copland, L., Kamp, U., Owen, L.A., Seong, Y.B. & Shroder, J.F. (2010) Climate change and mountain topographic evolution in the Central Karakoram, Pakistan. *Annals of the Association of American Geographers*, 100, 772–793.

Bishop, M.P., Bonk, R., Kamp, U. Jr & Shroder, J.F. (2012a) Terrain analysis and data modeling for alpine glacier modeling. *Polar Geography*, 24, 257–276.

Bishop, M.P., James, L.A., Shroder, J.F. & Walsh, S.J. (2012b) Geospatial technologies and digital geomorphological mapping: Concepts, issues, and research. *Geomorphology*, 137, 5–26.

Bishop, T.F.A. & Minasny, B. (2005) Environmental soil–terrain modeling: The predictive potential and uncertainty. In: S. Grunwald (ed.) *Environmental Soil–Landscape Modeling: Geographic Information Technologies and Pedometrics*, pp. 185–213. Boca Raton, FL: CRC Press.

Blaszczynski, J.S. (1997) Landform characterization with geographic information systems. *Photogrammetric Engineering and Remote Sensing*, 63, 183–191.

Boast, R.R. & Shelito, R.G. (1989) *Soil Survey of Madison County Area, Montana*. Washington, DC: Soil Conservation Service, US Department of Agriculture.

Bogaart, P.W. & Troch, P.A. (2006) Curvature distribution within hillslopes and catchments and its effect on the hydrologic response. *Hydrology and Earth System Sciences*, 10, 925–936.

Böhner, J. & Antonić, O. (2009) Land-surface parameters specific to topo-climatology. In: T. Hengl & H.I. Reuter (eds) *Geomorphometry: Concepts, Software, Applications*, pp. 195–226. Amsterdam, Netherlands: Elsevier.

Böhner, J., Köthe, R., Conrad, O., Gross, J., Ringeler, A. & Selige, T. (2002) Soil regionalization by means of terrain analysis and process parameterization. In: E. Micheli, E. Nachtergaele & L. Montanarella (eds) *Soil Classification 2001, EUR 20398 EN*, pp. 213–222. Ispra, Italy: European Soil Bureau, Joint Research Centre.

Böhner, J., McCloy, K.R. & Strobl, J. (eds) (2006) *SAGA: Analysis and Modeling Applications*. Göttingen, Germany: Göttinger Geographische Abhandlungen.

Bolstad, P.V. & Lillesand, T.M. (1992) Improved classification of forest vegetation in northern Wisconsin through a rule-based combination of soils, terrain, and Landsat TM data. *Forest Science*, 38, 5–20.

Bolstad, P.V. & Stowe, T. (1994) An evaluation of DEM accuracy: Elevation, slope and aspect. *Photogrammetric Engineering and Remote Sensing*, 60, 1327–1332.

Bontemps, S., Defourny, P., Van Bogaert, E., Arino, E., Kalogirou, V. & Perez, J. (2011) *GLOBCOVER 2009: Products Description and Validation Report*. Brussels, Belgium: European Space Agency and University College London.

Borga, M., Dalla Fontana, C. & Cazorzi, F. (2002) Analysis of topographic and climatic control on rainfall-triggered shallow landsliding using a quasi-dynamic wetness index. *Journal of Hydrology*, 268, 56–71.

Brabyn, L.W. (1997) Classification of macro landforms using GIS. *ITC Journal*, 1, 26–40.

Brabyn, L.W. (1998) GIS analysis of macro landforms. In: *Proceedings of the 10th Colloquium of the Spatial Information Research Centre, University of Otago*, pp. 35–48. Dunedin, New Zealand.

Bradley, A.P. (1997) The use of the area under the TOC curve in the evaluation of machine learning algorithms. *Pattern Recognition*, 30, 1145–1159.

Brändli, M. (1996) Hierarchical models for the definition and extraction of terrain features. In: P.A. Burrough & A.W. Frank (eds) *Geographic Objects with Indeterminate Boundaries*, pp. 257–270. London, UK: Taylor & Francis.

Brooks, R.T. & Colburn, E.A. (2011) Extent and channel morphology of unmapped headwater stream segments of the Quabbin watershed, Massachusetts. *Journal of the American Water Resources Association*, 47, 158–168.

Brown, D.G. (1998) Classification and boundary vagueness in mapping pre-settlement forest types. *International Journal of Geographical Information Science*, 12, 105–129.

Brown, D.G. & Bara, T.J. (1994) Recognition and reduction of systematic error in elevation and derivative surfaces from 7.5-minute DEMs. *Photogrammetric Engineering and Remote Sensing*, 60, 189–194.

Bruy, A. (2015) *QGIS by Example*. Birmingham, UK: Packt Publishing.

Buchanan, B.P., Easton, Z.M., Schneider, R. & Walter, M.T. (2012) Incorporating variable source area hydrology into a spatially distributed direct runoff model. *Journal of the American Water Resources Association*, 48, 43–60.

Buchanan, B.P., Archibald, J.A., Easton, Z.M., Shaw, S.B., Schneider, R.L. & Walter, M.T. (2013) A phosphorous index that combines critical source areas and transport pathways using a travel time approach. *Journal of Hydrology*, 486, 123–135.

Buchanan, B.P., Fleming, M., Schneider, R.L., Richards, B.K., Archibald, J., Qiu, Z. & Walter, M.T. (2014) Evaluating topographic wetness indices across central New York agricultural landscapes. *Hydrology and Earth Systems Science*, 18, 3279–3299.

Bunn, A.G., Waggoner, L.A. & Graumlich, L.J. (2005) Topographic mediation of growth in high elevation foxtail pine (*Pinus balfournia* Grev. et Balf.) forests in the Sierra Nevada, USA. *Global Ecology and Biogeography*, 14, 103–114.

Burbank, D., Leland, J., Fielding, E., Anderson, R.S., Brozovik, N., Reid, M.R. & Duncan, C. (1996) Bedrock incision, rock uplift and threshold hillslopes in the northwestern Himalaya. *Nature*, 379, 505–510.

Burns, W.J., Coe, J.A., Kay, B.S. & Ma, L. (2010) Analysis of elevation changes detected from multi-temporal LiDAR surveys in forested landslide terrain in western Oregon. *Environmental and Engineering Geoscience*, 16, 315–341.

Burrough, P.A. (1996) Natural objects with indeterminate boundaries. In: P.A. Burrough & A.U. Frank (eds) *Geographic Objects with Indeterminate Boundaries*, pp. 3–28. London, UK: Taylor & Francis.

Burrough, P.A. & McDonnell, R. (1998) *Principles of Geographical Information Science*. New York, NY: Oxford University Press.

Burrough, P.A., van Gaans, P.F.M. & MacMillan, R.A. (2000) High-resolution landform classification using fuzzy k-means. *Fuzzy Sets and Systems*, 113, 37–52.

Burrough, P.A., Wilson, J.P., van Gaans, P.F.M. & Hansen, A.J. (2001) Fuzzy k-means classification of topo-climatic data as an aid to forest mapping in the Greater Yellowstone Area, USA. *Landscape Ecology*, 16, 523–546.

Burt, T.P. & Butcher D.P. (1985) Topographic controls of soil moisture distribution. *Journal of Soil Science*, 36, 469–486.

Buttenfield, B.P. & McMaster, R.B. (eds) (1991) *Map Generalization: Making Rules for Knowledge Representation*. New York, NY: John Wiley & Sons, Inc.

Buttenfield, B.P., Stanislawaski, L.V. & Brewer, C.A. (2011) Adapting generalization tools to physiographic diversity for the USGS National Hydrography Dataset. *Cartography and Geographic Information Science*, 38, 289–301.

Buttenfield, B.P., Ghandehari, M., Leyk, S., Stanislawski, L.V. & Brantley, M.E. (2016) Measuring distance "as the horse runs": Cross-scale comparison of terrain-based metrics. In: *Proceedings of the 9th International Conference on Geographic Information Science*, Montreal, Quebec.

Buytaert, W. (2011) Topmodel: Implementation of the Hydrological Model TOPMODEL in R, Version 0.7.2-2. Retrieved from http://cran.r-project.org/web/packages/topmodel/index.html

Byun, J. & Seong, Y.B. (2015) An algorithm to extract more accurate stream longitudinal profiles from unfilled DEMs. *Geomorphology*, 242, 38–48.

Carabajal, C. & Harding, D. (2006) SRTM C-Band and ICESat laser altimetry elevation comparisons as a function of tree cover and relief. *Photogrammetric Engineering and Remote Sensing*, 72, 287–298.

Carara, A., Bitelli, G. & Carla, R. (1997) Comparison of techniques for generating digital terrain models from contour lines. *International Journal of Geographical Information Science*, 11, 451–473.

Carlisle, B.H. (2000) The highs and lows of DEM error: Developing a spatially distributed DEM error model. In: *Proceedings of the 5th International Conference on Geocomputation*, pp. 23–25. Greenwich, UK.

Cavalli, M., Tarolli, P., Marchi, L. & Dalla Fontana, G. (2008) The effectiveness of airborne LiDAR data in the recognition of channel bed morphology. *Catena*, 73, 249–260.

Cavazzi, S., Corstanje, R., Mayr, T., Hannam, J. & Fealy, R. (2013) Are fine resolution digital elevation models always the best choice in digital soil mapping? *Geoderma*, 195–196, 111–121.

Chairat, S. & Delleur, J.W. (1993) Effects of topographic index distribution on predicted runoff using GRASS. *Water Resources Bulletin*, 29, 1029–1034.

Chan, Y.C., Chen, Y.G., Shih, T.Y. & Huang, C. (2007) Characterizing the Hsincheng active fault in northern Taiwan using airborne LiDAR data: Detailed geomorphic features and their structural implications. *Journal of Asian Earth Sciences*, 31, 303–316.

Chang, K. (2007) *Introduction to Geographic Information Systems*, 4th edn. New York, NY: McGraw-Hill.

Charnpratheep, K., Zhou, Q. & Garner, B. (1997) Preliminary landfill site screening using fuzzy geographical information systems. *Waste Management and Research*, 15, 197–215.

Chen, J., Chen, J., Liao, A., Cao, X., Chen, L., Chen, X., He, C., Han, G., Peng, S., Lu, M., Zhang, W., Tong, X. & Mills, J. (2015) Global land cover mapping at 30 m resolution: A POK-based operational approach. *ISPRS Journal of Photogrammetry and Remote Sensing*, 103, 7–27.

Chen, Y. & Zhou, Q. (2013) A scale-adaptive DEM for multi-scale terrain analysis. *International Journal of Geographical Information Science*, 27, 1329–1348.

Chen, Z. & Guevara, J. (1987) Systematic selection of very important points (VIP) from digital terrain model for constructing triangular irregular networks. In: *Proceedings of the Auto Carto XIII Conference*, pp. 50–56. Baltimore, MD.

Cheng, T. & Molenaar, M. (1999a) Objects with fuzzy spatial extent. *Photogrammetric Engineering and Remote Sensing*, 63, 403–414.

Cheng, T. & Molenaar, M. (1999b) Diachronic analysis of fuzzy objects. *GeoInformatica*, 3, 337–356.

Chinh, L., Iseri, H., Hiramatsu, K., Harada, M. & Mori, M. (2013) Simulation of rainfall runoff and pollution load for Chikugo River basin in Japan using a GIS-based distributed parameter model. *Paddy and Water Environment*, 11, 97–112.

Chirico, G.B., Western, A.W., Grayson, R.B. & Günter, B. (2005) On the definition of the flow width for calculating specific catchment area patterns from gridded elevation data. *Hydrological Processes*, 19, 2539–2556.

Chorowicz, J., Ichoku, C., Riazanoff, S. & Kim, Y.-J. (1992) A combined algorithm for automated drainage network extraction. *Water Resources Research*, 28, 1293–1302.

Chow, F.C. (1984) Summed-area tables for texture mapping. *Computer Graphics*, 18, 207–212.

Chow, T.E. & Hodgson, M.E. (2009) Effects of LiDAR post-spacing and DEM resolution to mean slope estimation. *International Journal of Geographic Information Science*, 23, 1277–1295.

Chow, T.-Y., Lin, W.-T., Lin, C.-Y., Chow, W.-C. & Huang, P.-H. (2004) Application of the PROMETHEE technique to determine depression outlet locations and flow directions in DEMs. *Journal of Hydrology*, 287, 49–61.

Christensen, T.R., Jackowicz-Korczyński, M., Aurela, M., Crill, P., Heliasz, M., Mastepanov, M. & Friborg, T. (2012) Monitoring the multi-year carbon balance of a subarctic Palsa Mire with micrometeorological techniques. *Ambio*, 41, 207–217.

Cimmery, V. (2010a) *SAGA User Guide, Updated for SAGA Version 2.0.5: Volume 1, An Introduction to the Graphical User Interface*. Retrieved from http://priede.bf.lu.lv/ftp/pub/TIS/gis_paketes/SAGA/SAGA_User_Guide_Vol1_Cimmery_version_2.0.5_20100823.pdf

Cimmery, V. (2010b) *SAGA User Guide, Updated for SAGA Version 2.0.5: Volume 2, "How To" Information on Many SAGA Modules, Functions, and GIS Applications*. Retrieved from http://priede.bf.lu.lv/ftp/pub/GIS/gis_paketes/SAGA/SAGA_User_Guide_Vol2_Cimmery_version_2.0.5_20101209.pdf

Claessens, L., Heuvelink, G.B.M., Schoorl, J.M. & Veldkamp, A. (2005) DEM resolution effects on shallow landslide hazard and soil redistribution modeling. *Earth Surface Processes and Landforms*, 30, 461–477.

Claessens, L., Lowe, D.J., Hayward, B.W., Schaap, B.F., Schoorl, J.M. & Veldkamp, A. (2006) Reconstructing high-magnitude/low-frequency landslide events on soil redistribution modeling and a Late Holocene sediment record from New Zealand. *Geomorphology*, 74, 29–49.

Clarke, K.C. (1988) Scale-based simulation of topographic relief. *American Cartographer*, 15, 173–181.

Clarke, K.C. & Lee, S.J. (2007) Spatial resolution and algorithm choice as modifiers of downslope flow computed from digital elevation models. *Cartography and Geographic Information Science*, 34, 215–230.

Clarke, K.C. & Romero, B.E. (2016) On the topology of topography: A review. *Cartography and Geographic Information Science*, 44, 271–282.

Clarke, K.C. & Schweizer, D.M. (1991) Measuring the fractal dimension of natural surfaces using a robust fractal estimator. *Cartography and Geographic Information Systems*, 18, 37–47.

Clubb, F.J., Mudd, S.M., Milodowski, D.T., Hurst, M.D. & Slater, L.J. (2014) Objective extraction of channel heads from high-resolution topographic data. *Water Resources Research*, 50, 4283–4304.

Cobby, D.M., Mason, D.C. & Davenport, I.J. (2001) Image processing of airborne scanning laser altimetry data for improved river flood modeling. *ISPRS Journal of Photogrammetry and Remote Sensing*, 56, 121–138.

Coe, M.T. (1998) A linked global model of terrestrial hydrologic processes: Simulation of modern rivers, lakes, and wetlands. *Journal of Geophysical Research: Atmospheres*, 103, 8885–8899.

Coe, M.T., Costa, M.H. & Soares-Filho, B.S. (2009) The influence of historical and potential future deforestation on the stream flow of the Amazon River: Land surface processes and atmospheric feedbacks. *Journal of Hydrology*, 369, 165–174.

Collins, S.H. & Moon, G.C. (1981) Algorithms for dense digital terrain models. *Photogrammetric Engineering and Remote Sensing*, 47, 71–76.

Conacher, A.J. & Dalrymple, J.B. (1977) The nine-unit land surface model: An approach to pedogeomorphic research. *Geoderma*, 18, 1–154.

Conrad, O., Bechtel, B., Bock, M., Dietrich, H., Fischer, E., Gerlitz, L., Wehberg, J., Wichmann, V. & Böhner, J. (2015) System for automated geoscientific analyses (SAGA) v. 2.1.4. *Geoscientific Model Development*, 8, 1991–2007.

Corbett, J.D. & Carter, S.E. (1996) Using GIS to enhance agricultural planning: The example of inter-seasonal rainfall variability in Zimbabwe. *Transactions in GIS*, 1, 207–218.

Costa-Cabral, M. & Burges, S.J. (1994) Digital Elevation Model Networks (DEMON), a model of flow over hillslopes for computation of contributing and dispersal areas. *Water Resources Research*, 30, 1681–1692.

Coughlan, J.C. & Running, S.W. (1996) Biophysical aggregations of a forested landscape using an ecological diagnostic system. *Transactions in GIS*, 1, 25–39.

Cova, T.J. & Goodchild, M.F. (2002) Extending geographical representation to include fields of spatial objects. *International Journal of Geographical Information Science*, 16, 509–532.

Csatho, B., Schenk, T., Kyle, P., Wilson, T. & Krabill, W.B. (2008) Airborne laser swath mapping of the summit of Erebus volcano, Antarctica: Applications to geological mapping of a volcano. *Journal of Volcanology and Geothermal Research*, 177, 531–548.

Culling, W. (1960) Analytical theory of erosion. *Journal of Geology*, 68, 336–344.

Culling, W. (1963) Soil creep and the development of hillside slopes. *Journal of Geology*, 71, 127–161.

Culling, W. (1965) Theory of erosion on soil-covered slopes. *Journal of Geology*, 73, 230–254.

Dadson, S.J. & Bell, V.A. (2010) *Comparison of Grid-2-Grid and TRIP Runoff Routing Schemes*. Wallingford, UK: Centre for Ecology and Hydrology Report.

Dadson, S.J., Ashpole, I., Harris, P., Davies, H.N., Clark, D.B., Blyth, E. & Taylor, C.M. (2010) Wetland inundation dynamics in a model of land surface climate: Evaluation in the Niger inland delta region. *Journal of Geophysical Research: Atmospheres*, 115, D23114.

Dadson, S.J., Bell, V.A. & Jones, R.G. (2011) Evaluation of a grid-based river flow model configured for use in a regional climate model. *Journal of Hydrology*, 411, 238–250.

Daly, C., Gibson, W.P., Taylor, G.H., Johnson, G.L. & Pasteris, P. (2002) A knowledge-based approach to the statistical mapping of climate. *Climatic Research*, 22, 99–113.

Danielson, J.J. & Gesch, D.B. (2011) *Global Multi-resolution Terrain Elevation Data 2010 (GMTED2010)*. Washington, DC: US Geological Survey Open-File Report 2011–1073.

Darnell, A.R., Tate, N.J. & Brunsdon, C. (2008) Improving user assessment of error implications in digital elevation models. *Computers, Environment and Urban Systems*, 32, 268–277.

Davies, K.W., Bates, J.D. & Miller, R.F. (2007) Environmental and vegetation characteristics of the *Artemisia tridentata* spp. *wyomingensis* alliance. *Journal of Arid Environments*, 70, 478–494.

Davies, K.W., Petersen, S.L., Johnson, D.D., Davis, D.B., Madsen, M.D., Zvirzdin, D.L. & Bates, J.D. (2010) Estimating juniper cover from National Agriculture Imagery Program (NAIP) imagery and evaluating relationships between potential cover and environmental variables. *Rangeland Ecology and Management*, 63, 630–637.

Day, T. & Miller, J.-P. (1988) Quality assessment of digital elevation models produced by automatic stereo matches from SPOT image pairs. *International Archives of Photogrammetry and Remote Sensing*, 27, 148–159.

DeBruin, S. (2000) Querying probabilistic land cover data using fuzzy set theory. *International Journal of Geographical Information Science*, 14, 359–372.

de Ferranti, J. (2014) *Digital Elevation Data*. Retrieved from http://viewfinderpanoramas.org/dem3.html

de Floriani, L., Falcidieno, B. & Penovi, C. (1984) A hierarchical structure for surface approximation. *Computers and Graphics*, 8, 183–193.

DeHaemer, M. Jr & Zyda, M.J. (1991) Simplification of objects rendered by polygonal approximations. *Computers and Graphics*, 15, 175–184.

Dehn, M., Gärtner, H. & Dikau, R. (2001) Principles of semantic modeling of landform structures. *Computers and Geosciences*, 27, 1005–1010.

Deng, Y.X. (2007) New trends in digital terrain analysis: Landform definition, representation, and classification. *Progress in Physical Geography*, 31, 405–419.

Deng, Y.X. & Wilson, J.P. (2006) The role of attribute selection in GIS representations of the biophysical environment. *Annals of the Association of American Geographers*, 96, 47–63.

Deng, Y.X. & Wilson, J.P. (2008) Multi-scale and multi-criteria mapping of mountain peaks as fuzzy entities. *International Journal of Geographical Information Science*, 22, 205–218.

Deng, Y.X., Wilson, J.P. & Sheng, J. (2006) The sensitivity of fuzzy landform classification to variable attribute weights. *Earth Surface Processes and Landforms*, 31, 1452–1462.

Deng, Y.X., Wilson, J.P. & Bauer, B.O. (2007) DEM resolution dependencies of terrain attributes across a landscape. *International Journal of Geographical Information Science*, 21, 187–213.

Deng, Y.X., Wilson, J.P. & Gallant, J.C. (2008) Terrain analysis. In: J.P. Wilson & A.S. Fotheringham (eds) *The Handbook of Geographic Information Science*, pp. 417–435. Oxford, UK: Blackwell.

De Rose, R.C. & Basher, L.R. (2011) Measurement of river bank and cliff erosion from sequential LiDAR and historical aerial photography. *Geomorphology*, 126, 132–147.

Desmet, P.J.J. (1997) Effects of interpolation errors on the analysis of DEMs. *Earth Surface Processes and Landforms*, 22, 563–580.

Desmet, P.J.J. & Govers, G. (1996a) A GIS procedure for the automated calculation of the USLE LS factor on topographically complex landscape units. *Journal of Soil and Water Conservation*, 51, 427–433.

Desmet, P.J.J. & Govers, G. (1996b) Comparison of routing algorithms for digital elevation models and their implications for predicting ephemeral gullies. *International Journal of Geographical Information Systems*, 10, 311–331.

Deza, M.M. & Deza, E. (2014) *Encylcopedia of Distances*, 3rd edn. Berlin, Germany: Springer.

Dietrich, W., Wilson, C., Montgomery, D., McKean, J. & Bauer, R. (1992) Erosion thresholds and land surface morphology. *Geology*, 20, 675–679.

Dietrich, W., Wilson, C., Montgomery, D. & McKean, J. (1993) Analysis of erosion thresholds, channel networks, and landscape morphology using a digital terrain model. *Journal of Geology*, 101, 259–278.

Dikau, R. (1989) The application of a digital relief model to landform analysis. In: J. Raper (ed.) *Three-dimensional Applications of Geographical Information Systems*, pp. 77–90. London, UK: Taylor & Francis.

Dikau, R., Brabb, E.E. & Mark, R.M. (1991) *Landform Classification of New Mexico by Computer*. Washington, DC: US Geological Survey Open File Report No. 91–364.

Dikau, R., Brabb, E.E., Mark, R.M. & Pike, R.J. (1995) Morphometric landform analysis of New Mexico. *Zeitschrift für Geomorphologie, Suppl-Bd*, 101, 109–126.

Dingman, J.R., Sweet, L.C., McCullough, I., Davis, F.W., Flint, A., Franklin, J. & Flint, L.E. (2013) Cross-scale modeling of surface temperature and tree seedling establishment in mountain landscapes. *Ecological Processes*, 2, 30.

Dobos, E. & Hengl, T. (2009) Soil mapping applications. In: T. Hengl & H.I. Reuter (eds) *Geomorphometry: Concepts, Software, Applications*, pp. 461–480. Amsterdam, Netherlands: Elsevier.

Dobos, E., Daroussin, J. & Montanarella, L. (2005) *An SRTM-based Procedure to Delineate SOTER Terrain Units on 1:1 and 1:5 Million Scales, EUR 21571 EN*. Luxembourg: Office for Official Publications of the European Communities.

Dobrowski, S.Z., Safford, H.D., Cheng, Y.B. & Ustin, S.L. (2008) Mapping mountain vegetation using species distribution modeling, image-based texture analysis, and object-based classification. *Applied Vegetation Science*, 11, 499–508.

D'Oleire-Oltmanns, S., Eisank, C., Drăguţ, L. & Blaschke, T. (2013) An object-based workflow to extract landforms at multiple scales from two distinct data types. *IEEE Geoscience and Remote Sensing Letters*, 10, 947–951.

Douglas, D.H. (1986) Experiments to locate ridges and channels and create a new type of digital elevation model. *Cartographica*, 23, 29–61.

Douglas, D.H. & Peucker, T.K. (1973) Algorithms for the reduction in the number of points required to represent a digitized line or its caricature. *Canadian Cartographer*, 10, 112–122.

Dozier, J., Bruno, J. & Downey, P. (1981) A faster solution to the horizon problem. *Computers and Geosciences*, 7, 145–151.

Drăguţ, L. & Blaschke, T. (2006) Automated classification of landform elements using object-based image analysis. *Geomorphology*, 81, 330–344.

Drăguţ, L. & Blaschke, T. (2008) Terrain segmentation and classification using SRTM data. In: Q. Zhou, B. Lees & G.-A. Tang (eds) *Advances in Digital Terrain Analysis*, pp. 141–158. Berlin, Germany: Springer.

Drăguţ, L. & Eisank, C. (2011) Object representations at multiple scales from digital elevation models. *Geomorphology*, 129, 183–189.

Drăguţ, L. & Eisank, C. (2012) Automated object-based classification of topography from SRTM data. *Geomorphology*, 141, 21–33.

Drăguţ, L., Eisank, C., Strasser, T. & Blaschke, T. (2009) A comparison of methods to incorporate scale in geomorphometry. In: R. Purves, S. Gruber, R. Straumann & T. Hengl (eds) *Proceedings of Geomorphometry 2009*, pp. 133–139. Zurich, Switzerland: University of Zurich.

Drăguţ, L., Tiede, D. & Levick, S.R. (2010) ESP: A tool to estimate scale parameter for multiresolution image segmentation of remotely sensed data. *International Journal of Geographical Information Science*, 24, 859–871.

Drăguţ, L., Eisank, C. & Strasser, T. (2011) Local variance for multi-scale analysis in geomorphometry. *Geomorphology*, 130, 162–172.

Duan, J. & Grant, G.E. (2000) Shallow landslide delineation for steep forest watersheds based on topographic attributes and probability analysis. In: J.P. Wilson & J.C. Gallant (eds) *Terrain Analysis: Principles and Applications*, pp. 311–330. New York, NY: John Wiley & Sons, Inc.

Duan, X., Li, L., Zhu, H. & Ying, S. (2017) A high-fidelity multiresolution digital elevation model for Earth systems. *Geoscientific Model Development*, 10, 239–253.

Dubayah, R. (1992) Estimating net solar radiation using Landsat Thematic Mapper and digital elevation data. *Water Resources Research*, 28, 2469–2484.

Dubayah, R. & Loechel, S. (1997) Modeling topographic solar radiation using GOES data. *Journal of Applied Meteorology*, 36, 141–154.

Dubayah, R. & Rich, P.M. (1995) Topographic solar radiation models for GIS. *International Journal of Geographical Information Systems*, 9, 405–419.

Dubayah, R. & van Katwijk, V. (1992) The topographic distribution of annual incoming solar radiation in the Rio Grande river basin. *Geophysical Research Letters*, 19, 2231–2234.

Ducharne, A. (2009) Reducing scale dependence in TOPMODEL using a dimensionless topographic index. *Hydrology and Earth System Science*, 13, 2399–2412.

Ducharne, A., Koster, R.D., Suarez, M.J. & Kumar, P. (1999) A catchment-based land surface model for GCMs and the framework for its evaluation. *Physics and Chemistry of the Earth*, 24, 769–773.

Dunn, M. & Hickey, R. (1998) The effect of slope algorithms on slope estimates within a GIS. *Cartography*, 27, 9–15.

Dymond, J.R., Derose, R.C. & Harmsworth, G.R. (1995) Automated mapping of land components from digital elevation data. *Earth Surface Processes and Landforms*, 20, 131–137.

Easton, Z.M., Fuka, D.R., Walter, M.T., Cowan, D.M., Scheidermann, E.M. & Steenhuis, T.S. (2008) Re-conceptualizing the soil and water assessment tool (SWAT) model to predict runoff from variable source areas. *Journal of Hydrology*, 348, 279–291.

Eckert, S., Kellenberger, T. & Itten, K. (2005) Accuracy assessment of automatically derived digital elevation models from ASTER data in mountainous terrain. *International Journal of Remote Sensing*, 26, 1943–1957.

Endreny, T.A. & Wood E.F. (2001) Representing elevation uncertainty in runoff modeling and flowpath mapping. *Hydrological Processes*, 15, 2223–2236.

Endreny, T.A. & Wood, E.F. (2003) Maximizing spatial congruence of observed and DEM-delineated overland flow networks. *International Journal of Geographical Information Science*, 17, 699–713.

Erbs, D.G., Klein, S.A. & Duffie, J.A. (1982) Estimation of the diffuse radiation fraction for hourly, daily and monthly average global radiation. *Solar Energy*, 28, 293–302.

Erskine, R.H., Green, T.R., Ramirez, J.A. & MacDonald, L.H. (2006) Comparison of grid-based algorithms for computing upslope contributing area. *Water Resources Research*, 42, W09416.

Estomell, J., Ruiz, L.A., Velzquez-Mart, B. & Hemosilla, T. (2011) Analysis of the factors affecting LiDAR DTM accuracy in a steep shrub area. *International Journal of Digital Earth*, 4, 521–538.

Evans, I.S. (1972) General geomorphometry, derivatives of altitude, and descriptive statistics. In: R.J. Chorley (ed.) *Spatial Analysis in Geomorphology*, pp. 17–90. London, UK: Harper & Row.

Evans, I.S. (1979) *An Integrated System of Terrain Analysis and Slope Mapping*. Durham, UK: University of Durham, Final report on Grant DA-ERO-591-73-G0040.

Evans, I.S. (1980) An integrated system of terrain analysis and slope mapping. *Zeitschrift für Geomorphologie NF, Suppl-Bd*, 36, 274–295.

Evans, I.S. (1987) The morphometry of specific landforms. In: V. Gardiner (ed.) *International Geomorphology 1986*, Vol. 2, pp. 105–124. Chichester, UK: John Wiley & Sons Ltd.

Evans, I.S. (2012) Geomorphology and landform mapping: What is a landform? *Geomorphology*, 137, 94–106.

Evans, I.S. (2013) Land surface derivatives: History, calculation, and further development. In: *Proceedings of the 3rd International Conference on Geomorphometry*, Nanjing, China.

Evans, I.S., Hengl, T. & Gorsevski, P. (2009) Applications in geomorphology. In: T. Hengl & H.I. Reuter (eds) *Geomorphometry: Concepts, Software, Applications*, pp. 497–525. Amsterdam, Netherlands: Elsevier.

Evans, J.S. (2003) CTI.aml Compound Topographic Index AML script. Unpublished report.

Evans, J.S. & Cushman, S.A. (2009) Gradient modeling of conifer species using random forests. *Landscape Ecology*, 24, 673–683.

Evans, J.S., Oakleaf, J., Cushman, S.A. & Theobald, D. (2014) *An ArcGIS Toolbox for Surface Gradient and Geomorphometric Modeling (Version 2.0-0)*. Retrieved from http://evansmurphy.wixsite.com/evansspatial/arcgis-gradient-metrics-toolbox

Fairfield, J. & Leymarie, P. (1991) Drainage networks from grid digital elevation models. *Water Resources Research*, 27, 709–717.

Farr, T.G. & Kobrick, M. (2000) Shuttle radar topography mission produces a wealth of data. *EOS, Transactions of the American Geophysical Union*, 81, 83–85.

Farr, T.G., Rosen, P.A., Caro, E., Crippen, R., Duren, R., Hensley, S. et al. (2007) The Shuttle Radar Topography Mission. *Reviews of Geophysics*, 45, RG2004/2007.

Fei, L. & He, J. (2009) A three-dimensional Douglas–Peucker algorithm and its application to automated generalization of DEMs. *International Journal of Geographical Information Science*, 23, 703–718.

Fenneman, N.M. (1938) *Physiography of the Eastern United States*. New York, NY: McGraw-Hill.

FGDC (1998) *Geospatial Positional Accuracy Standards, Part 3: National Standard for Spatial Data Accuracy*. Retrieved from http://www.fgdc.gov/standards/projects/FGDC-standards-projects/accuracy/part3/chapter3

FGDC (2017) *ISO Geospatial Metadata Standards*. Retrieved from https://www.fgdc.gov/metadata/iso-standards

Fiddes, J. & Gruber, S. (2014) TopoSCALE v. 1.0: Downscaling gridded climate data in complex terrain. *Geoscientific Model Development*, 7, 387–405.

Filin, S. (2003) Recovery of systematic biases in laser altimetry data using natural surfaces. *Photogrammetric Engineering and Remote Sensing*, 69, 1235–1242.

Fisher, P.F. (1991) First experiments in viewshed uncertainty: The accuracy of the viewshed area. *Photogrammetric Engineering and Remote Sensing*, 57, 1321–1327.

Fisher, P.F. (1992) First experiments in viewshed uncertainty: Simulating fuzzy viewsheds. *Photogrammetric Engineering and Remote Sensing*, 58, 345–352.

Fisher, P.F. (1993) Algorithm and implementation uncertainty in viewshed analysis. *International Journal of Geographical Information Systems*, 7, 331–347.

Fisher, P.F. (1995) An exploration of probable viewsheds in landscape planning. *Environment and Planning B*, 22, 527–546.

Fisher, P.F. (1996) Reconsideration of the viewshed function in terrain modeling. *Geographical Systems*, 3, 33–58.

Fisher, P.F. (1997) The pixel: A snare and a delusion. *International Journal of Remote Sensing*, 18, 679–685.

Fisher, P.F. (1998) Improved modeling of elevation error with geostatistics. *Geoinformatica*, 2, 215–233.

Fisher, P.F. (2000a) Fuzzy modeling. In: S. Openshaw, R. Abrahart & T. Harris (eds) *Geocomputing*, pp. 161–186. London, UK: Taylor & Francis.

Fisher, P.F. (2000b) Sorties paradox and vague geographies. *Fuzzy Sets and Systems*, 113, 7–18.

Fisher, P.F. & Pathriana, S. (1994) The evaluation of fuzzy membership of land cover classes in the suburban zone. *Remote Sensing of Environment*, 34, 121–132.

Fisher, P.F. & Tate, N.J. (2006) Causes and consequences of error in digital elevation models. *Progress in Physical Geography*, 30, 467–489.

Fisher, P.F. & Wood, J. (1998) What is a mountain? Or the Englishman who went up a Boolean geographical concept and realized it was fuzzy. *Geography*, 83, 247–256.

Fisher, P.F., Wood, J. & Cheng, T. (2004) Where is Helvellyn? Fuzziness of multi-scale landscape morphometry. *Transactions of the Institute of British Geographers NS*, 29, 106–128.

Fisher, S.G. & Welter, J.R. (2005) Flowpaths as integrators of heterogeneity in streams and landscapes. In: G.M. Lovett, C.G. Jones, M.G. Turner & K.C. Weathers (eds) *Ecosystem Function in Heterogeneous Landscapes*, pp. 311–321. New York, NY: Springer.

Fitzgerald, R.W. & Lees, B.G. (1992) The application of neural networks to the floristic classification of remote sensing and GIS data in complex terrain. In: *Proceedings of the 17th International Society of Photogrammetry and Remote Sensing Congress*. Washington, DC: ASPRS.

Fitzgerald, R.W. & Lees, B.G. (1994) Spatial context and scale relationships in raster data for thematic mapping in natural systems. In: T. Waugh & R. Hedley (eds) *Advances in GIS Research*, pp. 462–476. London, UK: Taylor & Francis.

Fleming, P.M. (1987) *Notes on a Radiation Index for Use in Studies of Aspect Effects on Radiation Climate*. Canberra, Australia: Commonwealth Scientific and Industrial Research Organization, Division of Water Resources Research, Institute of Biological Resources Technical Memorandum.

Florinsky, I.V. (1998) Accuracy of local topographic variables derived from digital elevation models. *International Journal of Geographical Information Science*, 12, 47–62.

Florinsky, I.V. (2002) Errors of signal processing in digital terrain modeling. *International Journal of Geographical Information Science*, 16, 475–501.

Florinsky, I.V. (2012) *Digital Terrain Analysis in Soil Science and Geology*. Oxford, UK: Elsevier.

Florinsky, I.V. (2017) Spheroidal equal angular DEMs: The specificity of morphometric treatment. *Transactions in GIS*, 21 (in press).

Florinsky, I.V. & Kuryakova, G.A. (2000) Determination of grid size for digital terrain modeling in landscape investigations: Exemplified by soil moisture distribution at a micro-scale. *International Journal of Geographical Information Science*, 14, 815–832.

Foody, G.M. (1996) Fuzzy modeling of vegetation from remotely sensed imagery. *Ecological Modeling*, 85, 3–12.

Fornaciai, A., Behncke, B., Favalli, M., Neri, M., Tarquini, S. & Boschi, E. (2010) Detecting short-term evolution of Etnean scoria cones: A LiDAR-based approach. *Bulletin of Volcanology*, 72, 1209–1222.

Foster, G.R. & Wischmeier, W.H. (1974) Evaluating irregular slopes for soil loss prediction. *Transactions of the American Society of Agricultural Engineers*, 17, 305–309.

Fowler, R.J. & Little, J.J. (1979) Automatic extraction of irregular network digital terrain models. *Computer Graphics*, 13, 199–207.

Franklin, J. (1995) Predictive vegetation mapping: Geospatial mapping of biospatial patterns in relation to environmental gradients. *Progress in Physical Geography*, 19, 474–499.

Franklin, J. (1998) Predicting the distribution of shrub species in southern California from climate and terrain-derived variables. *Journal of Vegetation Science*, 9, 733–748.

Franklin, W.R. & Ray, C. (1994) Higher isn't necessarily better: Visibility algorithms and experiments. In: R.G. Healey & T.C. Waugh (eds) *Advances in GIS: Proceedings of the 6th Symposium on Spatial Data Handling*, pp. 751–770. London: Taylor & Francis.

Freeman, G.T. (1991) Calculating catchment area with divergent flow based on a regular grid. *Computers and Geosciences*, 17, 413–422.

Fried, J.S., Brown, D.G., Zweifler, M.O. & Gold, M.A. (2000) Mapping contributing areas for stormwater discharge to streams using terrain analysis. In: J.P. Wilson & J.C. Gallant (eds) *Terrain Analysis: Principles and Applications*, pp. 183–204. New York, NY: John Wiley & Sons, Inc.

Fu, P. & Rich, P.M. (2002) A geometric solar radiation model with applications in agriculture and forestry. *Computers and Electronics in Agriculture*, 37, 25–35.

Gallant, A.L., Douglas, D.B. & Hoffer, R.M. (2005) Automated mapping of Hammond's landforms. *IEEE Geoscience and Remote Sensing Letters*, 2, 384–388.

Gallant, J.C. (2011) Adaptive smoothing for noisy DEMs. In: *Proceedings of the 2nd International Geomorphometry Conference*, Redlands, CA.

Gallant, J.C. & Dowling, T.I. (2003) A multi-resolution index of valley bottom flatness for mapping depositional areas. *Water Resources Research*, 39, 1347–1360.

Gallant, J.C. & Hutchinson, M.F. (1997) Scale dependence in terrain analysis. *Mathematics and Computers in Simulation*, 43, 313–321.

Gallant, J.C. & Hutchinson, M.F. (2011) A differential equation for specific catchment area. *Water Resources Research*, 47, W05535.

Gallant, J.C. & Read, A.M. (2016) A near-global bare-earth DEM from SRTM. *ISPRS International Archives of the Photogrammetry, Remote Sensing and Spatial Information Sciences*, 41, 137–141.

Gallant, J.C. & Wilson, J.P. (1996) TAPES-G: A grid-based terrain analysis program for the environmental sciences. *Computers and Geosciences*, 22, 713–722.

Gallant, J.C. & Wilson, J.P. (2000) Primary topographic attributes. In: J.P. Wilson & J.C. Gallant (eds) *Terrain Analysis: Principles and Applications*, pp. 51–85. New York, NY: John Wiley & Sons, Inc.

Gallant, J.C., Hutchinson, M.F. & Wilson, J.P. (2000) Future directions for terrain analysis. In: J.P. Wilson & J.C. Gallant (eds) *Terrain Analysis: Principles and Applications*, pp. 423–427. New York, NY: John Wiley & Sons, Inc.

Gallant, J.C., Read, A.M. & Dowling, T.I. (2012) Removal of tree offsets from SRTM and other digital surface models. In: *Proceedings of the 22nd Congress of the International Society for Photogrammetry and Remote Sensing*, Melbourne, Australia.

Garbrecht, J. & Martz, L.W. (1994) Grid size dependency of parameters extracted from digital elevation models. *Computers and Geosciences*, 20, 85–87.

Garbrecht, J. & Martz, L.W. (1997) The assignment of drainage direction over flat surfaces in raster digital elevation models. *Journal of Hydrology*, 193, 204–213.

Garbrecht, J. & Starks, P. (1995) Note on the use of USGS Level 1 7.5-minute DEM coverages for landscape drainage analyses. *Photogrammetric Engineering and Remote Sensing*, 61, 519–522.

Garland, M. & Heckbert, P.S. (1995) *Fast Polygonal Approximation of Terrains and Height Fields*. Pittsburgh, PA: Carnegie Mellon University Technical Report No. CMU-CS-95-181.

Gates, D.M. (1980) *Biophysical Ecology*. New York, NY: Springer.

Gedney, N. & Cox, P.M. (2003) The sensitivity of global climate model simulations to the representation of soil moisture heterogeneity. *Journal of Hydrometeorology*, 4, 1265–1275.

Gerçek, D., Toprak, V. & Strobl, J. (2011) Object-based classification of landforms based on their local geometry and geomorphometric context. *International Journal of Geographical Information Science*, 25, 1011–1023.

Gesch, D.B. (2007) The National Elevation Dataset. In: D. Maune (ed.) *Digital Elevation Model Technologies and Applications: The DEM User's Manual*, pp. 99–118. Bethesda, MD: American Society of Photogrammetry and Remote Sensing.

Gesch, D.B., Oimoen, M., Greenlee, S., Nelson, C., Steuck, M. & Tyler, D. (2002) The national elevation dataset. *Photogrammetric Engineering and Remote Sensing*, 68, 5–11.

Gesch, D.B., Oimoen, M., Zhang, Z., Danielson, J. & Meyer, D. (2011) *Validation of the ASTER Global Digital Elevation Model (GDEM) Version 2 over the Conterminous United States*. Sioux Falls, SD: US Geological Survey, Earth Resources Science Center.

Gesch, D.B., Oimoen, M.J. & Evans, G.A. (2014) *Accuracy Assessment of the U.S. Geological Survey National Elevation Dataset, and Comparison with Other Large-Area Elevation Datasets: SRTM and ASTER*. Reston, VA: US Geological Survey Open-File Report 2014-1008.

Gessler, P.E., Moore, I.D., McKenzie, N.J. & Ryan, P.J. (1995) Soil–landscape modeling and spatial prediction of soil attributes. *International Journal of Geographical Information Systems*, 9, 421–432.

Gessler, P.E., McKenzie, N.J. & Hutchinson, M.F. (1996) Progress in soil–landscape modeling and spatial prediction of soil attributes for environmental models. In: *Proceedings of the 3rd International Conference Integrating GIS and Environmental Modeling*, Santa Fe, NM.

Gessler, P.E., Chadwick, O.A., Chamron, F., Holmes, K. & Althouse, L. (2000) Modeling soil–landscape and ecosystem properties using terrain attributes. *Soil Science Society of America Journal*, 64, 2046–2056.

Gessler, P., Pike, R., MacMillan, R.A., Hengl, T. & Reuter, H.I. (2009) The future of geomorphometry. In: T. Hengl & H.I. Reuter (eds) *Geomorphometry: Concepts, Software, Applications*, pp. 31–63. Amsterdam, Netherlands: Elsevier.

Gironás, J., Niemann, J.D., Roesner, L.A., Rodriguez, F. & Andrieu, H. (2010) Evaluation of methods for representing terrain in storm-water modeling. *Journal of Hydraulic Engineering*, 15, 1–14.

Glenn, N.F., Streutker, D.R., Chadwick, D.J., Thackray, G.D. & Dorsch, S.J. (2006) Analysis of LiDAR-derived topographic information for characterizing and differentiating landslide morphology and activity. *Geomorphology*, 73, 131–148.

Gold, A.J., Groffman, P.M., Addy, K., Kellogg, D.Q., Stolt, M. & Rosenblatt, A.E. (2001) Landscape attributes as controls on ground water nitrate removal capacity of riprarian zones. *Journal of the American Water Resources Assoication*, 37, 1457–1464.

Goldberg, D.E. (1989) *Genetic Algorithms in Search, Optimization, and Machine Learning*. Reading, MA: Addison-Wesley.

Gonga-Saholiariliva, N., Gunnell, Y., Petit, C. & Mering, C. (2011) Techniques for quantifying the accuracy of gridded models and for mapping uncertainty in digital terrain analysis. *Progress in Physical Geography*, 35, 739–764.

Goodchild, M.F. (1980) Fractals and the accuracy of geographical measures. *Mathematical Geology*, 12, 85–98.

Goodchild, M.F. (2001) Metrics of scale in remote sensing and GIS. *International Journal of Applied Earth Observation and Geoinformation*, 3, 114–120.

Goodchild, M.F. (2011) Scale in GIS: An overview. *Geomorphology*, 130, 5–9.

Goodchild, M.F. & Proctor, J. (1997) Scale in a digital geographic world. *Geographical and Environmental Modeling*, 1, 5–23.

Goodchild, M.F., Yuan, M. & Cova, T.J. (2007) Towards a general theory of geographic representation in GIS. *International Journal of Geographical Information Science*, 21, 239–260.

Goovaerts, P. (1997) *Geostatistics for Natural Resources Evaluation*. New York, NY: Oxford University Press.

Gorte, B.G.H. & Koolhoven, W. (1990) Interpolation between isolines based on the Borgefors distance trasnform. *ITC Journal*, 1, 245–247.

Grabs, T., Seibert, J., Bishop, K. & Laudon, H. (2009) Modeling spatial patterns of saturated areas: A comparison of the topographic wetness index and a dynamic distributed model. *Journal of Hydrology*, 373, 15–23.

Graham, L. (2005) The LAS 1.1 standard. *Photogrammetric Engineering and Remote Sensing*, 71, 777–781.

Graser, A. (2016) *Learning QGIS*, 3rd edn. Birmingham, UK: Packt Publishing.

Grayson, R.B., Blöschl, G., Barling, R.D. & Moore, I.D. (1993) Process, scale, and constraints to hydrological modeling in GIS. In: K. Kovar & H.P. Nachtnebel (eds) *Application of Geographic Information Systems in Hydrology and Water Resources: Proceedings of the HydroGIS Conference Held in Vienna, April 1993*, pp. 83–92. Wallingford, UK: International Association of Hydrological Sciences Publication No. 211.

Grimaldi, S., Nardi, F., Di Benedetto, F., Istanbulluoglu, E. & Bras, R.L. (2007) A physically-based method for removing pits in digital elevation models. *Advances in Water Resources*, 30, 2151–2158.

Grohmann, C.H., Smith, M.J. & Riccomini, C. (2011) Multiscale analysis of topographic surface roughness in the Midland Valley, Scotland. *IEEE Transactions on Geoscience and Remote Sensing*, 49, 1200–1213.

Gruber, A., Wessel, B., Huber, M. & Roth, A. (2013) Operational TanDEM-X DEM calibration and first validation results. *ISPRS Journal of Photogrammetry and Remote Sensing*, 73, 39–49.

Gruber, S. (2007) MTD: A mass-conserving algorithm to parameterize gravitational transport and deposition processes using digital elevation models. *Water Resources Research*, 43, W06412.

Gruber, S. & Peckham, S. (2009) Land-surface parameters and objects in hydrology. In: T. Hengl & H.I. Reuter (eds) *Geomorphometry: Concepts, Software, Applications*, pp. 171–194. Amsterdam, Netherlands: Elsevier.

Gruber, S., Huggel, C. & Pike, R. (2009) Modeling mass movements and landslide susceptibility. In: T. Hengl & H.I. Reuter (eds) *Geomorphometry: Concepts, Software, Applications*, pp. 527–549. Amsterdam, Netherlands: Elsevier.

Guisan, A., Theurillat, J. & Kienast, F. (1998) Predicting the potential distribution of plant species in an alpine environment. *Journal of Vegetation Science*, 9, 65–74.

Güntner, A., Seibert, J. & Uhlenbrook, S. (2004) Modeling spatial patterns of saturated areas: An evaluation of different terrain indices. *Water Resources Research*, 40, W05114.

Gupta, V.K., Waymire, E.C. & Wang, C.T. (1980) A representation of an instantaneous unit hydrograph from geomorphology. *Water Resources Research*, 16, 855–862.

Gustavsson, M. & Kolstrup, E. (2009) New geomorphological mapping system used at different scales in a Swedish glaciated area. *Geomorphology*, 110, 37–44.

Guth, P.L. (1995) Slope and aspect calculations on DEMs. *Zeitschrift für Geomorphologie NF, Suppl-Bd*, 101, 31–52.

Guth, P.L. (2009) Geomorphometry in MicroDEM. In: T. Hengl & H.I. Reuter (eds) *Geomorphometry: Concepts, Software, Applications*, pp. 351–366. Amsterdam, Netherlands: Elsevier.

Guth, P.L. (2010) Slope, reflectance, and viewshed algorithms for arc-second digital elevation models. In: *Proceedings of the Annual American Society of Photogrammetry and Remote Sensing Conference*, San Diego, CA.

Guth, P.L. (2013) The Giga revolution in geomorphometry: Gigabytes of RAM, gigabyte-sized data sets, and gigabit internet access. In: *Proceedings of the 3rd International Conference on Geomorphometry*, Nanjing, China.

Guzzetti, F. & Reichenbach, P. (1994) Toward the definition of topographic divisions for Italy. *Geomorphology*, 11, 57–75.

Guzzetti, F., Reichenbach, P., Cardinali, M., Galli, M. & Ardizzone, F. (2005) Probabilistic landslide hazard assessment at the basin scale. *Geomorphology*, 72, 272–299.

Gyasi-Agyai, Y.G., Wilgoose, G. & de Troch, F.P. (1995) Effects of vertical resolution and map scale of digital elevation models on geomorphological parameters used in hydrology. *Hydrological Processes*, 9, 363–382.

Hammond, E.H. (1954) Small-scale continental landform maps. *Annals of the Association of American Geographers*, 44, 33–42.

Hammond, E.H. (1964) Analysis of properties in land form geography: An application to broad-scale land form mapping. *Annals of the Association of American Geographers*, 54, 11–19.

Hammond, E.H. (1965) What is a landform? Some further comments. *Professional Geographer*, 17, 12–13.

Hancock, G.R. (2006) The impact of different gridding methods on catchment geomorphology and soil erosion over long timescales using a landscape evolution model. *Earth Surface Processes and Landforms*, 31, 10–35.

Hancock, G.R. (2008) The impact of depression removal on catchment geomorphology, soil erosion, and landscape evolution. *Earth Surface Processes and Landforms*, 33, 459–474.

Hancock, G.R. & Evans, K.G. (2006) Channel head location and characteristics using digital elevation models. *Earth Surface Processes and Landforms*, 31, 809–824.

Hansen, M.C., Potapov, P.V., Moore, R., Hancher, M., Turubanova, S.A., Tyukavina, A. et al. (2013) High-resolution global maps of 21st-century forest cover change. *Science*, 342, 850–853.

Harding, D.J., Bufton, J.L. & Frawley, J. (1994) Satellite laser altimetry of terrestrial topography: Vertical accuracy as a function of surface slope, roughness, and cloud cover. *IEEE Transactions on Geoscience and Remote Sensing*, 32, 329–339.

Hasan, A., Pilesjö, P. & Persson, A. (2013a) Drainage area estimate in practice: How to tackle artifacts in real world data. In: *Proceedings of the GIS Ostrava 2012: Surface Models for Geosciences Conference*. Ostrava, Czech Republic.

Hasan, A., Pilesjö, P. & Persson, A. (2013b) On generating digital elevation models from LiDAR data: Resolution versus accuracy and topographic wetness indices in northern peatlands. *Geodesy and Cartography*, 38, 57–69.

Hastings, D.A. & Dunbar, P.K. (1998) Development and assessment of the Global Land One-km Base Elevation digital elevation model (GLOBE). *ISPRS Archives*, 32, 218–221.

Hebeler, F. & Purves, R.S. (2009) The influence of elevation uncertainty on derivation of topographic indices. *Geomorphology*, 111, 4–16.

Heckbert, P.S. & Garland, M. (1997) *Survey of Polygonal Surface Simplification Algorithms*. Pittsburgh, PA: Carnegie Mellon University, School of Computer Science Technical Report.

Heller, M. (1990) Triangulation algorithms for adaptive terrain modeling. In: *Proceedings of the 4th International Symposium on Spatial Data Handling*, Vol. 1, pp. 163–174. Zurich, Switzerland.

Hengl, T. (2006) Finding the right pixel size. *Computers and Geosciences*, 32, 1283–1298.

Hengl, T. & Evans, I.S. (2009) Mathematical and digital models of the land surface. In: T. Hengl & H.I. Reuter (eds) *Geomorphometry: Concepts, Software, Applications*, pp. 31–63. Amsterdam, Netherlands: Elsevier.

Hengl, T. & MacMillan, R.A. (2009) Geomorphometry: A key to landscape mapping and modeling. In: T. Hengl & H.I. Reuter (eds) *Geomorphometry: Concepts, Software, Applications*, pp. 433–460. Amsterdam, Netherlands: Elsevier.

Hengl, T. & Reuter, H.I. (eds) (2009) *Geomorphometry: Concepts, Software, Applications*. Amsterdam, Netherlands: Elsevier.

Hengl, T., Gruber, S. & Shrestha, D.P. (2003) Digital terrain analysis in ILWIS. Unpublished lecture notes, International Institute for Geo-Information Science and Earth Observation (ITC), Enschede, Netherlands.

Hengl, T., Gruber, S. & Shrestha, D.P. (2004) Reduction of errors in digital terrain parameters used in soil-landscape modeling. *International Journal of Applied Earth Observation and Geoinformation*, 5, 97–112.

Hengl, T., Maathuis, B.H.P. & Wang, L. (2009) Geomorphometry in ILWIS. In: T. Hengl & H.I. Reuter (eds) *Geomorphometry: Concepts, Software, Applications*, pp. 309–331. Amsterdam, Netherlands: Elsevier.

Hengl, T., Heuvelink, G.B.M. & van Loon, E.E. (2010) On the uncertainty of stream networks derived from elevation data: The error propagation approach. *Hydrology and Earth System Sciences*, 14, 1153–1165.

Heritage, G.L. & Milan, D.J. (2009) Terrestrial laser scanning of grain roughness in a gravel bed river. *Geomorphology*, 113, 4–11.

Heritage G.L., Milan, D.J., Large, A.R.G. & Fuller, I.C. (2009) Influence of survey strategy and interpolation model on DEM quality. *Geomorphology*, 112, 334–344.

Hernández Encinas, A., Hernández Encinas, L., Hoya White, S., Martin del Rey, A. & Rodriquez Sanchez, G. (2007) Simulation of forest fire fronts using cellular automata. *Advances in Engineering Software*, 38, 372–378

Hetrick, W.A., Rich, P.M., Barnes, F.J. & Weiss, S.B. (1993a) GIS-based solar radiation flux models. In: *Proceedings of the ASPRS-ACSM Annual Convention*, Vol. 3, pp. 132–143. New Orleans, LA: ACSM.

Hetrick, W.A., Rich, P.M. & Weiss, S.B. (1993b) Modeling insolation on complex surfaces. In: *Proceedings of the 13th Esri International User Conference*, Vol. 2, pp. 447–458. Palm Springs, CA: Esri.

Hickey, R. (2000) Slope angle and slope length solutions for GIS. *Cartography*, 29, 1–8.

Hirano, A., Welch, R. & Lang, H. (2003) Mapping from ASTER stereo image data: DEM validation and accuracy assessment. *ISPRS Journal of Photogrammetry and Remote Sensing*, 57, 356–370.

Hirt, C., Filmer, M.S. & Featherstone, W.E. (2010) Comparison and validation of the recent freely available ASTER GDEM ver1, SRTM ver4.1, and GEODATA DEM-9S ver3 digital elevation models over Australia. *Australian Journal of Earth Sciences*, 57, 337–347.

Hjerdt, K.N., McDonnell, J.J., Seibert, J. & Rodhe, A. (2004) A new topographic index to quantify downslope controls on local drainage. *Water Resources Research*, 40, W05602.

Hobson, R.D. (1972) Surface roughness in topography: A quantitative approach. In: R.J. Chorley (ed.) *Spatial Analysis in Geomorphology*, pp. 221–245. London, UK: Harper & Row.

Hodgson, M.E. (1995) What cell size does the computed slope/aspect angle represent? *Photogrammetric Engineering and Remote Sensing*, 61, 513–517.

Hodgson, M.E. & Alexander, B.E. (1990) Use of historic maps in GIS analyses. In: *Proceedings of the ASPRS-ACSM Annual Convention*, pp. 109–116. Denver, CO: ACSM.

Hodgson, M.E. & Bresnahan, P. (2004) Accuracy of airborne LiDAR-derived elevation: Empirical assessment and error budget. *Photogrammetric Engineering and Remote Sensing*, 70, 331–333.

Hodgson, M.E., Jensen, J.R., Schmidt, L., Schill, S. & Davis, B. (2003) An evaluation of LiDAR- and IFSAR-derived digital elevation models in leaf-on conditions with USGS Level 1 and Level 2 DEMs. *Remote Sensing of Environment*, 84, 295–308.

Hodgson, M.E., Jensen, J.R., Raber, G., Tullis, J., Davis, B., Schuckman, K. & Thompson, G. (2005) An evaluation of LiDAR-derived elevation and terrain slope in leaf-off conditions. *Photogrammetric Engineering and Remote Sensing*, 71, 817–823

Hoechstetter, S., Thinh, N.X. & Walz, U. (2006) 3D indices for the analysis of spatial patterns of landscape structure. In: *Proceedings of the 12th International Conference on GIS and Sustainable Development*, pp. 108–118. Berlin, Germany.

Hoechstetter, S., Walz, U., Dang, L.H. & Thinh, N.X. (2008) Effects of topography and surface roughness in analyses of landscape structure: A proposal to modify the existing set of landscape metrics. *Landscape Online*, 1, 1–14.

Hofierka, J. (1997) Direct solar radiation within an open GIS environment. In: *Proceedings of the 1997 Joint European GI Conference*, pp. 575–584. Vienna, Austria.

Hofierka, J., Mitaášová, H. & Neteler, M. (2009) Geomorphometry in GRASS GIS. In: T. Hengl & H.I. Reuter (eds) *Geomorphometry: Concepts, Software, Applications*, pp. 387–410. Amsterdam, Netherlands: Elsevier.

Hofton, M., Dubayah, R., Blair, J.B. & Rabine, D. (2006) Validation of SRTM elevations over vegetated and non-vegetated terrain using medium footprinting LiDAR. *Photogrammetric Engineering and Remote Sensing*, 72, 279–285.

Höhle, J. & Höhle, M. (2009) Accuracy assessment of digital elevation models by means of robust statistical methods. *ISPRS Journal of Photogrammetry and Remote Sensing*, 64, 398–406.

Hollenhorst, T., Host, G. & Johnson, L. (2008) Scaling issues in mapping riparian zones with remote sensing data: Quantifying errors and sources of uncertainty. In: J. Wu, K.B. Jones, H. Li & O.J. Loucks (eds) *Scaling and Uncertainty Issues in Ecology*, pp. 275–295. Berlin, Germany: Springer.

Hollis, J.M., Hannam, J. & Bellamy, P.H. (2012) Empirically-derived pedotransfer functions for predicting bulk density in European soils. *European Journal of Soil Science*, 63, 96–109.

Holmes, K.W., Chadwick, O.A. & Kyriakidis, P.C. (2000) Error in a USGS 30 m digital elevation model and its impact on digital terrain modeling. *Journal of Hydrology*, 233, 154–173.

Holmgren, P. (1994) Multiple flow direction algorithms for runoff modeling in grid based elevation models: An empirical evaluation. *Hydrological Processes*, 8, 327–334.

Homer, C.G., Dewitz, J.A., Yang, L., Jin, S., Danielson, P., Xian, G., Coulston, J., Herold, N., Wickham, J. & Megown, K. (2015) Completion of the 2011 National Land Cover Database for the conterminous United States: Representing a decade of land cover change information. *Photogrammetric Engineering and Remote Sensing*, 81, 345–354.

Horn, B.K.P. (1981) Hill shading and the reflectance map. *Proceedings of the Institute of Electrical and Electronic Engineers*, 69, 14–47.

Horton, R.E. (1932) Drainage basin characteristics. *Transactions of the American Geophysical Union*, 14, 350–361.

Horsburgh, J.S., Morsy, M.M., Castronova, A.M., Goodall, J.L., Gan, T., Yi, H., Stealey, M.J. & Tarboton, D.G. (2016) Hydroshare: Sharing diverse environmental data types and models as social objects with application to the hydrology domain. *Journal of the American Water Resources Association*, 52, 873–889.

Howard, A.D. (1990) Role of hypsometry and planform in basin hydrologic response. *Hydrological Processes*, 4, 373–385.

Howard, A.D. (1994) A detachment-limited model of drainage basin evolution. *Water Resources Research*, 30, 2261–2286.

Howard, A. & Kerby, G. (1983) Channel changes in badlands. *Geological Society of America Bulletin*, 94, 739–752.

Hrvatin, M. & Perko, D. (2009) Suitability of Hammond's method for determining landform units in Slovenia. *Acta Geographica Slovenica*, 49, 343–366.

Hu, P., Liu, X. & Hu, H. (2009a) Isomorphism in digital elevation models and its implication to interpolation functions. *Photogrammetric Engineering and Remote Sensing*, 75, 713–721.

Hu, P., Liu, X. & Hu, H. (2009b) Accuracy assessment of digital elevation models based on approximation theory. *Photogrammetric Engineering and Remote Sensing*, 75, 49–56.

Huggett, R. (1975) Soil landscape systems: A model of soil genesis. *Geoderma*, 13, 1–22.

Hughes, M., Lastra, A.A. & Saxe, E. (1996) Simplification of global-illumination meshes. *Computer Graphics Forum*, 15, 339–345.

Hungerford, R.D., Nemani, R.R., Running, S.W. & Coughlan, J.C. (1989) *MTCLIM: A Mountain Microclimate Simulation Model*. Ogden, UT: US Department of Agriculture, Forest Service, Intermountain Research Station Research Paper No. INT-414.

Hunter, G.J. & Goodchild, M.F. (1997) Modeling the uncertainty of slope and aspect estimates derived from spatial databases. *Geographical Analysis*, 29, 35–49.

Hutchinson, M.F. (1988) Calculation of hydrologically sound digital elevation models. In: *Proceedings of the 3rd International Symposium on Spatial Data Handling*, pp. 117–133. Sydney, Australia.

Hutchinson, M.F. (1989) A new procedure for gridding elevation and stream line data with automatic removal of spurious pits. *Journal of Hydrology*, 106, 211–232.

Hutchinson, M.F. (1995) Interpolating mean rainfall using thin plate smoothing splines. *International Journal of Geographical Information Science*, 9, 385–403.

Hutchinson, M.F. (1996) A locally adaptive approach to the interpolation of digital elevation models. In: *Proceedings of the 3rd International Conference on Integrating GIS and Environmental Modeling*, Santa Fe, NM.

Hutchinson, M.F. (2000) Optimizing the degree of data smoothing for locally adaptive finite element bivariate smoothing splines. *ANZIAM Journal*, 42, C774–C796.

Hutchinson, M.F. (2008) Adding the Z-dimension. In: J.P. Wilson & A.S. Fotheringham (eds) *The Handbook of Geographic Information Science*, pp. 144–168. Oxford, UK: Blackwell.

Hutchinson, M.F. (2011) ANUDEM Version 5.3. Retrieved from http://fennerschool.anu.edu.au/research/products/anudem-vrsn-53

Hutchinson, M.F. & Gallant, J.C. (2000) Digital elevation models and representation of terrain shape. In: J.P. Wilson & J.C. Gallant (eds) *Terrain Analysis: Principles and Applications*, pp. 29–50. New York, NY: John Wiley & Sons, Inc.

Hutchinson, M.F., Stein J.A., Stein J.L. & Xu T. (2009) Locally adaptive gridding of noisy high resolution topographic data. In: R.S. Anderson, R.D. Braddock & L.T.H. Newham (eds) *Eighteenth World IMACS Congress and MODSIM09 International Congress on Modeling and Simulation*, pp. 2493–2499. Canberra, Australia: Modelling and Simulation Society of Australia and New Zealand and International Association for Mathematics and Computers in Simulation.

Hutchinson, M.F., Stein, J.L., Gallant, J.C. & Dowling, T.I. (2013) New methods for incorporating and analyzing drainage structure in digital elevation models. In: *Proceedings of the 3rd International Conference on Geomorphometry*. Nanjing, China. Available at http://geomorphometry.org/Hutchinson2013

Idso, S.B. (1969) Atmospheric attenuation of solar radiation. *Journal of the Atmospheric Sciences*, 26, 1088–1095.

Iqbal, J., Read, J.J., Thomasson, A.J. & Jenkins, J.N. (2005) Relationships between soil-landscape and dryland cotton lint yield. *Soil Science Society of America Journal*, 69, 872–882.

Irvin, B.J., Ventura, S.J. & Slater, B.K. (1997) Fuzzy and isodata classification of landform elements from digital terrain data in Pleasant Valley, Wisconsin. *Geoderma*, 77, 137–154.

Issacson, D.L. & Ripple, W.J. (1991) Comparison of 7.5 minute and 1 degree digital elevation models. *Photogrammetric Engineering and Remote Sensing*, 56, 1523–1527.

Istanbulluoghu, E., Tarboton, D.G., Pack, R.T. & Luce, C. (2002) A probabilistic approach to channel initiation. *Water Resources Research*, 38, WR000782.

Iverson, L.R., Scott, C.T., Dale, M. & Prasad, A.M. (1996) Develoment of an integrated moisture index for predicting species composition. In: M. Kohl & G.Z. Gertner (eds) *Caring for the Forest: Research in a Changing World, Statistics, Mathematics, and Computers*, pp. 101–116. Birmensdorf, Switzerland: Swiss Federal Institute for Forest, Snow and Landscape Research.

Iverson, L.R., Dale, M.E., Scott, C.T. & Prasad, A. (1997) A GIS-derived integrated moisture index to predict forest composition and productivity of Ohio forests (USA). *Landscape Ecology*, 12, 331–348.

Iverson, L.R., Prasad, A.M. & Rebbeck, J. (2004) A comparison of the integrated moisture index and the topographic wetness index as related to two years of soil moisture monitoring in Zaleski State Forest, Ohio. In: D.A. Yaussy, D.M. Hix, R.P. Long & P.C. Goebel (eds) *Proceedings of the 14th Central Hardwood Forest Conference*, pp. 515–517. Newtown Square, PA: US Department of Agriculture, Forest Service, Northeastern Research Station General Technical Report No. NE-316.

Iverson, R.M., Schilling, S.P. & Vallance, J.W. (2003) Objective delineation of lahar-inundation hazard zones. *Geological Society of America Bulletin*, 110, 972–984.

Iwahashi, J. & Pike, R.J. (2007) Automated classifications of topography from DEMs by an unsupervised nested-means algorithm and a three-part geometric signature. *Geomorphology*, 86, 409–440.

Jaboyedoff, M., Oppikofer, T., Abellán, A., Derron, M.-H., Loye, A., Metzger, R. & Pedrazzini, A. (2012) Use of LiDAR in landslide investigations: A review. *Natural Hazards*, 61, 5–28.

Jacoby, B.S., Peterson, E.W. & Dogwiler, T. (2011) Identifying the stream erosion potential of cave levels in Carter Cave State Resort Park, Kentucky. *Journal of Geographic Information Systems*, 3, 323–333.

James, L.A., Watson, D.G. & Hansen, W.F. (2007) Using LiDAR to map gullies and headwater streams under forest canopy: South Carolina, USA. *Catena*, 71, 132–144.

James, L.A., Hodgson, M.E., Ghoshal, S. & Latiolais, M.M. (2012) Geomorphic change detection using historic maps and DEM differencing: The temporal dimension of geospatial analysis. *Geomorphology*, 137, 181–198.

Jana, R.B. & Mohanty, B.P. (2011) Enhancing PTFs with remotely sensed data for multi-scale soil water retention estimates. *Journal of Hydrology*, 399, 201–211.

Jarvis, A., Rubiano, J., Nelson, A., Farrow, A. & Mulligan, M. (2004) *Practical Use of SRTM Data in the Tropics: Comparisons with Digital Elevation Models Generated from Cartographic Data*. Cali, Colombia: International Centre for Tropical Agriculture Working Document No. 198.

Jarvis, A., Reuter, H., Nelson, A. & Guevara, E. (2006) Void-filled seamless SRTM data (Version 3). Washington, DC: CGIAR-CSI Consortium for Spatial Information.

Jarvis, C.H. & Stuart, N. (2001) A comparison among strategies for interpolating maximum and minimum daily air temperatures. *Journal of Applied Meteorology*, 40, 1075–1084.

Jasiewicz, J. & Stepinski, T.F. (2013) Geomorphons: A pattern recognition approach to classification and mapping of landforms. *Geomorphology*, 182, 147–156.

Jasiewicz, J., Netzel, P. & Stepinski, T.F. (2015) GeoPAT: A toolbox for pattern-based information retrieval from large geospatial databases. *Computers and Geosciences*, 80, 62–73.

Jefferson, A.J. & McGee, R.W. (2013) Channel network extent in the context of historical land use, flow generation processes, and landscape evoluition in the North Carolina Piedmont. *Earth Surfaces Processes and Landforms*, 38, 601–613.

Jellema, A., Stobbelaar, D.-J., Groot, J.C.J. & Rossing, W.A.H. (2009) Landscape character assessment using region growing techniques in geographical information systems. *Journal of Environmental Management*, 90, S161–S174.

Jencso, K.G. & McGlynn, B.L. (2011) Hierarchical controls on runoff generation: Topographically driven hydrologic connectivity, geology, and vegetation. *Water Resources Research*, 47, W11527.

Jencso, K.G., McGlynn, B.L., Gooseff, M.N., Wondzell, S.M., Bencala, K.E. & Marshall, L.A. (2009) Hydrologic connectivity between landscapes and streams: Transferring reach- and plot-scale understanding to the catchment scale. *Water Resources Research*, 45, W07225.

Jenness, J.S. (2004) Calculating landscape surface area from gridded elevation models. *Wildlife Society Bulletin*, 32, 829–839.

Jenson, S.K. & Domingue, J.O. (1988) Extracting topographic attributes from digital elevation data for geographical information system analysis. *Photogrammetric Engineering and Remote Sensing*, 54, 1593–1600.

Jersey, J.K. (1993) *Assessing vegetation patterns and hydrologic characteristics in a semi-arid environment using a geographic information system and terrain-based models*. Unpublished MS thesis, Montana State University, Bozeman, MT.

Jessop, D.E., Kelfoun, K., Labazuy, P., Mangeney, A., Roche, O., Tillier, J.-L., Trouillet, M. & Thibault, G. (2012) LiDAR derived morphology of the 1993 Lascar pyroclastic flow deposits, and implication for flow dynamics and rheology. *Journal of Volcanology and Geothermal Research*, 245–246, 81–97.

Jiang, H. & Eastman, J. R (1996) Application of fuzzy measures in multi-criteria evaluation in GIS. *International Journal of Geographical Information Science*, 14, 173–184.

Jiang, R.-Q. & Tang, G.-A. (2015) A method of depression filling with consideration of local micro-relief features. In: *Proceedings of the 4th International Conference on Geomorphometry*, Poznań, Poland.

Jobin, T., Prasannakumar, V. & Vineetha, P. (2015) Suitability of spaceborne digital elevation models of different scales in topographic analysis: An example from Kerala, India. *Environmental Earth Sciences*, 73, 1245–1263.

Jones, A.F., Brewer, P.A., Johnstone, E. & Macklin, M.G. (2007) High resolution interpretative geomorphological mapping of river valley environments using airborne LiDAR data. *Earth Surface Processes and Landforms*, 21, 1574–1592.

Jones, J.A. (1986) Some limitations to the a/s index for predicting basin-wide patterns of soil water drainage. *Zeitschrift für Geomorphologie*, 60, 7–20.

Jones, J.A. (1987) The initiation of natural drainage networks. *Progress in Physical Geography*, 11, 205–245.

Jones, K.H. (1998) A comparison of algorithms used to compute hill slope as a property of the DEM. *Computers and Geosciences*, 24, 315–323.

Jones, N.L., Wright, S.G. & Maidment, D.R. (1990) Watershed delineation with triangle-based terrain models. *Journal of Hydraulic Engineering*, 116, 1232–1251.

Junk, W., Piedade, M., Schöngart, J., Cohn-Haft, M., Adeney, J.M. & Wittmann, F. (2011) A classification of major naturally-occurring Amazonian Lowland wetlands. *Wetlands*, 31, 623–640.

Kalbermatten, M., Van De Ville, D., Turberg, P., Tuia, D. & Joost, S. (2012) Multiscale analysis of geomorphological and geological features in high resolution digital elevation models using the wavelet transform. *Geomorphology*, 138, 352–363.

Karagulle, D., Frye, C., Sayre, R., Breyer, S., Aniello, P., Vaughan, R. & Wright, D. (2017) Modeling global Hammond landform regions from 250 m elevation data. *Transactions in GIS*, 21, 1040–1060.

Kasai, M., Ikeda, M., Asahina, T. & Fujisawa, K. (2009) LiDAR-derived DEM evaluation of deep-seated landslides in a steep and rocky region of Japan. *Geomorphology*, 113, 57–69.

Katzenbeisser, R. (2003) On the calibration of LiDAR sensors. In: H.-G. Maas, G. Vosselman & A. Streilein (eds) *3-D Reconstruction from Airborne Laserscanner and InSAR Data*, pp. 59–64. Enschede, Netherlands: Institute of Photogrammetry and Remote Sensing, Faculty of GeoInformation Science and Earth Observation, University of Twente.

Kelly, R.E., McConnell, P.R. & Mildenberger, S.J. (1978) The Gestalt photomapping system. *Photogrammetric Engineering and Remote Sensing*, 43, 1407–1417.

Kennelly, P.J. (2008) Terrain maps displaying hill-shading with curvature. *Geomorphology*, 102, 567–577.

Kenny, F. & Matthews, B. (2005) A methodology for aligning raster flow direction data with photogrammetrically mapped hydrology. *Computers and Geosciences*, 31, 768–779.

Kenny, F., Matthews, B. & Todd, K. (2008) Routing overland flow through sinks and flats in interpolated raster terrain surfaces. *Computers and Geosciences*, 34, 1417–1430.

Kereszturi, G., Procter, J., Cronin, S.J., Németh, K., Bebbington, M. & Lindsay, J. (2012) LiDAR-based quantification of lava flow susceptibility in the City of Auckland (New Zealand). *Remote Sensing of Environment*, 125, 198–213.

Kheir, R.B., Wilson, J.P. & Deng, Y.X. (2007) Use of terrain variables for mapping gully erosion susceptibility in Lebanon. *Earth Surface Processes and Landforms*, 32, 1770–1782.

Kidner, D.B. (2003) Higher-order interpolation of regular grid digital elevation models. *International Journal of Remote Sensing*, 24, 2981–2987.

Kienzle, S. (2004) The effect of DEM raster resolution on first order, second order, and compound terrain derivatives. *Transactions in GIS*, 8, 83–111.

Kim, S. & Lee, H. (2004) A digital elevation analysis: A spatially distributed flow apportioning algorithm. *Hydrological Processes*, 18, 1777–1794.

Kirby, M.J. (1976) Tests of the random network model, and its application to basin hydrology. *Earth Surface Processes*, 1, 197–212.

Kleinen, T., Brovkin, V. & Schuldt, R.J. (2012) A dynamic model of wetland extent and peat accumulation: Results for the Holocene. *Biogeosciences*, 9, 235–248.

Klinkenberg, B. (1994) A review of methods used to determine the fractal dimension of linear features. *Mathematical Geology*, 26, 23–46.

Klir, G.J. & Yuan, B. (1995) *Fuzzy Sets and Fuzzy Logic: Theory and Applications*. Upper Saddle Creek, NJ: Prentice-Hall.

Kondratyev, K.Y. (1969) *Radiation Regime of Inclined Surfaces*. Geneva, Switzerland: World Meterological Organization Technical Note No. 152.

Konecny, G., Lohmann, P., Engel, H. & Kruck, E. (1987) Evaluation of SPOT imagery on analytical instruments. *Photogrammetric Engineering and Remote Sensing*, 53, 1223–1230.

Koons, P., Upton, P. & Barker, A.D. (2012) The influence of mechanical properties on the link between tectonic and topographic evolution. *Geomorphology*, 137, 168–180.

Kopecký, M. & Čížková, Š. (2010) Using topographic wetness index in vegetation ecology: Does the algorithm matter? *Applied Vegetation Science*, 13, 450–459.

Kopecký, M. & Vojta, J. (2009) Land use legacies in post-agricultural forests in the Doupovské Mountains, Czech Republic. *Applied Vegetation Science*, 12, 251–260.

Köthe, R. & Lehmeier, F. (1993) SAGA: Ein Programmsystem zur Automatischen Relief-Analyse. *Zeitschrift für Angewandte Geographie*, 4, 11–21.

Krebs, P., Stocker, M., Pezzatti, G.B. & Conedera, M. (2015) An alternative approach to transverse and profile terrain curvature. *International Journal of Geographical Information Science*, 29, 643–656.

Kreznor, W.R., Olson, K.R., Banwart, W.L. & Johnson, D.L. (1989) Soil landscape and erosion relationships in a northwest Illinois watershed. *Soil Science Society of America Journal*, 53, 1763–1771.

Krieger, T., Curtis, W. & Haase, J. (2010) *Global Validation of the ASTER Global Digital Elevation Model (GDEM) Version 2*. Springfield, VA: National Geospatial-Intelligence Agency.

Kumar, L., Skidmore, A.K. & Knowles, E. (1997) Modelling topographic variation in solar radiation in a GIS environment. *International Journal of Geographical Information Science*, 11, 475–497.

Kumar, S.V., Peters-Lidard, C.D., Santanello, J., Harrison, K., Liu, Y. & Shaw, M. (2012) Land surface Verification Toolkit (LVT): A generalized framework for land surface model evaluation. *Geoscientific Model Development*, 5, 869–886.

Kumler, M.P. (1994) An intensive comparison of triangulated irregular networks and digital elevation models. *Cartographica*, 31, 2, 1–99.

Kyriakidis, P.C., Shortridge, A.M. & Goodchild, M.F. (1999) Geostatistics for conflation and accuracy assessment of digital elevation models. *International Journal of Geographical Information Science*, 13, 677–707.

La Barbera, P. & Rosso, R. (1989) On the fractal dimension of stream networks. *Water Resources Research*, 25, 735–741.

Lagacherie, P., Moussa, R., Cormary, D. & Molenat, J. (1993) Effects of DEM data source and sampling pattern on topographic parameters and on a topography-based hydrological model. In: K. Kovar & H.P. Nachtnebel (eds) *Application of Geographic Information Systems in Hydrology and Water Resources: Proceedings of the HydroGIS Conference held in Vienna, April 1993*, pp. 191–199. Wallingford, UK: International Association of Hydrological Sciences Publication No. 211.

LaLonde, T., Shortridge, A. & Messina, J. (2010) The influence of land cover on Shuttle Radar Topography Mission (SRTM) elevations in low-relief areas. *Transactions in GIS*, 14, 461–479.

Lane, S.N., Brookes, C.J., Kirkby, M.J. & Holden, J. (2004) A network-index-based version of TOPMODEL for use with high-resolution digital topographic data. *Hydrological Processes*, 18, 191–201.

Lane, S.N., Westaway, R.M. & Hicks, D.M. (2003) Estimation of erosion and deposition volumes in a large gravel-bed, braided river using synoptic remote sensing. *Earth Surface Processes and Landforms*, 28, 249–271.

Lane, S.N., Reaney, S.M. & Heathwaite, A.L. (2009) Representation of landscape hydrological connectivity using a topographically-driven surface flow index. *Water Resources Research*, 45, W08423.

Lanni, C., McDonnell, J.J. & Rigon, R. (2011) On the relative role of upslope and downslope topography for describing water flow path and storage dynamics: A theoretical analysis. *Hydrological Processes*, 25, 3909–3923.

Lassueur, T., Joost, S. & Randin, C.F. (2006) Very high resolution digital elevation models: Do they improve models of plant species distribution? *Ecological Modelling*, 198, 139–153.

Lawhead, J. (2015) *QGIS Python Programming Cookbook*. Birmingham, UK: Packt Publishing.

Lea, N.L. (1992) An aspect driven kinematic routing algorithm. In: A.J. Parsons & A.D. Abrahams (eds) *Overland Flow: Hydraulics and Erosion Mechanics*, pp. 147–175. London, UK: Chapman & Hall.

Leathwick, J.R. (1995) Climatic relationships of some New Zealand forest tree species. *Journal of Vegetation Science*, 6, 237–248.

Lee, D.T. & Schachter, B.J. (1980) Two algorithms for constructing a Delaunay triangulation. *International Journal of Parallel Programming*, 9, 219–242.

Lee, I.-S., Chang, H.-C. & Ge, L. (2005) GPS campaigns for validation of InSAR derived DEMs. *Journal of Global Positioning Systems*, 4, 82–87.

Lee, J. (1991) Comparison of existing methods for building triangular irregular network models of terrain from grid digital elevation models. *International Journal of Geographical Information Systems*, 5, 267–285.

Lee, R. (1978) *Forest Microclimatology*. New York, NY: Columbia University Press.

Leempoel, K., Parisod, C., Geiser, C., Daprà, L., Vittoz, P. & Joost, S. (2015) Very high-resolution digital elevation models: Are multi-scale derived variables ecologically relevant? *Methods in Ecology and Evolution*, 6, 1373–1383.

Lees, B.G. (1999) The Kioloa GLCTS Pathfinder Site. Retrieved from https://fennerschool-associated.anu.edu.au/pathfinder/

Lees, B.G. & Ritman, K. (1991) Decision-tree and rule induction approach to integration of remotely sensed and GIS data in mapping vegetation in disturbed or hilly environments. *Environmental Management*, 15, 823–831.

Legleiter, C.J. (2012) Remote measurement of river morphology via fusion of LiDAR topography and spectrally based bathymetry. *Earth Surface Processes and Landforms*, 37, 499–518.

Lehner, B. (2013) *HydroSHEDS Technical Documentation (Version 1.2)*. Washington, DC: World Wildlife Fund.

Lehner, B. & Döll, P. (2004) Development and validation of a global database of lakes, reservoirs and wetlands. *Journal of Hydrology*, 296, 1–22.

Lehner, B., Verdin, K. & Jarvis, A. (2008) New global hydrography derived from spaceborne elevation data. *Eos, Transactions of the American Geophysical Union*, 89, 93–94.

Leij, F.J., Romano, N., Palladino, M. & Schaap, M.G. (2004) Topographical attributes to predict soil hydraulic properties along a hillslope terrace. *Water Resources Research*, 40, 1–15.

Leitão, J.P., Prodanović, D & Maksimović, Č. (2016) Improving merge methods for grid-based digital elevation models. *Computers and Geosciences*, 88, 115–131.

Lemmens, M.J.P.M. (1978) A survey on stereo matching techniques. *International Archives of Photogrammetry and Remote Sensing*, 27, 11–23.

Lewis, G. & Holden, N.M. (2012) A comparison of grid-based computation methods of topographic wetness index derived from digital elevation model data. *Biosystems Engineering Research Review*, 17, 103.

Lewis, L.A., Verstraeten, G. & Zhu, H. (2005) RUSLE applied in a GIS framework: Calculating the LS factor and deriving homogeneous patches for estimating soil loss. *International Journal of Geographical Information Science*, 19, 809–829.

Li, J., Taylor, G. & Kidner, D.B. (2005) Accuracy and reliability of map-matched GPS coordinates: The dependence on terrain model resolution and interpolation algorithm. *Computers and Geosciences*, 31, 241–251.

Li, Z. (2008) Multi-scale digital terrain modeling and analysis. In: Q. Zhou, B. Lees & G. Tang (eds) *Advances in Digital Terrain Analysis*, pp. 59–83. Berlin, Germany: Springer.

Li, Z., Zhu, Q. & Gold, C. (2005) *Digital Terrain Modeling: Principles and Methodology*. Boca Raton, FL: CRC Press.

Liang, C. & Mackay D.S. (2000) A general model of watershed extraction and representation using globally optimal flow paths and upslope contributing areas. *International Journal of Geographical Information Science*, 4, 337–358.

Likens, G.E., Bormann, F.H., Pierce, R.S., Eaton, J.S. & Johnson, N.M. (1977) *Biogeochemistry of a Forested Ecosystem*. New York, NY: Springer.

Lin, K., Zhang, Q. & Chen, X. (2010) An evaluation of impacts of DEM resolution and parameter correlation on TOPMODEL modeling uncertainty. *Journal of Hydrology*, 394, 370–383.

Lin, S., Jing, C., Coles, N., Chaplot, V., Moore, N. & Wu, J. (2013) Evaluating DEM source and resolution uncertainties in the Soil and Water Assessment Tool. *Stochastic Environmental Research and Risk Assessment*, 27, 209–221.

Lin, Z., Kaneda, H., Mukoyama, S., Asada, N. & Chiba, T. (2013) Detection of subtle tectonic-geomorphic features in densely forested mountains by very high-resolution airborne LiDAR survey. *Geomorphology*, 182, 104–115.

Linacre, E. (1992) *Climate Data and Resources: A Reference and Guide*. London, UK: Routledge.

Lindsay, J.B. (2005) The terrain analysis system: A tool for hydro-geomorphic applications. *Hydrological Processes*, 19, 1123–1130.

Lindsay, J.B. (2006) Sensitivity of channel mapping techniques to uncertainty in digital elevation data. *International Journal of Geographical Information Science*, 20, 669–692.

Lindsay, J.B. (2009) Geomorphometry in TAS GIS. In: T. Hengl & H.I. Reuter (eds) *Geomorphometry: Concepts, Software, Applications*, pp. 367–386. Amsterdam, Netherlands: Elsevier.

Lindsay, J.B. (2014) The Whitebox Geospatial Analysis Tools project and open-access GIS. In: *Proceedings of the GIS Research UK 22nd Annual Conference*, Glasgow, UK.

Lindsay, J.B. (2016a) Efficient hybrid breaching-filling sink removal methods for flow path enforcement in digital elevation models. *Hydrological Processes*, 30, 846–857.

Lindsay, J.B. (2016b) The practice of DEM stream burning revisited. *Earth Surface Processes and Landforms*, 41, 658–668.

Lindsay, J.B. (2016c) Whitebox GAT: A case study in geomorphometric analysis. *Computers and Geosciences*, 95, 75–84.

Lindsay, J.B. & Creed, I.F. (2005a) Removal of artifact depressions from digital elevation models: Towards a minimum impact approach. *Hydrological Processes*, 19, 3113–3126.

Lindsay, J.B. & Creed, I.F. (2005b) Sensitivity of digital landscapes to artifact depressions in remotely-sensed DEMs. *Photogrammetric Engineering and Remote Sensing*, 71, 1029–1036.

Lindsay, J.B. & Creed, I.F. (2006) Distinguishing between artifact and real depressions in digital elevation data. *Computers and Geosciences*, 32, 1194–1204.

Lindsay, J.B. & Dhun, K. (2015) Modelling surface drainage patterns in altered landscapes using LiDAR. *International Journal of Geographical Information Science*, 29, 397–411.

Lindsay, J.B. & Evans, M.G. (2006) The influence of elevation error on the morphometrics of channel networks extracted from DEMs and the implications for hydrological modeling. *Hydrological Processes*, 22, 1588–1603.

Lindsay, J.B. & Seibert, J. (2013) Measuring the significance of a divide to local drainage patterns. *International Journal of Geographical Information Science*, 27, 1453–1468.

Lindsay, J.B., Rothwell, J.J. & Davies, H. (2008) Mapping outlet points used for watershed delineation onto DEM-derived stream networks. *Water Resources Research*, 44, W08442.

Lindsay, J.B., Cockburn, J. & Russell, H. (2015) An integral image approach to performing multi-scale topographic position analysis. *Geomorphology*, 245, 51–61.

List, R.J. (1968) *Smithsonian Meteorological Tables*. Washington, DC: Smithsonian Miscellaneous Collections No. 114.

Liu, B.Y.H. & Jordan, R.C. (1960) The interrelationship and characteristic distribution of direct, diffuse and total solar radiation. *Solar Energy*, 4, 1–19.

Liu, T. (1992) Fractal structure and properties of stream networks. *Water Resources Research*, 28, 2981–2988.

Liu, X. (2008) Airborne LiDAR for DEM generation: Some critical issues. *Progress in Physical Geography*, 32, 31–49.

Liu, X. & Bian, L. (2008) Accuracy assessment of DEM slope algorithms related to spatial autocorrelation of DEM errors. In: Q. Zhou, B. Lees & G.A. Tang (eds) *Advances in Digital Terrain Analysis*, pp. 307–322. Berlin, Germany: Springer Lecture Notes in Geoinformation and Cartography.

Liu, X.-H., Hu, P., Hu, H. & Sherba, J. (2012) Approximation theory applied to DEM vertical accuracy assessment. *Transactions in GIS*, 16, 393–410.

Liu, X.-H., Hu, H. & Hu, P. (2015) The "M" in digital elevation models. *Cartography and Geographic Information Science*, 29, 235–243.

Lloyd, C.D. (2005) Assessing the effect of integrating elevation data into the estimation of monthly precipitation in Great Britain. *Journal of Hydrology*, 308, 128–150.

Lloyd, C.D. & Atkinson, P.D. (2006) Deriving ground surface digital elevation models from LiDAR with geostatistics. *International Journal of Geographical Information Science*, 20, 535–563.

Lohani, B. & Mason, D.C. (2001) Application of airborne scanning laser altimetry to the study of tidal channel geomorphology. *ISPRS Journal of Photogrammetry and Remote Sensing*, 56, 100–120.

Lucieer, A. & Stein, A. (2005) Texture-based landform segmentation of LiDAR imagery. *International Journal of Applied Earth Observation and Geoinformation*, 6, 261–270.

Luo, W. (2000) Quantifying groundwater-sapping landforms with a hypsometric technique. *Journal of Geophysical Research*, 105, 1685–1694.

Lyon, S.W., Walter, M.T., Grard-Marchant, P. & Steenhuis, T.S. (2004) Using a topographic index to distribute variable source area runoff predicted with the SCS curve-number equation. *Hydrological Processes*, 18, 2757–2771.

Maathuis, B.H.P. & Wang, L. (2006) Digital elevation model based hydro-processing. *Geocarto International*, 21, 21–26.

McBratney, J.A.B. & Odeh, I.O.A. (1997) Application of fuzzy sets in soil science: Fuzzy logic, fuzzy measurement, and fuzzy decisions. *Geoderma*, 77, 85–113.

McCool, D.K., Brown, L.C., Foster, G.R., Mutchler, C.K. & Meyer, L.D. (1987) Revised slope steepness factor for the Universal Soil Loss Equation. *Transactions of the American Society of Agricultural Engineers*, 30, 1387–1396.

McCool, D.K., Foster, G.R., Mutchler, C.K. & Meyer, L.D. (1989) Revised slope length factor for the Universal Soil Loss Equation. *Transactions of the American Society of Agricultural Engineers*, 32, 1571–1576.

McCool, D.K., Foster, G.R. & Weesies, G.A. (1997) Slope-length and steepness factors (LS). In: K.G. Renard, G.R. Foster, G.A. Weeisies, D.K. McCool & D.C. Yoder (eds) *Predicting Soil Erosion by Water: A Guide to Conservation Planning with the Revised Universal Soil Loss Equation (RUSLE)*, pp. 101–142. Washington, DC: US Department of Agriculture, Agriculture Handbook No. 703.

McCune, B. & Keon, D. (2002) Equations for potential annual direct incident radiation and heat load index. *Journal of Vegetation Science*, 13, 603–606.

MacDonald, R.I. & Urban, D.L. (2004) Forest edges and tree growth rates in the North Carolina Piedmont. *Ecology*, 85, 2258–2266.

McGlynn, B.L. & Seifert, J. (2003) Distributed assessment of contributing area and riparian buffering along stream networks. *Water Resources Research*, 39, W1082.

McGuire, A.D., Christensen, T.R., Hayes, D., Heroult, A., Euskirchen, E., Kimball, J.S. et al. (2012) An assessment of the carbon balance of Arctic tundra: Comparisons among observations, process models, and atmospheric inversions. *Biogeosciences*, 9, 3185–3204.

Machguth, H., Paul, F., Hoelzle, M. & Haeberli, W. (2006). Distributed glacier mass balance modeling as an important component of modern multi-level glacier monitoring. *Annals of Glaciology*, 43, 335–343.

Mackaness, W.A., Ruas, A. & Sarjakoski, L.T. (eds) (2007) *Generalisation of Geographic Information: Cartographic Modelling and Applications*. Oxford, UK: Elsevier.

Mackay, D.S. & Band, L.E. (1998) Extraction and representation of nested catchment areas from digital elevation models in lake-dominated topography. *Water Resources Research*, 34, 897–901.

Mackay, D.S., Samanta, S., Ahl, D.E., Ewers, B.E., Gower, S.T. & Burrows, S.N. (2003) Automated parameterization of land surface process models using fuzzy logic. *Transactions in GIS*, 7, 139–153.

McKean, J., Nagel, D., Tonina, D., Bailey, P., Wright, C.W., Bohn, C. & Nayegandhi, A. (2009) Remote sensing of channels and riparian zones with a narrow-beam aquatic-terrestrial LiDAR. *Remote Sensing*, 1, 1065–1096.

McKenzie, G., Raubal, M., Janowicz, K. & Flanagin, A. (2016) Provenance and credibility in spatial and platial data. *Journal of Spatial Information Science*, 13, 101–102.

McKenzie, N.J., Gessler, P.E., Ryan, P.J. & O'Connell, D. (2000) The role of terrain analysis in soil mapping. In: J.P. Wilson & J.C. Gallant (eds) *Terrain Analysis: Principles and Applications*, pp. 245–266. New York, NY: John Wiley & Sons, Inc.

Mackey, B.G. (1996) The role of GIS and environmental modeling in the conservation of biodiversity. In: *Proceedings of the Third International Conference on Integrating GIS and Environmental Modeling*, Santa Fe, NM.

Mackey, B.G., Mullen, I.C., Baldwin, K.A., Gallant, J.C., Sims, R.A. & McKenney, D.W. (2000) Towards a spatial model of boreal forest ecosystems: The role of digital terrain analysis. In: J.P. Wilson & J.C. Gallant (eds) *Terrain Analysis: Principles and Applications*, pp. 391–422. New York, NY: John Wiley & Sons, Inc.

MacMillan, R.A. & Pettapiece, W.W. (1997) *Soil Landscape Models: Automated Landscape Characterization and Generation of Soil-landscape Models*. Lethbridge, Canada: Agriculture and Agri-Food Canada Research Branch Research Report No. 1.

MacMillan, R.A. & Shary, P.A. (2009) Landforms and landform elements in geomorphometry. In: T. Hengl & H.I. Reuter (eds) *Geomorphometry: Concepts, Software, Applications*, pp. 227–254. Amsterdam, Netherlands: Elsevier.

MacMillan, R.A., Pettapiece, W.W., Nolan, S.C. & Goddard, T.W. (2000) A generic procedure for automatically segmenting landforms into landform elements using DEMs, heuristic rules and fuzzy logic. *Fuzzy Sets and Systems*, 113, 81–109.

Mandelbrot, B.B. (1977) *Fractals: Form, Chance, and Dimension*. San Francisco, CA: Freeman.

Marjerison, R.D., Dahlke, H., Easton, Z.M., Seifert, S. & Walter, M.T. (2011) A phosphorus index transport factor based on variable source area hydrology for New York State. *Journal of Soil and Water Conservation*, 66, 149–157.

Mark, D.M. (1975) Computer analysis of topography: A comparison of terrain storage methods. *Geografiska Annaler*, 57A, 179–188.

Mark, D.M. (1978) Concepts of "data structure" for digital terrain models. In: *Proceedings of the Digital Terrain Models Symposium*, pp. 24–31. St. Louis, MO.

Mark, D.M., Dozier, J. & Frew, J. (1984) Automated basin delineation from digital elevation data. *Geoprocessing*, 2, 299–311.

Marsden, L.E. (1960) How the national map accuracy standards were developed. *Surveying and Mapping*, 20, 427–439.

Marthews, T.R., Dadson, S.J., Lehner, B., Abele, S. & Gedney, N. (2015) High-resolution global topographic index values for use in large-scale hydrological modeling. *Hydrology and Earth System Sciences*, 19, 91–104.

Martinez, C., Hancock, G.R., Kalma, J.D., Wells, T. & Boland, L. (2010) An assessment of digital elevation models and their ability to capture geomorphic and hydrologic properties at the catchment scale. *International Journal of Remote Sensing*, 31, 6239–6257.

Martz, L.W. & de Jong, E. (1988) CATCH: A Fortran program for measuring catchment area from digital elevation models. *Computers and Geosciences*, 14, 627–640.

Martz, L.W. & Garbrecht, J. (1992) Numerical definition of drainage network and subcatchment areas from digital elevation models. *Computers and Geosciences*, 18, 747–761.

Martz, L.W. & Garbrecht, J. (1993) Automated extraction of drainage network and watershed data from digital elevation models. *Journal of the American Water Resources Association*, 29, 901–908.

Martz, L.W. & Garbrecht, J. (1999) An outlet breaching algorithm for the treatment of closed depressions in a raster DEM. *Computers and Geosciences*, 25, 835–844.

Melton, M.A. (1965) The geomorphic and paleoclimatic significance of alluvial deposits in southern Arizona. *Journal of Geology*, 73, 1–38.

Menke, K., Smith, R. Jr, Pirelli, L. & Van Hoesen, J. (2015) *Mastering QGIS*. Birmingham, UK: Packt Publishing.

Meybeck, M., Green, P. & Vorosmarty, C.J. (2001) A new typology for mountains and other relief classes: an application to global continental water resources and population distribution. *Mountain Research and Development*, 21, 34–45.

Meyer, T.H. (2004) The discontinuous nature of kriging interpolation for digital terrain modeling. *Cartography and Geographic Information Science*, 31, 209–216.

Miklánek, P. (1993) The estimation of energy income in grid points over the basin using a simple digital elevation model. *Annales Geophysicae*, 11, 296–312.

Milne, G. (1935) Some suggested units of classification and mapping particularly for East Africa soils. *Soil Research*, 4, 183–198.

Minár, J. & Evans, I.S. (2008) Elementary forms for land surface segmentation: The theoretical basis of terrain analysis and geomorphological mapping. *Geomorphology*, 95, 236–259.

Ming, D.-P., Li, J., Wang, J. & Zhang, M. (2015) Scale parameter selection by spatial statistics for GeOBIA: Using mean-shift based multi-scale segmentation as an example. *ISPRS Journal of Photogrammetry and Remote Sensing*, 106, 28–41.

Mitas, L. & Mitášová, H. (1998) Distributed soil erosion simulation for effective erosion prevention. *Water Resources Research*, 34, 505–516.

Mitas, L. & Mitášová, H. (1999) Spatial interpolation. In: P.A. Longley, M.F. Goodchild, D.J. Maguire & D.W. Rhind (eds) *Geographical Information Systems: Principles, Techniques, Management and Applications*, pp. 481–492. New York, NY: John Wiley & Sons, Inc.

Mitášová, H. & Hofierka, J. (1993) Interpolation by regularized spline with tension: II. Application to terrain modeling and surface geometry analysis. *Mathematical Geology*, 25, 657–669.

Mitášová, H. & Mitas, L. (1993) Interpolation by regularized spline with tension: I. Theory and implementation. *Mathematical Geology*, 25, 641–655.

Mitášová, H., Mitas, L., Brown, W.M., Gerdes, D.P., Kosinovsky, I. & Baker, T. (1995) Modeling spatially and temporally distributed phenomena: New methods and tools for GRASS GIS. *International Journal of Geographical Information Systems*, 9, 433–446.

Mitášová, H., Hofierka, J., Zlocha, M. & Iverson, R.I. (1996) Modeling topographic potential for erosion and deposition using GIS. *International Journal of Geographical Information Systems*, 10, 629–641.

Möller, M., Volk, M., Friedrich, K. & Lymburner, L. (2008) Placing soil-genesis and transport processes into a landscape context: A multiscale terrain-analysis approach. *Journal of Plant Nutrition and Soil Science*, 171, 419–430.

Molnar, P. & England, P. (1990) Late Cenozoic uplift of mountain ranges and climate change: Chicken or egg? *Nature*, 346, 29–34.

Momm, H.G., Bingner, R.L., Wells, R.R., Rigby, J.R. Jr & Dabney, S.M. (2013) Effect of topographic characteristics on compound topographic index for identification of gully channel initiation locations. *Transactions of the American Society of Agricultural and Biological Engineering*, 56, 523–537.

Montgomery, D.R. & Dietrich, W.E. (1988) Where do channels begin? *Nature*, 336, 232–234.

Montgomery, D.R. & Dietrich, W.E. (1989) Source areas, drainage density, and channel initiation. *Water Resources Research*, 25, 1907–1918.

Montgomery, D.R. & Dietrich, W.E. (1992) Channel initiation and the problem of landscape scale. *Science*, 255, 826–830.

Montgomery, D.R. & Dietrich, W.E. (1994) A physically-based model for the topographic control on shallow landsliding. *Water Resources Research*, 30, 1153–1171.

Montgomery, D.R. & Foufoula-Georgiou, E. (1993) Channel network source representation using digital elevation models. *Water Resources Research*, 29, 3925–3934.

Montgomery, D.R., Sullivan, K. & Greenberg, H.M. (1998) Regional test of a model for shallow landsliding. *Hydrological Processes*, 12, 943–955.

Moody, A. & Meentemeyer, R.K. (2001) Environmental factors influencing spatial patterns of woody plant diversity in chaparral, Santa Ynez Mountains, California. *Journal of Vegetation Science*, 12, 41–52.

Moore, D.M., Lees, B.G. & Davey, S.M. (1991) A new method for predicting vegetation distributions using decision tree analysis in a geographic information system. *Environmental Management*, 15, 59–71.

Moore, I.D. (1996) Hydrologic modeling and GIS. In: M.F. Goodchild, L.T. Steyaert, B.O. Parks, C. Johnston, D. Maidment, M. Crane & S. Glendinning (eds) *GIS and Environmental Modeling: Progress and Research Issues*, pp. 143–148. Fort Collins, CO: GIS World Books.

Moore, I.D. & Burch, G.J. (1986a) Sediment transport capacity of sheet and rill flow: Application of unit stream power theory. *Water Resources Research*, 22, 1350–1360.

Moore, I.D. & Burch, G.J. (1986b) Physical basis of the length–slope factor in the Universal Soil Loss Equation. *Soil Science Society of America*, 50, 1294–1298.

Moore, I.D. & Burch, G.J. (1986c) Modeling erosion and deposition: Topographic effects. *Transactions of the American Society of Agricultural Engineers*, 29, 1624–1630, 1640.

Moore, I.D. & Grayson, R.B. (1991) Terrain-based catchment partitioning and runoff prediction using vector elevation data. *Water Resources Research*, 27, 1177–1191.

Moore, I.D. & Nieber, J.L. (1989) Landscape assessment of soil erosion and non-point source pollution. *Journal of the Minnesota Academy of Science*, 55, 18–25.

Moore, I.D. & Wilson, J.P. (1992) Length–slope factors for the Revised Universal Soil Loss Equation: Simplified method of estimation. *Journal of Soil and Water Conservation*, 47, 423–428.

Moore, I.D. & Wilson, J.P. (1994) Reply to "Comment on length-slope factors for the Revised Universal Soil Loss Equation: Simplified method of estimation" by G.R. Foster. *Journal of Soil and Water Conservation*, 49, 174–180.

Moore, I.D., Burch, G.J. & MacKenzie, D.H. (1988a) Topographic effects on the distribution of surface soil water and the location of ephemeral gullies. *Transactions of the American Society of Agricultural Engineers*, 31, 1098–1117.

Moore, I.D., O'Loughlin, E.M. & Burch, G.J. (1988b) A countour-based topographic model for hydrological and ecological applications. *Earth Surface Processes and Landforms*, 13, 305–320.

Moore, I.D., Grayson, R.B. & Ladson, A.R. (1991) Digital terrain modeling: A review of hydrological, geomorphological, and biological applications. *Hydrological Processes*, 5, 3–30.

Moore, I.D., Gessler, P.E., Nielsen, G.A. & Petersen, G.A. (1993a) Soil attribute prediction using terrain analysis. *Soil Science Society of America Journal*, 57, 443–452.

Moore, I.D., Gessler, P.E., Nielsen, G.A. & Peterson, G.A. (1993b) Terrain analysis for soil specific crop management. In: P.C. Robert, R.H. Rust & W.E. Larson (eds) *Soil Specific Crop Management*, pp. 27–55. Madison, WI: Soil Science Society of America.

Moore, I.D., Lewis, A. & Gallant, J.C. (1993c) Terrain attributes: Estimation methods and scale effects. In: A.J. Jakeman, M.B. Beck & M.J. McAleer (eds) *Modeling Change in Environmental Systems*, pp. 189–214. New York, NY: John Wiley & Sons, Inc.

Moore, I.D., Norton, T.W. & Williams, J.E. (1993d) Modeling environmental heterogeneity in forested landscapes. *Journal of Hydrology*, 150, 717–747.

Moore, I.D., Turner, A.K., Wilson, J.P., Jensen, S.K. & Band, L.B. (1993e) GIS and land surface-subsurface modeling. In: M.F. Goodchild, B.O. Parks & L.T. Steyaert (eds) *Geographic Information Systems and Environmental Modeling*, pp. 196–230. Oxford, UK: Oxford University Press.

Moore, R.B. & Dewald, T.G. (2016) The road to NHDPlus: Advancements in digital stream networks and associated catchments. *Journal of the American Water Resources Association*, 52, 890–900.

Moretti, G. & Orlandini, S. (2008) Automatic delineation of drainage basins from contour elevation data using skeleton construction techniques. *Water Resources Research*, 44, W05403.

Morgan, J. & Lesh, A. (2005) Developing landform maps using Esri's Model Builder. In: *Proceedings of the Esri International User Conference*, San Diego, CA.

Morris, D. & Heerdegen, R. (1988) Automatically derived catchment boundaries and channel networks and their hydrological applications. *Geomorphology*, 1, 131–141.

Mukherjee, S., Mukherjee, S., Garg, R.D., Bhardwaj, A. & Raju, P.L.N. (2013) Evaluation of topographic index in relation to terrain roughness and DEM grid spacing. *Journal of Earth System Science*, 122, 869–886.

Mulligan, M. & Wainwright, J. (2013) Modelling and model building, In: J. Wainwright & M. Mulligan (eds) *Environmental Modelling: Finding Simplicity in Complexity*, 2nd edn, pp. 7–26. Chichester, UK: John Wiley & Sons Ltd.

Murphy, M.A., Evans, J.S. & Storfer, A. (2010) Quantifying *Bufo boreas* connectivity in Yellowstone National Park with landscape genetics. *Ecology*, 91, 252–261.

National Imagery and Mapping Agency (2000) *Performance Specification Digital Terrain Elevation Dataset (DTED)*. Washington, DC: National Imagery and Mapping Agency Report No. MIL-PRF-89020B.

National Water Center (2016) *National Water Model: Improving NOAA's Water Prediction Services*. Retrieved from http://water.noaa.gov/documents/wrn-national-water-model.pdf

Natural Resources Institute, University of Minnesota Duluth (2016) *Critical Lands Project*. Retrieved from https://beaver.nrri.umn.edu/EcolRank/water-quality/

Nelson, A., Reuter, H.I. & Gessler, P. (2009) DEM production methods and sources. In: T. Hengl & H.I. Reuter (eds) *Geomorphometry: Concepts, Software, Applications*, pp. 65–85. Amsterdam, Netherlands: Elsevier.

Nelson, E.J., Jones, N.L. & Miller, A.W. (1994) Algorithm for precise drainage-basin delineation. *Journal of Hydraulic Engineering*, 120, 298–312.

Nelson, E.J., Jones, N.L. & Berrett, R.J. (1999) Adaptive tessellation method for creating TINs from GIS data. *Journal of Hydraulic Engineering*, 41, 2–9.

Neri, M., Mazzarini, F., Tarquini, S., Bisson, M., Isola, I., Behncke, B. & Pareschi, M.T. (2008) The changing face of Mount Etna's summit area documented with LiDAR technology. *Geophysical Research Letters*, 35, L09305.

Neteler, M. & Mitášová, H. (2008) *Open Source GIS: A GRASS GIS Approach*, 3rd edn. New York, NY: Springer.

Netzel, P., Jasiewicz, J. & Stepinski, T.F. (2016) TerraEx: A GeoWeb app for world-wide content-based search and distribution of elevation and landforms data. In: *Proceedings of the 9th International Conference on Geographic Information Science*, Montreal, Quebec. Available at http://escholarship.org/uc/item/0dk1t0vc

Nguyen, T.M. & Wilson, J.P. (2010) Sensitivity of the quasi-dynamic topographic wetness index to choice of DEM resolution, flow routing algorithm and soil variability. In: *Proceedings of the 9th International Symposium on Spatial Accuracy Assessment in Natural Resources and the Environmental Sciences*, Leicester, UK.

Nico, G., Leva, D., Antonello, G. & Tarchi, D. (2004) Ground-based SAR interferometry for terrain mapping: Theory and sensitivity analysis. *IEEE Transactions on Geoscience and Remote Sensing*, 42, 1344–1350.

Nico, G., Leva, D., Fortuny, J., Antonello, G. & Tarchi, D. (2005a) Generating digital terrain models by a ground-based synthetic aperture radar interferometer. *IEEE Transactions on Geoscience and Remote Sensing*, 43, 45–49.

Nico, G., Rutigliano, P., Benedetto, C. & Vespe, F. (2005b) Terrain modeling by kinematical GPS survey. *Natural Hazards and Earth System Sciences*, 5, 293–299.

Nikora, V.I., Sapozhnikov, V.B. & Noever, D.A. (1993) Fractal geometry of individual river channels and its computer simulation. *Water Resources Research*, 29, 3561–3568.

Norouzi, M., Ayoubi, S., Jalalian, A., Khademi, H. & Dehghani, A.A. (2010) Predicting rainfed wheat quality and quantity by artificial neural network using terrain and soil characteristics. *Acta Agriculturae Scandinavica, Section B, Soil and Plant Science*, 60, 341–352.

Notebaert, B., Verstraeten, G., Govers, G. & Poesen, J. (2009) Qualitative and quantitative applications of LiDAR imagery in fluvial geomorphology. *Earth Surface Processes and Landforms*, 34, 217–231.

O'Callaghan, J.F. & Mark, D.M. (1984) The extraction of drainage networks from digital elevation data. *Computer Vision, Graphics and Image Processing*, 28, 323–344.

Oimoen, M.J. (2000) An effective filter for removal of artifacts in U.S. Geological Survey 7.5-minute digital elevation models. In: *Proceedings of the 14th International Conference on Applied Geologic Remote Sensing*, pp. 311–319. Las Vegas, NV.

Oksanen, J. & Sarjakoski, T. (2005) Error propagation analysis of DEM-based drainage basin delineation. *International Journal of Remote Sensing*, 26, 3085–3102.

Oksanen, J. & Sarjakoski, T. (2006) Uncovering the statistical and spatial scales of fine toposcale DEM error. *International Journal of Geographical Information Science*, 20, 345–369.

Olaya, V. (2009) Basic land-surface parameters. In: T. Hengl & H.I. Reuter (eds) *Geomorphometry: Concepts, Software, Applications*, pp. 141–169. Amsterdam, Netherlands: Elsevier.

Olaya, V. & Conrad, O. (2009) Geomorphometry in SAGA. In: T. Hengl & H.I. Reuter (eds) *Geomorphometry: Concepts, Software, Applications*, pp. 293–308. Amsterdam, Netherlands: Elsevier.

Olefeldt, D., Roulet, N., Giesler, R. & Persson, A. (2013) Total waterborne carbon export and DOC composition from ten nested subarctic peatland catchments: Importance of peat-land cover, groundwater influence, and inter-annual variability of precipitation patterns. *Hydrological Processes*, 27, 2280–2294.

O'Loughlin, E.M. (1986) Prediction of surface saturation zones in natural catchments by topographic analysis. *Water Resources Research*, 22, 794–804.

Olsson, L. & Pilesjö, P. (2002) Approaches to spatially distributed hydrological modeling in a GIS environment. In: A. Skidmore (ed.) *Environmental Modeling with GIS and Remote Sensing*, pp. 166–199. London, UK: Taylor & Francis.

O'Neil, G. & Shortridge, A. (2013) Quantifying local flow direction uncertainty. *International Journal of Geographical Information Science*, 27, 1292–1311.

Onstad, C.A. & Brakensiek, D.L. (1968) Watershed simulation by stream path analogy. *Water Resources Research*, 4, 965–971.

Orlandini, S. & Moretti, G. (2009) Determination of surface flows from gridded elevation data. *Water Resources Research*, 45, W03417.

Orlandini, S., Moretti, G., Franchini, M., Aldighieri, B. & Testa, B. (2003) Path-based methods for the determination of non-dispersive drainage directions in grid-based elevation models. *Water Resources Research*, 39, W1144.

Orlandini, S., Moretti, G., Corticelli, M.A., Santangelo, P.E., Capra, A., Rivola, R. & Albertson, J.D. (2012) Evaluation of flow direction methods against field observations of overland flow dispersion. *Water Resources Research*, 48, W10523.

Ortega, L. & Rueda, A. (2010) Parallel drainage network computation on CUDA. *Computers and Geosciences*, 36, 171–178.

OSGF, Open Source Geospatial Foundation (2011) Geospatial Data Abstraction Library (Version 1.9.0), Translator Library. Retrieved from http://www.gdal.org/

Pain, C.F. (2005) Size does matter: Relationships between image pixel size and landscape process scales. In: A. Zerger & R.M. Argent (eds) *Proceedings of the International Congress on Modeling and Simulation (MODSIM 2005)*, pp. 1430–1436. Melbourne, Australia: Modeling and Simulation Society of Australia and New Zealand.

Palomar-Vázquez, J. & Pardo-Pascual, J. (2008) Automated spot heights generalization in trail maps. *International Journal of Geographical Information Science*, 22, 91–110.

Pan, F., Peters-Lidard, C., Sale, M. & King, A. (2004) A comparison of geographical information system-based algorithms for computing the TOPMODEL topographic index. *Water Resources Research*, 40, W06303.

Panuska, J.C., Moore, I.D. & Kramer, L.A. (1991) Terrain analysis: Integration into the Agricultural Nonpoint Source Pollution Model. *Journal of Soil and Water Conservation*, 46, 59–64.

Park, S.J., McSweeney, K. & Lowery, B. (2001) Identification of the spatial distribution of soils using a process-based terrain characterization. *Geoderma*, 103, 249–272.

Passalacqua, P., Do Trung, T., Foufoula-Georgiou, E., Sapiro, G. & Dietrich, W.E. (2010) A geometric framework for channel network extraction from LiDAR: Nonlinear diffusion and geodesic paths. *Journal of Geophysical Research*, 115, F01002.

Peckham, R.J. & Jordan, G. (eds) (2007) *Digital Terrain Modeling: Development and Applications in a Policy Support Environment*. Berlin, Germany: Springer Lecture Notes in Geoinformation and Cartography.

Peckham, S.D. (1995) *Self-similarity in the three-dimensional geometry and dynamics of large river basins*. Unpublished PhD dissertation, University of Colorado at Boulder, Boulder, CO.

Peckham, S.D. (1998) Efficient extraction of river networks and hydrologic measurements from digital elevation data. In: O.E. Barndorff-Nielsen, V.K. Gupta, V. Pérez-Abreu & E. Waymire (eds) *Stochastic Methods in Hydrology: Rain, Landforms, and Floods*, pp. 173–203. Singapore: World Scientific.

Peckham, S.D. (2009) Geomorphometry in RiverTools. In: T Hengl & H.I. Reuter (eds) *Geomorphometry: Concepts, Software, Applications*, pp. 411–430. Amsterdam, Netherlands: Elsevier.

Peckham, S.D. (2013) Mathematical surfaces for which specific and total contributing area can be computed: Testing contributing area algorithms. In: *Proceedings of the 3rd International Conference on Geomorphometry*, Nanjing, China.

Peckham, S.D. & Gupta, V.K. (1999) A reformulation of Horton's laws for large river networks in terms of statistical self-similarity. *Water Resources Research*, 35, 2763–2777.

Pei, T., Qin, C.-Z., Zhu, A.-X., Yang, L., Luo, M., Li, B. & Zhou, C. (2010) Mapping soil organic matter using the topographic wetness index: A comparative study based on different flow-direction algorithms and kriging methods. *Ecological Indicators*, 10, 610–619.

Pelletier, J.D. (2013) A robust, two-parameter method for the extraction of drainage networks from high-resolution digital elevation models (DEMs): Evaluation using synthetic and real-world DEMs. *Water Resources Research*, 49, 75–89.

Pennock, D.J., Zebarth, B.J. & de Jong, E. (1987) Landform classification and soil distribution in hummocky terrain, Saskatchewan, Canada. *Geoderma*, 40, 297–315.

Pennock, D.J., Anderson, D.W. & de Jong, E. (1994) Landscape-scale changes in indicators of soil quality due to cultivation in Saskatchewan, Canada. *Geoderma*, 64, 1–19.

Perron, J.T. & Royden, L. (2013) An integral approach to bedrock river process analysis. *Earth Surface Processes and Landforms*, 38, 570–576.

Perron, J.T., Dietrich, W. & Kirchner, J. (2008) Controls on the spacing of first-order valleys. *Journal of Geophysical Research*, 113, F04016.

Perron, J.T., Kirchner, J. & Dietrich, W. (2009) Formation of evenly spaced ridges and valleys. *Nature*, 460, 502–505.

Perron, J.T., Richardson, P., Ferrier, K. & Lapôtre, M. (2012) The root of branching river networks. *Nature*, 492, 100–103.

Persson, A., Pilesjö, P. & Eklundh, L. (2005) Spatial influence of topographical factors on yield. In: J. Stafford (ed.) *Precision Agriculture*, pp. 341–357. Berlin, Germany: Springer.

Persson, A., Hasan, A., Tang, J. & Pilesjö, P. (2012) Modeling flow routing in permafrost landscapes with TWI: An evaluation against site-specific wetness measurements. *Transactions in GIS*, 16, 703–713.

Petrasova, A., Harmon, B., Petras, V. & Mitášová, H. (2015) *Tangible Modeling with Open Source GIS*. Berlin, Germany: Springer.

Petrescu, A.M.R., van Huissteden, J., Jackowicz-Korczyński, M., Yurova, A., Christensen, T.R., Crill, P.M., Bäckstrand, K. & Maximov, T.C. (2008) Modeling $CH_4$ emissions from Arctic wetlands: Effects of hydrological parameterization. *Biogeosciences*, 5, 111–121.

Peucker, T.K. & Douglas, D.H. (1975) Detection of surface-specific points by local parallel processing of discrete terrain elevation data. *Computer Graphics and Image Processing*, 4, 375–387.

Peucker, T.K., Fowler, R.J., Little, J.J. & Mark, D.M. (1978) The triangulated irregular network. In: *Proceedings of the Auto Carto III Conference*, San Francisco, CA.

Phillips, J.D. (1990) A saturation-based model of relative wetness for wetland identification. *Water Resources Bulletin*, 26, 333–342.

Pike, R.J. (1988) The geometric signature: Quantifying landslide-terrain types from digital elevation models. *Mathematical Geology*, 20, 491–511.

Pike, R.J. (1995) Geomorphometry: Progress, practice, and prospect. *Zeitschrift für Geomorphologie*, 101, 221–238.

Pike, R.J. (2000) Geomorphometry: Diversity in quantitative surface analysis. *Progress in Physical Geography*, 24, 1–20.

Pike, R.J. & Wilson, S.E. (1971) Elevation relief ratio, hypsometric integral, and geomorphic area altitude analysis. *Bulletin of the Geological Society of America*, 82, 1079–1084.

Pike, R.J., Evans, I.S. & Hengl, T. (2009) Geomorphometry: A brief guide. In T. Hengl & H.I. Reuter (eds) *Geomorphometry: Concepts, Software, Applications*, pp. 3–30. Amsterdam, Netherlands: Elsevier.

Pilesjö, P. (2008) An integrated raster-TIN surface flow algorithm. In: Q. Zhou, B. Lees & G. Tang (eds) *Advances in Digital Terrain Analysis*, pp. 237–255. Berlin, Germany: Springer.

Pilesjö, P. & Hasan, A. (2014) A triangular form-based multiple flow algorithm to estimate overland flow distribution and accumulation on a digital elevation model. *Transactions in GIS*, 18, 108–124.

Pilesjö, P. & Zhou, Q. (1997) Theoretical estimation of flow accumulation from a grid-based digital elevation model. In: *Proceedings of the GIS, AM/FM Asia and Geoinformatics Conference*, pp. 447–456. Taipei, Taiwan.

Pilesjö, P., Zhou, Q. & Harrie, L. (1998) Estimating flow distribution over digital elevation models using a form-based algorithm. *Annals of GIS*, 4, 44–51.

Pilesjö, P., Persson, A. & Harrie, L. (2006) Digital elevation data for estimation of potential wetness in rugged fields: Comparison of two different methods. *Agricultural Water Management*, 79, 225–247.

Pilouk, M. & Tempfli, K. (1992) A digital image processing approach to creating DTMs from digitized contours. *International Archives of Photogrammetry and Remote Sensing*, 29(B4), 956–961.

Pirotti, F. & Tarolli, P. (2010) Suitability of LiDAR point density and derived landform curvature maps for channel network extraction. *Hydrological Processes*, 24, 1187–1197.

Planchon, O. & Darboux, F. (2001) A fast, simple and versatile algorithm to fill the depressions of digital elevation models. *Catena*, 46, 159–176.

Podobnikar, T. (2005) Production of integrated digital terrain model from multiple datasets of different quality. *International Journal of Geographical Information Science*, 19, 69–89.

Pourali, S.H., Arrowsmith, C., Chrisman, N., Matkan, A.A. & Mitchell, D. (2016) Topography wetness index application to flood-risk-based land use planning. *Applied Spatial Analysis and Policy*, 9, 39–54.

Prigent, C., Papa, F., Aires, F., Rossow, W.B. & Matthews, E. (2007) Global inundation dynamics inferred from multiple satellite observations, 1993–2000. *Journal of Geophysical Research: Atmosphere*, 112, D12107.

Qi, F. (2004) Knowledge discovery from area-class resource maps: Data preprocessing for noise reduction. *Transactions in GIS*, 8, 297–308.

Qi, F. & Zhu, A.-X. (2003) Knowledge discovery from soil maps using inductive learning. *International Journal of Geographical Information Science*, 17, 771–785.

Qi, F., Zhu, A.-X., Harrower, M. & Burt, J.E. (2006) Fuzzy soil mapping based on prototype category theory. *Geoderma*, 136, 774–787.

Qin, C.-Z. & Zhan, L. (2012) Parallelizing flow-accumulation calculations on graphics processing units: From iterative DEM preprocessing algorithm to recursive multiple-flow-direction algorithm. *Computers and Geosciences*, 43, 7–16.

Qin, C., Zhu, A.-X., Pei, T., Li, B., Zhou, C. & Yang, L. (2007) An adaptive approach to selecting the flow partition exponent for multiple flow direction algorithms. *International Journal of Geographical Information Science*, 21, 443–458.

Qin, C.-Z., Zhu, A.-Z., Shi, X., Li, B.-L., Pei, T. & Zhou, C.-H. (2009) Quantification of spatial gradation of slope positions. *Geomorphology*, 110, 152–161.

Qin, C.-Z., Zhu, A.-X., Pei, T., Li, B.-L., Scholten, T., Behrens, T. & Zhou, C.-H. (2011) An approach to computing topographic wetness index based on maximum downslope gradient. *Precision Agriculture*, 12, 32–43.

Qin, C.-Z., Zhu, A.-X., Qiu, W.-L., Lu, Y.-J., Li, B.-L. & Pei, T. (2012) Mapping soil organic matter in small low-relief catchments using fuzzy slope position information. *Geoderma*, 171–172, 64–74.

Qin, C.-Z., Bao, L.-L., Zhu, A.-X., Hu, X.-M. & Qin B. (2013a) Artificial surfaces simulating complex terrain types for evaluating grid-based flow direction algorithms. *International Journal of Geographical Information Science*, 27, 1055–1072.

Qin, C.-Z., Bao, L.-L., Zhu, A.-X., Wang, R.-X. & Hu, X.-M. (2013b) Uncertainty due to DEM error in landslide susceptibility mapping. *International Journal of Geographical Information Science*, 27, 1364–1380.

Qin, C.-Z., Zhan, L.-J. & Zhu, A.-X. (2014a) How to apply the Geospatial Data Abstraction Library (GDAL) properly to parallel geospatial raster I/O? *Transactions in GIS*, 18, 950–957.

Qin, C.-Z., Zhan, L.-J., Zhu, A.-X. & Zhou, C.-H. (2014b) A strategy for raster-based geocomputation under different parallel computing platforms. *International Journal of Geographical Information Science*, 28, 2127–2144.

Qin, C.-Z., Wu, X.-W., Jiang, J.-C. & Zhu, A.-X. (2016) Case-based formalization and reasoning method for knowledge in digital terrain analysis: Application to extracting drainage networks. *Hydrology and Earth System Sciences*, 20, 3379–3392.

Qin, C.-Z., Ai, B.-B., Zhu, A.-Z. & Liu, J.-Z. (2017) An efficient method for applying a differential equation to deriving the spatial distribution of specific catchment area from gridded digital elevation models. *Computers and Geosciences*, 100, 94–102.

Quinn, P.F., Beven, K.J., Chevallier, P. & Planchon, O. (1991) The prediction of hillslope paths for distributed hydrological modeling using a digital terrain model. *Hydrological Processes*, 5, 59–79.

Quinn, P.F., Beven, K.J. & Lamb, R. (1995) The ln(a/tan b) index: How to calculate it and how to use it within the TOPMODEL framework. *Hydrological Processes*, 9, 161–182.

Quinn, T., Zhu, A.-X. & Burt, J.E. (2005) Effects of detailed soil spatial information on watershed modeling across different model scales. *International Journal of Applied Earth Observation and Geoinformation*, 7, 324–338.

Quiquet, A., Archibald, A.T., Friend, A.D., Chappellaz, J., Levine, J.G., Stone, E.J., Telford, P.J. & Pyle, J.A. (2015) The relative importance of methane sources and sinks over the Last Interglacial period and into the last glaciation. *Quaternary Science Reviews*, 112, 1–16.

Raaflaub, L.D. & Collins, M.J. (2006) The effect of error in gridded digital elevation models on the estimation of topographic parameters. *Environmental Modeling and Software*, 21, 710–732.

Raber, G.T., Jensen, J.R., Schill, S.R. & Schuckman, K. (2002) Creation of digital terrain models using an adaptive LiDAR vegetation point removal process. *Photogrammetric Engineering and Remote Sensing*, 68, 1307–1315.

Raber, G.T., Jensen, J.R., Hodgson, M.E., Tullis, J.A., Davis, B.A. & Berglund, J. (2007) Data impact of LiDAR nominal post-spacing on DEM accuracy and flood zone delineation. *Photogrammetric Engineering and Remote Sensing*, 73, 793–804.

Rabus, B., Eineder, M., Roth, A. & Bamler, R. (2003) The shuttle radar topography mission: A new class of digital elevation models acquired by spaceborne radar. *ISPRS Journal of Photogrammetry and Remote Sensing*, 57, 241–262.

Rawls, W.J. (1983) Estimating soil bulk density from particle size analysis and organic matter. *Soil Science*, 135, 123–125.

Reaney, S.M., Lane, S.N., Heathwaite, A.L. & Dugdale, L.J. (2011) Risk-based modeling of diffuse land use impacts from rural landscapes upon salmonid fry abundance. *Ecological Modeling*, 222, 1016–1029.

Regnauld, N. & Mackaness, W.A. (2006) Creating a hydrographic network from its cartographic representation: A case study using Ordnance Survey MasterMap data. *International Journal of Geographical Information Science*, 20, 611–631

Renard, K.G., Foster, G.R., Weesies, G.A. & Porter, J.P. (1991) Revised universal soil loss equation. *Journal of Soil and Water Conservation*, 46, 30–33.

Renslow, M. (2012) *Manual of Airborne Topographic Lidar*. Bethesda, MD: American Society of Photogrammetry and Remote Sensing.

Reuter, H.I. (2004) *Spatial Crop and Soil Landscape Processes Under Special Consideration of Relief Information in a Loess Landscape*. Osnabruck, Germany: Der Andere Verlag.

Reuter, H.I. (2009) ArcGIS Geomorphometry Toolbox. Retrieved from http://www.ai-relief.org

Reuter, H.I. & Nelson, A. (2009) Geomorphometry in ESRI packages. In: T. Hengl & H.I. Reuter (eds) *Geomorphometry: Concepts, Software, Applications*, pp. 269–292. Amsterdam, Netherlands: Elsevier.

Reuter, H.I., Kersebaum, K.C. & Wendroth, O. (2005) Modeling of solar radiation influenced by topographic shading: Evaluation and application for precision farming. *Physics and Chemistry of the Earth*, 30, 139–149.

Reuter, H.I., Nelson, A. & Jarvis, A. (2007) An evaluation of void-filling interpolation methods for SRTM data. *International Journal of Geographical Information Science*, 21, 983–1008.

Reuter, H.I., Hengl, T., Gessler, P. & Soille, P. (2009) Preparation of DEMs for geomorphometric analysis. In: T. Hengl & H.I. Reuter (eds) *Geomorphometry: Concepts, Software, Applications*, pp. 87–120. Amsterdam, Netherlands: Elsevier.

Rexer, M. & Hirt, C. (2014) Comparison of free high resolution digital elevation data sets (ASTER GDEM2, SRTM v2.1/v4.1) and the validation against accurate heights from the Australian National Gravity Database. *Australian Journal of Earth Sciences*, 61, 213–226.

Rieger, W. (1992) Automated river line and catchment area extraction from DEM data. *International Archives of Photogrammetry and Remote Sensing*, 29(B4), 642–649.

Rigol-Sanchez, J.P., Stuart, N. & Pulido-Bosch, A. (2015) ArcGeomorphometry: A toolbox for geomorphometric characterization of DEMs in the ArcGIS environment. *Computers and Geosciences*, 85, 155–163.

Rigon, R., Rinaldo, A., Rodriquez-Iturbe, I., Bras, R.L. & Ijjasz-Vasquez, E. (1993) Optimal channel networks: A framework for the study of river basin morphology. *Water Resources Research*, 29, 1635–1646.

Riley, S.J., DeGloria, S.D. & Elliot, R. (1999) A terrain ruggedness index that quantifies topographic heterogeneity. *Intermountain Journal of Sciences*, 5, 23–27.

Ringeval, B., Decharme, B., Piao, S.L., Ciais, P., Papa, F., de Noblet-Ducoudré, N., Prigent, C., Friedlingstein, P., Gouttevin, I., Koven, C. & Ducharne, A. (2012) Modelling sub-grid wetland in the ORCHIDEE global land surface model: Evaluation against river discharges and remotely sensed data. *Geoscientific Model Development*, 5, 941–962.

Robinson, N. (1966) *Solar Radiation*. Amsterdam, Netherlands: Elsevier.

Robinson, V.B. (2003) A perspective on the fundamentals of fuzzy sets and their use in geographic information systems. *Transactions in GIS*, 7, 3–30.

Rodriguez, E., Morris, C.S., Belz, J.E., Chapin, E.C., Martin, J.M., Daffer, W. & Hensley, S. (2005) *An Assessment of the SRTM Topographic Products*. Pasadena, CA: Jet Propulsion Laboratory.

Rodriguez, E., Morris, C.S. & Belz, J.E. (2006) A global assessment of the SRTM performance. *Photogrammetric Engineering and Remote Sensing*, 72, 249–260.

Romano, N. & Chirico, G.B. (2004) The role of terrain analysis in using and developing pedotransfer functions. *Developments in Soil Science*, 30, 273–294.

Romstad, B. & Etzelmüller, B. (2009) Structuring the digital elevation model into landform elements through watershed segmentation of curvature. In: R. Purves, S. Gruber, R. Straumann & T. Hengl (eds) *Proceedings of Geomorphometry 2009*, pp. 55–60. Zurich, Switzerland: University of Zurich.

Romstad, B. & Etzelmüller, B. (2012) Mean-curvature watersheds: A simple method for segmentation of a digital elevation model into terrain units. *Geomorphology*, 139–140, 293–302.

Rost, S., Gerten, D., Bondeau, A., Lucht, A., Rohwer, J. & Schaphoff, S. (2008) Agricultural green and blue water consumption and its influence on the global water system. *Water Resources Research*, 44, W09405.

Rowbotham, D.N. & Dudycha, D. (1998) GIS modelling of slope stability in Phewa Tal watershed, Nepal. *Geomorphology*, 26, 151–170.

Ruhl, R.V. (1960) Elements of the soil landscape. In: *Proceedings of the 7th Congress of the International Society of Soil Science*, pp. 32–40. Madison, WI.

Ruhl, R.V. & Walker, P.H. (1968) Hillslope models and soil formation II: Open systems. In: *Proceedings of the 9th Congress of the International Soil Science Society*, pp. 551–560. Adelaide, Australia.

Ruiz, M. (1997) A causal analysis of error in viewsheds from USGS digital elevation models. *Transactions in GIS*, 2, 85–94.

Running, S.W. (1991) Computer simulation of regional evapotranspiration by integrating landscape biophysical attributes with satellite data. In: T.J. Schmugge & J. Andre (eds), *Land Surface Evaporation: Evaporation Measurement and Parameterization*, pp. 359–369. London, UK: Springer.

Running, S.W. & Thornton, P.E. (1996) Generating daily surfaces of temperature and precipitation over complex topography. In: M.F. Goodchild, L.T. Steyaert, B.O. Parks, C. Johnston, D.R. Maidment, M. Crane & S. Glendinning (eds) *GIS and Environmental Modeling: Progress and Research Issues*, pp. 93–98. Fort Collins, CO: GIS World Books.

Running, S.W., Nemani, R.R. & Hungerford, R.D. (1987) Extrapolation of synoptic meteorological data in mountainous terrain and its use for simulating forest evapotranspiration and photosynthesis. *Canadian Journal of Forest Research*, 17, 472–483.

Saalfeld, A. (1999) Topologically consistent line simplification with the Delaunay–Peucker algorithm. *Cartography and Geographic Information Science*, 26, 7–18.

Saunders, W.K. & Maidment, D.R. (1996) *A GIS Assessment of Nonpoint Source Pollution in the San Antonio-Nueces Coastal Basin*. Austin, TX: University of Texas Center for Research in Water Resources Report No. 96–1.

Saxton, K.E. & Rawls, W.J. (2006) Soil water characteristic estimates by texture and organic matter for hydrologic solutions. *Soil Science Society of America Agronomy Journal*, 70, 1569–1578.

Sayre, R., Bow, J., Josse, C., Sotomayor, L. & Touval, J. (2008) Terrestrial ecosystems of South America. In: J.C. Campbell, K.B. Jones, J.H. Smith & M.T. Koeppe (eds) *North America Land Cover Summit*, pp. 131–152. Washington, DC: Association of American Geographers.

Sayre, R., Comer, P., Warner, H. & Cress, J. (2009) *A New Map of Standardized Terrestrial Ecosystems of the Conterminous United States*. Washington, DC: US Geological Survey Professional Paper No. 1768.

Sayre, R., Comer, P., Hak, J., Josse, C., Bow, J., Warner, H. et al. (2013) *A New Map of Standardized Terrestrial Ecosystems of Africa*. Washington, DC: Association of American Geographers.

Sayre, R., Dangermond, J., Frye, C., Vaughan, R., Aniello, P., Breyer, S. et al. (2014) *A New Map of Global Ecological Land Units: An Ecophysiographic Stratification Approach*. Washington, DC: Association of American Geographers.

Scarlatos, L. & Pavlidis, T. (1992) Hierarchical triangulation using cartographic coherence. *Graphical Models and Image Processing*, 54, 147–161.

Schmidt, J. & Andrew, R. (2005) Multi-scale landform characterization. *Area*, 37, 341–350.

Schmidt, J. & Dikau, R. (1999) Extracting geomorphometric attributes and objects from digital elevation models: Semantics, methods, future needs. In: R. Dikau & H. Saurer (eds) *GIS for Earth Surface Systems: Analysis and Modeling of the Natural Environment*, pp. 153–173. Berlin, Germany: Schweizbart'sche Verlagbuchhandlung.

Schmidt, J. & Hewitt, A. (2004) Fuzzy land element classification from DTMs based on geometry and terrain position. *Geoderma*, 121, 243–256.

Schmidt, J., Merz, B. & Dikau, R. (1998) Morphological structure and hydrological process modelling. *Zeitschrift für Geomorphologie NF*, 112, 55–66.

Schmidt, J., Hennrich, K. & Dikau, R. (2000) Scales and similarities in runoff processes with respect to geomorphometry. *Hydrological Processes*, 20, 1963–1979.

Schmidt, J., Evans, I.S. & Brinkmann, J. (2003) Comparison of polynomial models for land surface curvature calculation. *International Journal of Geographical Information Science*, 17, 797–814.

Schneevoight, N.J., van der Linden, S., Thamm, H.-P. & Schrott, L. (2008) Detecting Alpine landforms from remotely sensed imagery. A pilot study in the Bavarian Alps. *Geomorphology*, 93, 104–119.

Schneiderman, E.M., Steenhuis, T.S., Thongs, D.J., Easton, Z.M., Zion, M.S., Neal, A.L., Mendoza, G.F. & Walter, M.T. (2007) Incorporating variable source area hydrology into a curve-number-based watershed model. *Hydrological Processes*, 21, 3420–3430.

Schoorl, J.M., Sonneveld, M.P.W. & Veldkamp, A. (2000) Three-dimensional landscape process modeling: The effect of DEM resolution. *Earth Surface Processes and Landforms*, 25, 1025–1034.

Schoorl, J.M., Veldkamp, A. & Bouma, J. (2002) Modeling soil and water redistribution in a dynamic landscape content. *Soil Science Society of America Journal*, 66, 1610–1619.

Schroeder, R., McDonald, K.C., Chapman, B.D., Jensen, K., Podest, E., Tessler, Z.D., Bohn, T.J. & Zimmermann, R. (2015) Development and evaluation of a multi-year fractional surface water data set derived from active/passive microwave remote sensing data. *Remote Sensing*, 7, 16688–16732.

Schröter, I., Paasche, H., Dietrich, P. & Wollschläger, U. (2015) Estimation of catchment-scale soil moisture patterns based on terrain data and sparse TDR measurements using a fuzzy c-means clustering approach. *Vadose Zone Journal*, 14, 11.

Seibert, J. & McGlynn, B.L. (2007) A new triangular flow direction algorithm for computing upslope areas from gridded digital elevation models. *Water Resources Research*, 43, W04501.

Seidl, M. & Dietrich, W. (1992) The problem of channel erosion into bedrock. *Catena Supplement*, 23, 101–124.

Selige, T., Böhner, J. & Ringeler, A. (2006) Processing of SRTM X-SAR data to correct interferometric elevation models for land surface process applications. In: J. Böhner, K.R. McCloy & J. Strobl (eds) *SAGA: Analyses and Modelling Applications*, pp. 97–104. Warsaw, Poland: Verlag Erich Goltze GmbH.

Seybold, C.A., Grossman, R.B. & Reinsch, T.G. (2005) Predicting cation exchange capacity for soil survey using linear models. *Soil Science Society of America Journal*, 69, 856–863.

Sharma, S.K., Mohanty, B.P. & Zhu, J. (2006) Including topography and vegetation attributes for developing pedotransfer functions. *Soil Science Society of America Journal*, 70, 1430–1440.

Shary, P.A. (1995) Land surface in gravity points classification by complete system of curvatures. *Mathematical Geology*, 27, 373–390.

Shary, P.A. & Stepanov, I.N. (1991) On the second derivative method in geology. *Doklady Academii Nauk SSSR*, 319, 456–460. (In Russian)

Shary, P.A., Sharaya, L.S. & Mitusov, A.V. (2002) Fundamental quantitative methods of land surface analysis. *Geoderma*, 107, 1–32.

Shary, P.A., Sharaya, L.S. & Mitusov, A.V. (2005) The problem of scale-specific scale-free approaches in geomorphometry. *Geografia Fisica e Dimanica Quaternaria*, 28, 81–101.

Shelef, E. & Hilley, G.E. (2013) Impact of flow routing on catchment area calculations, slope estimates, and numerical simulations of landscape development. *Journal of Geophysical Research: Earth Science*, 118, 2105–2123.

Sheng, J., Wilson, J.P., Chen, N., Devinny, J.S. & Sayre, J.M. (2007) Evaluating the quality of the National Hydrography Dataset for watershed assessments in metropolitan regions. *GIScience and Remote Sensing*, 44, 283–304.

Sheng, J., Wilson, J.P. & Lee, S. (2009) Comparison of land surface temperature (LST) modeled with a spatially distributed solar radiation model (SRAD) and remote sensing data. *Environmental Modelling and Software*, 24, 436–443.

Shi, W.Z. & Tian, Y. (2006) A hybrid interpolation method for the refinement of a regular grid digital elevation model. *International Journal of Geographical Information Science*, 20, 53–67.

Shi, W.Z., Li, Q. & Zhu C.Q. (2005) Estimating the propagation error of DEM from higher-order interpolation algorithms. *International Journal of Remote Sensing*, 26, 3069–3084.

Shi, W.Z., Wang, B. & Tian, Y. (2014) Accuracy analysis of digital elevation model relating to spatial resolution and terrain slope by bilinear interpolation. *Mathematical Geosciences*, 46, 445–481.

Shortridge, A.M. (2001) Characterizing uncertainty in digital elevation models. In: C.T. Hunsaker, M.F. Goodchild, M.A. Friedl & T.J. Case (eds) *Spatial Uncertainty in Ecology: Implications for Remote Sensing and GIS Applications*, pp. 238–257. New York, NY: Springer.

Shortridge, A.M. (2006) Shuttle Radar Topography Mission elevation data error and its relationship to land cover. *Cartography and Geographic Information Science*, 33, 65–75.

Sitch, S., Smith, B., Prentice, I.C., Arneth, A., Bondeau, A., Cramer, W., Kaplan, J.O., Levis, S., Lucht, W., Sykes, M.T., Thonicke, K. & Venevsky, S. (2003) Evaluation of ecosystem dynamics, plant geography and terrestrial carbon cycling in the LPJ dynamic global vegetation model. *Global Change Biology*, 9, 161–185.

Sithole, G. & Vosselman, G. (2004) Experimental comparison of filter algorithms for bare Earth extraction from airborne laser scanning point clouds. *ISPRS Journal of Photogrammetry and Remote Sensing*, 59, 85–101.

Sivapalan, M., Beven, K. & Wood, E.F. (1987) On hydrologic similarity: 2. A scaled model of storm runoff production. *Water Resources Research*, 23, 2266–2278.

Skidmore, A.K. (1989) A comparison of techniques for calculating gradient and aspect from a gridded digital elevation model. *International Journal of Geographical Information Systems*, 3, 323–334.

Skidmore, A.K., Ryan, P.J., Dawes, W., Short, D. & O'Loughlin, E. (1991) Use of an expert system to map forest soils from a geographical information system. *International Journal of Geographical Information Systems*, 5, 431–444.

Sklar, L. & Dietrich, W.E. (1998) River longitudinal profiles and bedrock incision models: Stream power and the influence of sediment supply. In: K.J. Tinkler & E.E. Wohl (eds) *Rivers Over Rock: Fluvial Processes in Bedrock Channels*, pp. 237–260. Washington, DC: American Geophysical Union.

Slater, J.A., Garvey, G., Johnston, C., Haase, J., Heady, B., Kroenung, G. & Little, J. (2006) The SRTM data "finishing" process and products. *Photogrammetric Engineering and Remote Sensing*, 72, 237–247.

Slater, J.A., Heady, B., Kroenung, G., Curtis, W., Haase, J., Hoegemann, D. et al. (2009) *Evaluation of the New ASTER Global Digital Elevation Model*. Springfield, VA: National Geospatial-Intelligence Agency.

Smith B. & Mark, D. (2003) Do mountains exist? Towards an ontology of landforms. *Environment and Planning B*, 30, 411–427.

Smith, B. & Varzi, A.C. (2000) Fiat and bona fide boundaries. *Philosophy and Phenomenological Research*, 60, 401–420.

Smith, B., Prentice, I.C. & Sykes, M.T. (2001) Representation of vegetation dynamics in the modelling of terrestrial ecosystems: Comparing two contrasting approaches within European climate space. *Global Ecology and Biogeography*, 10, 621–637.

Smith, M.P., Zhu, A.X., Burt, J.E. & Stiles, C. (2006) The effects of DEM resolution on and neighborhood size on digital soil survey. *Geoderma*, 137, 58–67.

Smith, T.R., Zhan, C. & Gao, P. (1990) A knowledge-based, two-step procedure for extracting channel networks from noisy DEM data. *Computers and Geosciences*, 16, 777–786.

Snyder, G.I. (2012) *The 3D Elevation Program: Summary of Program Direction*. Reston, VA: US Geological Survey Fact Sheet No. 2012–3089.

Soille, P. (2004) Optimal removal of spurious pits in grid digital elevation models. *Water Resources Research*, 40, W12509.

Soille, P. & Gratin, C. (1994) An efficient algorithm for drainage networks extraction on DEMs. *Journal of Visual Communication and Image Representation*, 5, 181–189.

Soille, P., Vogt, J. & Colombo, R. (2003) Carving and adaptive drainage enforcement of grid digital elevation models. *Water Resources Research*, 39, 1366–1378.

Sørensen, R. & Seibert, J. (2007) Effects of DEM resolution on the calculation of topographical indices: TWI and its components. *Journal of Hydrology*, 347, 79–89.

Sørensen, R., Zinko, U. & Seibert, J. (2006) On the calculation of the topographic wetness index: Evaluation of different methods based on field observations. *Hydrology and Earth System Sciences*, 10, 101–112.

Southard, D.A. (1991) Piecewise planar surface models from sampled data. In: N.M. Patrikalakis (ed.) *Scientific Visualization of Physical Phenomena*, pp. 667–680. Tokyo, Japan: Springer-Verlag.

Speight, J.G. (1968) Parametric description of land form. In: G.A. Stewart (ed.) *Land Evaluation: Papers of a CSIRO Symposium*, pp. 239–250. Melbourne, Australia: MacMillan.

Speight, J.G. (1974) A parametric approach to landform regions. In: E.H. Brown & R.S. Waters (eds) *Progress in Geomorphology*, pp. 213–230. Oxford, UK: Alden Press.

Speight, J.G. (1980) The role of topography in controlling throughflow generation: A discussion. *Earth Surface Processes*, 5, 187–191.

Speight, J.G. (1990) Landforms. In: R.C. MacDonald, R.F. Isbell, J.G. Speight, J. Walker & M.S. Hop (eds) *Australian Soil and Land Survey Field Handbook*, pp. 9–57. Melbourne, Australia: Inkata Press.

Srivastava, K.P. & Moore, I.D. (1989) Application of terrain analysis to land resource investigations of small catchments in the Caribbean. In: *Proceedings of the 20th International Conference of the Erosion Control Association*, pp. 229–242. Streamboat Springs, CO.

Stanislawski, L.V. (2009) Feature pruning by upstream drainage area to support automated generalization of the United States National Hydrography Dataset. *Computers, Environment and Urban Systems*, 33, 325–333.

Stanislawski, L.V. & Buttenfield, B.P. (2011) Hydrographic generalization tailored to dry mountainous regions. *Cartography and Geographic Information Science*, 38, 117–125.

Stanislawski, L.V., Buttenfield, B.P. & Doumbouya, A. (2015a) A rapid approach for automated comparison of independently derived stream networks. *Cartography and Geographic Information Science*, 42, 435–448.

Stanislawski, L.V., Falgout, J. & Buttenfield, B.P. (2015b) Automated extraction of natural drainage density patterns for the U.S. through high performance computing. *Cartographic Journal*, 52, 185–192.

Stehr, A., Debels, P., Romero, F. & Alcayaga, H. (2008) Hydrological modeling with SWAT under conditions of limited data availability. Evaluation of results from a Chilean case study. *Hydrological Sciences Journal*, 53, 588–601.

Stein, E.D., Dark, S., Longcore, T., Hall, N., Beland, M., Grossinger, R., Casanova, J. & Sutula, M. (2007) *Historical Ecology and Landscape Change of the San Gabriel River and Floodplain*. Costa Mesa, CA: Southern California Coastal Water Research Project Technical Report No. 499.

Stepinski, T.F. & Bagaria, C. (2009) Segmentation-based unsupervised terrain classification for generation of physiographic maps. *IEEE Geoscience and Remote Sensing Letters*, 6, 733–737.

Stocker, B.D., Spahni, R. & Joos, F. (2014) DYPTOP: A cost-efficient TOPMODEL implementation to simulate sub-grid spatio-temporal dynamics of global wetlands and peatlands. *Geoscientific Model Development*, 7, 3089–3110.

Stoker, J., Harding, D. & Parrish, J. (2008) The need for a national LiDAR dataset. *Photogrammetric Engineering and Remote Sensing*, 74, 1066–1068.

Strahler, A.N. (1952) Hypsometric (area–altitude) analysis of erosional topography. *Geological Society of America Bulletin*, 63, 1117–1142.

Strahler, A.N. (1957) Quantitative analysis of watershed geomorphology. *Transactions of the American Geophysical Union*, 38, 912–920.

Su, J. & Bork, E. (2006) Influence of vegetation slope and LiDAR sampling angle on DEM accuracy. *Photogrammetric Engineering and Remote Sensing*, 72, 1265–1274.

Sugarbaker, L.J., Constance, E.W., Heidemann, H.K., Jason, A.L., Lukas, V., Saghy, D.L. & Stoker, J.M. (2014) *The 3D Elevation Program Initiative: A Call for Action*. Reston, VA: US Geological Survey Circular No. 1399.

Sulebak, J.R. & Hjelle, Ø. (2003) Multi-resolution spline models and their applications in geomorphology. In: I.S. Evans, R. Dikau, R. Tokunaga, H. Ohmori & M. Hirano (eds) *Concepts and Modeling in Geomorphology: International Perspectives*, pp. 221–237. Tokyo, Japan: Terra Publications.

Šúri, M. & Hofierka, J. (2004) A new GIS-based solar radiation model and its application for photovoltaic assessments. *Transactions in GIS*, 8, 175–190.

Survila, K., Yildirim, A.A., Li, T., Liu, Y.Y., Tarboton, D.G. & Wang, S. (2016) A scalable high-performance topographic flow direction algorithm for hydrological information analysis. In: *Proceedings of the 5th Annual Extreme Science and Engineering Discovery Environment Conference*, Miami, FL.

Svenning, J.-C., Kinner, D.A., Stallard, R.F., Engelbrecht, B.M.J. & Wright, S.J. (2004) Ecological determinism in plant community structure across a tropical forest landscape. *Ecology*, 85, 2526–2538.

Tachikawa, T., Hato, M., Kaku, M. & Iwasaki, A. (2011a) Characteristics of ASTER GDEM Version 2. In: *Proceedings of the IEEE International Geoscience and Remote Sensing Symposium*, pp. 3657–3660. Vancouver, BC: IEEE.

Tachikawa, T., Kaku, M., Iwasaki, A., Gesch, D., Oimoen, M., Zhang, Z., Danielson, J.J., Krieger, T., Curtis, B., Haase, J., Abrams, M. & Carabajal, C. (2011b) *ASTER Global Digital Elevation Model Version 2: Summary of Validation Results*. Pasadena, CA: Joint Japan–US ASTER Science Team.

Tachikawa, Y., Shiba, M. & Takasao, T. (1994) Development of a basin geomorphic information system using a TIN-DEM data structure. *Water Resources Bulletin*, 30, 9–17.

Tadono, T., Shimada, M., Murakami, H. & Takaku, J. (2009) Calibration of PRISM and AVNIR-2 onboard ALOS "Daichi". *IEEE Transactions on Geoscience and Remote Sensing*, 47, 4042–4050.

Tadono, T., Ishida, H., Oda, F., Naito, S., Minakawa, K. & Iwamoto, H. (2014) Precise global DEM generation by ALOS PRISM. *ISPRS Annals of Photogrammetry, Remote Sensing and Spatial Information Sciences*, II-4, 71–76.

Tajchman, S.J. & Lacey, C.J. (1986) Bioclimatic factors in forest site potential. *Forest Ecology and Management*, 14, 211–218.

Tang, G., Shi, W. & Zhao, M. (2002) Evaluation of the accuracy of hydrologic data derived from DEMs of different spatial resolution. In: G.J. Hunter & K. Lowell (eds) *Proceedings of the 5th International Symposium on Spatial Accuracy Assessment in Natural Resources and Environmental Sciences*, pp. 201–213. Melbourne, Australia: RMIT University.

Tang, J., Pilesjö, P. & Persson, A. (2013) Estimating slope from raster data: A test of eight algorithms at different resolutions in flat and steep terrain. *Geodesy and Cartography*, 39, 41–52.

Tang, J., Pilesjö, P., Miller, P.A., Persson, A., Yang, Z., Hanna, E. & Callaghan, T.V. (2014) Incorporating topographic indices into dynamic ecosystem modeling using LPJ-GUESS. *Ecohydrology*, 7, 1147–1162.

Tang, J., Miller, L.A., Crill, P.M., Olin, S., and Pilesjö, P. (2015) Investigating the influence of two different flow routing algorithms on soil–water–vegetation interactions using the dynamic ecosystem model LPJ-GUESS. *Ecohydrology*, 8, 570–583.

Tarboton, D.G. (1997) A new method for the determination of flow directions and upslope areas in grid digital elevation models. *Water Resources Research*, 33, 309–319.

Tarboton, D.G. (2016) *Terrain Analysis Using Digital Elevation Models (TauDEM)*. Logan, UT: Utah Water Research Laboratory, Utah State University.

Tarboton, D.G., Bras, R.L. & Rodriguez-Iturbe, I. (1988) The fractal nature of river networks. *Water Resources Research*, 24, 1317–1322.

Tarboton, D.G., Bras, R.L. & Rodriguez-Iturbe, I. (1991) On the extraction of channel networks from digital elevation data. *Water Resources Research*, 33, 309–319.

Tarboton, D.G., Bras, R.L. & Rodriguez-Iturbe, I. (1992) A physical basis for drainage density. *Geomorphology*, 5, 59–76.

Tarboton, D.G., Idaszak, R., Horsburgh, J.S., Ames, D.P., Goodall, J.L., Band, L.E. et al. (2015a) *Clearing your desk! Software and data services for collaborative web-based GIS analysis*. Retrieved from https://agu.confex.com/agu/fm15/meetingapp.cgi/Paper/76460

Tarboton, D.G., Idaszak, R., Horsburgh, J.S., Ames, D., Goodall, J.L., Band, L. et al. (2015b) *HydroShare: Advancing hydrology through collaborative data and model sharing*. Retreived from https://www.hydroshare.org/resource/dc52e6aa93154521af08522de27ec276/

Tarolli, P. (2014) High-resolution topography for understanding Earth surface processes: Opportunities and challenges. *Geomorphology*, 216, 295–312.

Taverna, K., Urban, D.L. & MacDonald, R.I. (2005) Modeling landscape vegetation pattern in response to historic land-use: A hypothesis-driven approach for the North Carolina Piedmont, USA. *Landscape Ecology*, 20, 689–702.

Temme, A.J.A.M. & Veldkamp, A. (2009) Multi-process Late Quaternary landscape evolution modeling reveals lags in climate response over small spatial scales. *Earth Surface Processes and Landforms*, 34, 573–589.

Temme, A.J.A.M., Schoorl, J.M. & Veldkamp, A. (2006) An algorithm for dealing with depressions in dynamic landscape evolution models. *Computers and Geosciences*, 32, 452–461.

Temme, A.J.A.M., Heuvelink, G.B.M., Schoorl, J.M. & Claessens, L. (2009) Geostatistical simulation and error propagation in geomorphometry. In: T. Hengl & H.I. Reuter (eds) *Geomorphometry: Concepts, Software, Applications*, pp. 121–140. Amsterdam, Netherlands: Elsevier.

Temme, A.J.A.M., Claessens, L., Veldkamp, A. & Schoorl, J.M. (2011) Evaluating choices in multi-process landscape evolution models. *Geomorphology*, 125, 271–281.

Tesfa, T.K., Tarboton, D.G., Watson, D.W., Schreuders, K.A., Baker, M.E. & Wallace, R.M. (2011) Extraction of hydrological proximity measures from DEMs using parallel processing. *Environmental Modeling and Software*, 26, 1696–1709.

Thoma, D.P., Gupta, S.C., Bauer, M.E. & Kirchoff, C.E. (2005) Airborne laser scanning for riverbank erosion assessment. *Remote Sensing of Environment*, 95, 493–501.

Thompson, J.A., Bell, J.C. & Butler, C.A. (2001) Digital elevation model resolution: Effects on terrain attribute calculation and quantitative soil-landscape modeling. *Geoderma*, 100, 67–89.

Thorne, C.R., Zevenbergen, L.W., Grissinger, F.H. & Murphey, J.B. (1986) Ephemeral gullies as sources of sediment. In: *Proceedings of the 4th Federal Interagency Sedimentation Conference*, Vol. 1, pp. 3.152–3.161. Las Vegas, NV.

Thornton, P.E. & Running, S.W. (1999) An improved algorithm for estimating incident daily solar radiation from measurements of temperature, humidity, and precipitation. *Agricultural and Forest Meteorology*, 93, 211–228.

Thornton, P.E., Running, S.W. & White, M.A. (1997) Generating surfaces of daily meteorological variables over large regions of complex terrain. *Journal of Hydrology*, 190, 214–251.

Thornton, P.E., Hasenauer, H. & White, M.A. (2000) Simultaneous estimation of daily solar radiation and humidity from observed temperature and precipitation: An application over complex terrain in Austria. *Agricultural and Forest Meteorology*, 104, 255–271.

Tomer, M.D. & Anderson, J.L. (1995) Variation in soil water storage across a sand plain hillslope. *Soil Science Society of America Proceedings*, 54, 1091–1100.

Toutin, Th. & Cheng, P. (2000) Demystification of IKONOS. *Earth Observation Magazine*, 9, 17–21.

Trevisani, S. & Cavalli, M. (2016) Topography-based flow-directional roughness: potential and challenges. *Earth Surface Dynamics*, 4, 343–358.

Tribe, A. (1992) Automated recognition of valley lines and drainage networks from grid digital elevation models: A review and a new method. *Journal of Hydrology*, 139, 263–293.

Trimble, S.W. (1983) A sediment budget for Coon Creek basin in the Driftless Area, Wisconsin, 1853–1977. *American Journal of Science*, 283, 454–474.

Trimble, S.W. (1999) Decreased rates of alluvial sediment storage in the Coon Creek Basin, Wisconsin, 1975–93. *Science*, 285, 1244–1246.

Trimble, S.W. (2009) Fluvial processes, morphology and sediment budgets in the Coon Creek Basin, WI, USA, 1975–1993. *Geomorphology*, 108, 8–23.

Troeh, F.R. (1964) Landform paramters correlated to soil drainage. *Soil Science Society of America Proceedings*, 59, 808–812.

Troeh, F.R. (1965) Landform equations fitted to contour maps. *American Journal of Science*, 263, 616–627.

Troutman, B.M. & Karlinger, M.R. (1984) On the expected width function of topologically random channel networks. *Journal of Applied Probability*, 21, 836–849.

True, D. (2002) *Landforms of the Lower Mid-West*. Columbia, MO: University of Missouri MoRAP Map Series MS-2003-001.

Tseng, C.-M., Lin, C.-W., Stark, C.P., Liu, J.-K., Fei, L.-Y. & Hsieh, Y.-C. (2013) Application of a multi-temporal, LiDAR-derived, digital terrain model in a landslide-volume estimation. *Earth Surface Processes and Landforms*, 38, 1587–1601.

Tucker, G.E., Lancaster, S.T., Gasparini, N.M., Bras, R.L. & Rybarczyk, S.M. (2001) An object-oriented framework for distributed hydrologic and geomorphic modeling using triangulated irregular networks. *Computers and Geosciences*, 27, 959–973.

US Department of Agriculture, Natural Resources Conservation Service (2000) *1997 Natural Resources Inventory*. Washington, DC: US Department of Agriculture, Natural Resources Conservation Service.

US Department of Agriculture, Natural Resources Conservation Service (2009) *Soil Data Viewer*. Retrieved from https://www.nrcs.usda.gov/wps/portal/nrcs/detail/soils/survey/geo/?cid=nrcs142p2_053614

US Department of Agriculture, Natural Resources Conservation Service (2016) *Web Soil Survey*. Retrieved from http://websoilsurvey.nrcs.usda.gov/

Usery, E.L. (1996) A conceptual framework and fuzzy set implementation of geographic features. In: P.A. Burrough & A.U. Frank (eds) *Geographic Objects with Indeterminate Boundaries*, pp. 71–85. London, UK: Taylor & Francis.

US Geological Survey (1999) *Map Accuracy Standards Fact Sheet FS-171-99*. Retrieved from http://pubs.usgs.gov/fs/1999/0171/

US Geological Survey (2000) *HYDRO1k elevation derivative database*. Sioux Falls, SD: US Geological Survey Earth Resources Observation and Science (EROS) Center.

US Geological Survey (2015) *About NED*. Retrieved from http://ned.usgs.gov/about.html

Van Engelen, V.W.P. & Ting-Tiang, W. (1995) *Global and National Soils and Terrain Digital Databases Procedures Manual*. Wageningen, Netherlands: United Nations Food and Agriculture Organization, Land and Water Division, World Soil Resources Report No. 74.

Van Niel, K.P. & Austin, M.P. (2007) Predictive vegetation modeling for conservation: Impact of error propagation from digital elevation data. *Ecological Applications*, 17, 266–280.

Van Niel, K.P., Laffan, S.W. & Lees, B.G. (2004) Effect of error in the DEM on environmental variables for predictive vegetation modeling. *Journal of Vegetation Science*, 15, 747–756.

Van Remortel, R.D., Hamilton, M.E. & Hickey, R.J. (2001) Estimating the LS factor for RUSLE through iterative slope length processing of digital elevation data. *Cartography*, 30, 27–35.

Van Remortel, R.D., Maichle, R.W. & Hickey, R.J. (2004) Computing the LS factor for the Revised Universal Soil Loss Equation through array-based slope processing of digital elevation data using a C++ executable. *Computers and Geosciences*, 30, 1043–1053.

Vayssières, M.P., Plant, R.E. & Allen-Diaz, B.H. (2000) Classification trees: An alternative non-parametric approach for predicting species distributions. *Journal of Vegetation Science*, 11, 679–694.

Vaze, J., Teng, J. & Spencer, G. (2010) Impact of DEM accuracy and resolution on topographic indices. *Environmental Modeling and Software*, 25, 1086–1098.

Veatch, J.O. (1935) Graphic and quantitative comparisons of land types. *Journal of the American Society of Agronomy*, 27, 505–510.

Velleux, M.L., England, J.F. Jr & Julien, P.Y. (2008) TREX: Spatially distributed model to assess watershed contamination transport and fate. *Science of the Total Environment*, 404, 113–128.

Ventura, G., Vilardo, G., Terranova, C. & Sessa, E.B. (2011) Tracking and evolution of complex active landslides by multi-temporal airborne LiDAR data: The Montaguto landslide (Southern Italy). *Remote Sensing of Environment*, 115, 3237–3248.

Ventura, S.C. & Irvin, B.J. (2000) Automated landform classification for soil-landscape studies. In: J.P. Wilson & J.C. Gallant (eds) *Terrain Analysis: Principles and Applications*, pp. 276–294. New York, NY: John Wiley & Sons, Inc.

Verdin, K.L., Godt, J.W., Funk, C., Pedreros, D., Worstell, D. & Verdin J. (2007) *Development of a Global Slope Dataset for Estimation of Landslide Occurrence Resulting from Earthquakes*. Reston, VA: US Geological Survey Open-File Report No. 2007–1188.

Vieux, B.E. (1993) DEM aggregation and smoothing effects on surface runoff modeling. *Journal of Computing in Civil Engineering*, 7, 310–338.

Vieux, B.E. & Needham, S. (1993) Non-point pollution model sensitivity to grid-cell size. *Journal of Water Resources Planning and Management*, 119, 141–157.

Vivoni, E.R., Ivanov, Y.Y., Bras, R.L. & Entekhabi, D. (2004) Generation of triangulated irregular networks based on hydrologic similarity. *Journal of Hydraulic Engineering*, 9, 288–302.

Vivoni, E.R., Teles, V., Ivanov, V.Y., Bras, R.L. & Entekhabi, D. (2005) Embedding landscape processes into triangulated terrain models. *International Journal of Geographical Information Science*, 19, 249–457.

Wack, R. & Wimmer, A. (2002) Digital terrain models from airborne laser scanner data: A grid-based approach. *International Archives of Photogrammetry and Remote Sensing*, 35, 293–296.

Walker, J.P. & Willgoose, G.R. (1999) On the effect of digital elevation model accuracy on hydrology and geomorphology. *Water Resources Research*, 35, 2259–2266.

Walsh, S.J. (1989) User considerations in landscape characterization. In: M.F. Goodchild & S. Gopal (eds) *The Accuracy of Spatial Databases*, pp. 35–43. London, UK: Taylor & Francis.

Walsh, S.J., Lightfoot, D.R. & Butler, D.R. (1987) Recognition and assessment of error in geographic information systems. *Photogrammetric Engineering and Remote Sensing*, 53, 1423–1430.

Wang, H., Zhou, Y., Fu, J., Gao, J. & Wang, G. (2012) Maximum speedup ratio curve (MSC) in parallel conputing of the binary-tree-based drainage network. *Computers and Geosciences*, 38, 127–135.

Wang, L. & Liu, H. (2006) An efficient method for identifying and filling surface depressions in digital elevation models for hydrologic analysis and modeling. *International Journal of Geographical Information Science*, 20, 193–213.

Wang, S. (2010) A CyberGIS framework for the synthesis of cyberinfrastructure, GIS, and spatial analysis. *Annals of the Association of American Geographers*, 100, 535–557.

Wang, S., Anselin, L., Bhaduri, B., Crosby, C., Goodchild, M.F., Liu, Y. & Nyerges, T.L. (2013) CyberGIS software: A synthetic review and integration roadmap. *International Journal of Geographical Information Science*, 27, 2122–2145.

Wania, R., Ross, I. & Prentice, I.C. (2009a) Integrating peatlands and permafrost into a dynamic global vegetation model: 1. Evaluation and sensitivity of physical land surface processes. *Global Biogeochemical Cycles*, 23, GB3014.

Wania, R., Ross, I. & Prentice, I.C. (2009b) Integrating peatlands and permafrost into a dynamic global vegetation model: 2. Evaluation and sensitivity of vegetation and carbon cycle processes. *Global Biogeochemical Cycles*, 23, GB3015.

Wania, R., Ross, I. & Prentice, I.C. (2010) Implementation and evaluation of a new methane model within a dynamic global vegetation model: LPJ-WHyMe v1.3.1. *Geoscientific Model Development*, 3, 565–584.

Warren, S.D., Hohmann, M.G., Auerswald, K. & Mitášová, H. (2004) An evaluation of methods to determine slope using digital elevation data. *Catena*, 58, 215–233.

Webster, T.L. & Dias, G. (2006) An automated GIS procedure for comparing GPS and proximal LiDAR elevations. *Computers and Geosciences*, 32, 713–726.

Wechsler, S.P. (2007) Uncertainities associated with digital elevation models for hydrologic models: A review. *Hydrology and Earth Systems Science*, 11, 1481–1500.

Wechsler, S.P. & Knoll, C.N. (2006) Quantifying DEM uncertainty and its effect on topographic parameters. *Photogrammetric Engineering and Remote Sensing*, 72, 1081–1090.

Wehr, A. & Lohr, U. (1999) Airborne laser scanning: An introduction and overview. *ISPRS Journal of Photogrammetry and Remote Sensing*, 54, 68–82.

Weibel, R. (1992) Models and experiments for adaptive computer-assisted terrain generalization. *Cartography and Geographic Information Science*, 19, 133–153.

Weiss, A.D. (2001) Topographic position and landforms analysis. *Poster presented at the Esri International Users Conference*, San Diego, CA.

Weller, D.E., Jordan, R.E. & Correll, D.L. (1998) Heuristic models for material discharge from landscapes with riparian buffers. *Ecological Applications*, 8, 1159–1169.

Weller, D.E., Baker, M.E. & Jordan, R.E. (2011) Effects of riparian buffers on nitrate concentrations in watershed discharges: New models and management applications. *Ecological Applications*, 21, 1679–1695.

Western, A.W., Grayson, R.B., Blöschl, G., Willgoose, G.R. & McMahon, T.R. (1999) Observed spatial organization of soil moisture and its relation to terrain indices. *Water Resources Research*, 35, 797–810.

Wheaton, J.M., Brasington, J., Darby, S.E. & Sear, D. (2010) Accounting for uncertainty in DEMs from repeat topographic surveys: Improved sediment budgets. *Earth Surface Processes and Landforms*, 35, 136–156.

Whipple, K. & Tucker, G. (1999) Dynamics of the stream-power river incision model: Implications for height limits of mountain ranges, landscape response timescales, and research needs. *Journal of Geophysical Research*, 104, 17661–17674.

Wieczorek, G.F. & Snyder, J.B. (2009) Monitoring slope movements. In: R. Young & L. Norby (eds) *Geological Monitoring*, pp. 245–271. Boulder, CO: Geological Society of America.

Wilby, R.L. & Wigley, T.M. L. (1997) Downscaling general circulation model output: A review of methods and limitations. *Progress in Physical Geography*, 21, 530–548.

Wilson, D.J., Western, A.M. & Grayson, R.B. (2005) A terrain and data-based method for generating the spatial distribution of soil moisture. *Advances in Water Resources*, 28, 43–54.

Wilson, J.P. (1986) Estimating the topographic factor in the universal soil loss equation for watersheds. *Journal of Soil and Water Conservation*, 41, 179–184.

Wilson, J.P. (2012) Digital terrain modeling. *Geomorphology*, 137, 107–121.

Wilson, J.P. & Bishop, M.P. (2013) Geomorphometry. In: J.F. Shroder (ed.) *Treatise in Geomorphology. Volume 3, Remote Sensing and GIScience in Geomorphology*, pp. 162–186. San Diego, CA: Academic Press.

Wilson, J.P. & Burrough, P.A. (1999) Dynamic modeling, geostatistics, and fuzzy classification: New sneakers for a new geography? *Annals of the Association of American Geographers*, 89, 736–746.

Wilson, J.P. & Gallant, J.C. (1996) EROS: A grid-based program for estimating spatially distributed erosion indices. *Computers and Geosciences*, 22, 707–712.

Wilson, J.P. & Gallant, J.C. (eds) (2000a) *Terrain Analysis: Principles and Applications*. New York, NY: John Wiley & Sons, Inc.

Wilson, J.P. & Gallant, J.C. (2000b) Digital terrain analysis. In: J.P. Wilson & J.C. Gallant (eds) *Terrain Analysis: Principles and Applications*, pp. 1–27. New York, NY: John Wiley & Sons, Inc.

Wilson, J.P. & Gallant, J.C. (2000c) Secondary topographic attributes. In: J.P. Wilson & J.C. Gallant (eds) *Terrain Analysis: Principles and Applications*, pp. 51–85. New York, NY: John Wiley & Sons, Inc.

Wilson, J.P. & Lorang, M.S. (1999) Spatial models of soil erosion and GIS. In: A.S. Fotheringham & M. Wegener (eds) *Spatial Models and GIS: New Potential and New Models*, pp. 83–108. London, UK: Taylor & Francis.

Wilson, J.P., Repetto, P.L. & Snyder, R.D. (2000) Effect of data source, grid resolution, and flow-routing method on computed topographic attributes. In: J.P. Wilson & J.C. Gallant (eds) *Terrain Analysis: Principles and Applications*, pp. 133–161. New York, NY: John Wiley & Sons, Inc.

Wilson, J.P., Lam, C.S. & Deng, Y.X. (2007) Comparison of performance of flow routing algorithms used in geographic information systems. *Hydrological Processes*, 21, 1026–1044.

Wilson, J.P., Aggett, G.R., Deng, Y.X. & Lam, C.S. (2008) Water in the landscape: A review of contemporary flow routing algorithms. In: Q. Zhou, B.G. Lees & G.-A. Tang (eds) *Advances in Digital Terrain Analysis*, pp. 213–236. Berlin, Germany: Springer Lecture Notes in Geoinformation and Cartography.

Winchell, M.F., Jackson, S.H., Wadley, A.M. & Srinivasan, R. (2008) Extension and validation of a geographic information system-based method for calculating the Revised Universal Soil Loss Equation length–slope factor for erosion risk assessments in large watersheds. *Journal of Soil and Water Conservation*, 63, 105–111.

Wischmeier, W.H. & Smith, D.D. (1978) *Predicting Rainfall Erosion Losses*. Agriculture Handbook No. 537. Washington, DC: US Department of Agriculture.

Wisconsin Department of Natural Resources (2016) Erosion vulnerability assessment for agricultural lands. Retrieved from http://dnr.wi.gov/topic/nonpoint/evaal.html.

Wise, S.M. (1998) The effect of GIS interpolation errors on the use of digital elevation models in geomorphology. In: S.N. Lane, K.S. Richards & J.H. Chandler (eds) *Landform Monitoring, Modeling and Analysis*, pp. 139–164. Chichester, UK: John Wiley & Sons Ltd.

Wise, S.M. (2000a) GIS data modeling: Lessons from the analysis of DTMs. *International Journal of Geographical Information Science*, 14, 313–318.

Wise, S.M. (2000b) Assessing the quality of hydrological applications of digital elevation models derived from contours. *Hydrological Processes*, 14, 1909–1929.

Wise, S.M. (2010) Assessing the spatial characteristics of DEM interpolation error. In: *Proceedings of the 9th International Symposium on Spatial Data Assessment in Natural Resources and the Environmental Sciences*, pp. 117–119. Leicester, UK.

Wolf, A. (2011) Estimating the potential impact of vegetation on the water cycle requires accurate soil water parameter estimation. *Ecological Modeling*, 222, 2595–2605.

Wolock, D.M. & McCabe, G.J. (1995) Comparison of single and multiple flow direction algorithms for computing topographic parameters in TOPMODEL. *Water Resources Research*, 31, 1315–1324.

Wolock, D.M. & Price, C.V. (1994) Effects of digital elevation model and map scale and data resolution on a topography-based watershed model. *Water Resources Research*, 30, 3041–3052.

Wood, E.F., Sivapalan, M. & Beven, K.J. (1990) Similarity and scale in catchment storm response. *Reviews in Geophysics*, 28, 1–18.

Wood, E.F., Roundy, J.K., Troy, T.J., van Beek, L.P.H., Bierkens, M.F.P., Blyth, E. et al. (2011) Hyperresolution global land surface modeling: Meeting a grand challenge for monitoring Earth's terrestrial water. *Water Resources Research*, 47, W05301.

Wood, J. (1996a) Scale-based characterization of digital elevation models. In: D. Parker (ed.) *Innovations in GIS 3*, pp. 163–175. London, UK: Taylor & Francis.

Wood, J. (1996b) *The geomorphological characterization of digital elevation models*. Unpublished PhD dissertation, University of Leicester, Leicester, UK.

Wood, J. (1998) Modelling the continuity of surface form using digital elevation models. In: *Proceedings of the 8th International Symposium on Spatial Data Handling*, pp. 725–736. Burnaby, British Columbia.

Wood, J. (2009a) Overview of software packages used in geomorphometry. In: T. Hengl & H.I. Reuter (eds) *Geomorphometry: Concepts, Software, Applications*, pp. 257–267. Amsterdam, Netherlands: Elsevier.

Wood, J. (2009b) Geomorphometry in LandSerf. In: T. Hengl & H.I. Reuter (eds) *Geomorphometry: Concepts, Software, Applications*, pp. 333–349. Amsterdam, Netherlands: Elsevier.

Wood, R., Sivapalan, M. & Robinson, J. (1997) Modeling the spatial variability of surface runoff using a topographic index. *Water Resources Research*, 33, 1061–1073.

Woodcock, C.E. & Strahler, A.H. (1987) The factor of scale in remote sensing. *Remote Sensing of the Environment*, 21, 311–332.

Woodrow, K., Lindsay, J.B. & Berg, A.A. (2016) Evaluating DEM conditioning techniques, elevation source data, and grid resolution for field-scale hydrological parameter extraction. *Journal of Hydrology*, 540, 1022–1029.

Wright, D.J. & Wang, S. (2011) The emergence of spatial cyberinfrastructure. *Proceedings of the National Academy of Sciences USA*, 108, 5488–5491.

Wright, J.W., Moore, A.B. & Leonard, G.H. (2014) Flow direction algorithms in a hierarchical hexagonal surface model. *Journal of Spatial Science*, 59, 333–346.

Wu, S., Li, J. & Huang, G.H. (2008) A study on DEM-derived primary topographic attributes for hydrological applications: Sensitivity to elevation data resolution. *Applied Geography*, 28, 210–223.

Yang, B., Shi, W. & Li, Q. (2005) An integrated TIN and grid method for constructing multi-resolution digital terrain models. *International Journal of Geographical Information Science*, 19, 1019–1038.

Yi, L., Zhang, W.-C. & Yan, C.-A. (2017) Modified topographic index that incorporates the hydraulic and physical properties of soil. *Hydrology Research*, 48 (in press).

Yildirim, A.A., Watson, D., Tarboton, D.G. & Wallace, R.M. (2015) A virtual tile approach to raster-based calculations of large digital elevation models in a shared memory system. *Computers and Geosciences*, 82, 78–88.

Yildirim, A.A., Tarboton, D.G., Liu, Y.Y., Sazib, N.S. & Wang, S. (2016) Accelerating TauDEM for extracting hydrology information from a national-scale high resolution topographic dataset. In: *Proceedings of the 5th Annual Extreme Science and Engineering Discovery Environment Conference*, Miami, FL.

Ying, L.-X., Shen, Z.-H., Piao, S.-L. & Malanson, G.P. (2014) Terrestrial surface area increment: The effects of topography, DEM resolution, and algorithm. *Physical Geography*, 35, 297–312.

Yitayew, M., Pokrzywka, S.J. & Renard, K.G. (1999) Using GIS for facilitating erosion estimation. *Applied Engineering in Agriculture*, 15, 295–301.

Yokoyama, R., Shirasawa, M. & Pike, R.J. (2002) Visualizing topography by openness: A new application of image processing to digital elevation models. *Photogrammetric Engineering and Remote Sensing*, 68, 257–265.

Yong, B., Ren, L.-L., Chen, X., Zhang, Y., Zhang, W.-C., Fu, C.-B. & Niu, G.-Y. (2009) Development of a large-scale hydrological model TOPX and its coupling with Regional Integrated Environment Modeling System RIEMS. *Chinese Journal of Geophysics*, 52, 762–771.

Young, M. (1978) *Terrain Analysis: Program Documentation: Report No. 6 on Grant DA-ERO-591-73-G0040 – Statistical Characterization of Altitude Matrices by Computer*. Durham, UK: Department of Geography, University of Durham.

Yue, T.X., Du, Z.P., Song, D.J. & Gong, Y. (2007) A new method of surface modeling and its application to DEM construction. *Geomorphology*, 91, 161–172.

Zakerinejad, R. & Maerker, M. (2015) An integrated assessment of soil erosion dynamics with special emphasis on gully erosion in the Mazayjan basin, southwestern Iran. *Natural Hazards*, 79 (Suppl. 1), 25–50.

Zakšed, K. & Podobnikar, T. (2005) An effective DEM generalization with basic GIS operations. In: *Proceedings of the 8th ICA Workshop on Generalization and Multiple Representations*, A Coruńa, Spain.

Zandbergen, P.A. (2008) Applications of Shuttle radar topography mission elevation data. *Geography Compass*, 2, 1404–1431.

Zandbergen, P.A. (2012) *Python Scripting for ArcGIS*. Redlands, CA: Esri Press.

Zevenbergen, L.W. & Thorne, C.R. (1987) Quantitative analysis of land surface topography. *Earth Surface Processes and Landforms*, 12, 47–56.

Zhang, H., Han, W., Yang, Y., Yu, S., Li., S. & Zhao, X. (2013a) DEM-based extraction of LS factor: Integrate channel networks and convergence flow. In: *Proceedings of the 3rd International Conference on Geomorphometry*, Nanjing, China.

Zhang, H., Yang, Q., Li, R., Liu, Q., Moore, D., He, P., Ritsema, C.J. & Geissen, V. (2013b) Extension of a GIS procedure for calculating the RUSLE equation LS factor. *Computers and Geosciences*, 52, 177–188.

Zhang, K., Chen, S.-C., Whitman, D., Shyu, M.-L., Yan, J. & Zhang, C. (2003) A progressive morphological filter for removing non-ground measurements from airborne LiDAR data. *IEEE Transactions on Geoscience and Remote Sensing*, 41, 872–882.

Zhang, W., Miller, P.A., Smith, B., Wania, R., Koenigk, T. & Döscher, R. (2013) Tundra shrubification and tree-line advance amplify arctic climate warming: Results from an individual-based dynamic vegetation model. *Environmental Research Letters*, 8, 034023.

Zhang, W.H. & Montgomery, D.R. (1994) Digital elevation model grid size, landscape representation, and hydrological simulations. *Water Resources Research*, 30, 1019–1028.

Zhang, Z., Zimmermann, N.E., Kaplan, J.O. & Poulter, B. (2016) Modeling spatiotemporal dynamics of global wetlands: Comprehensive evaluation of a new sub-grid TOPMODEL parameterization and uncertainties. *Biogeosciences*, 13, 1387–1408.

Zhao, G.-J., Gao, J.-F., Tian, P. & Tian, K. (2009) Comparison of two different methods for determining flow direction in catchment hydrological modeling. *Water Science and Engineering*, 2(4), 1–15.

Zhao, J. (1995) *Physical Geography of China*. Beijing, China: Higher Education Press. (In Chinese)

Zhou, Q. & Chen, Y. (2011) Generalization of DEM for terrain analysis using a compound method. *ISPRS Journal of Photogrammetry and Remote Sensing*, 66, 38–45.

Zhou, Q. & Liu, X. (2002) Error assessment of grid-based flow routing algorithms used in hydrological models. *International Journal of Geographical Information Science*, 16, 819–842.

Zhou, Q. & Liu, X. (2004a) Analysis of errors of derived slope and aspect related to DEM data properties. *Computers and Geosciences*, 30, 369–378.

Zhou, Q. & Liu, X. (2004b) Error analysis on grid-based slope and aspect algorithms. *Photogrammetric Engineering and Remote Sensing*, 70, 957–962.

Zhou, Q., Liu, X. & Sun, Y. (2006) Terrain complexity and uncertainties in grid-based digital terrain analysis. *International Journal of Geographical Information Science*, 20, 1137–1147.

Zhou, Q., Lees, B.G. & Tang, G.-A. (eds) (2008) *Advances in Digital Terrain Analysis*. Berlin, Germany: Springer Lecture Notes in Geoinformation and Cartography.

Zhou, Q., Pilesjö, P. & Chen, Y. (2011) Estimating surface flow paths on a digital elevation model using a triangular facet network. *Water Resources Research*, 47, W07522.

Zhu, A.-X. (1997a) A similarity model for representing soil spatial information. *Geoderma*, 77, 217–242.

Zhu, A.X. (1997b) Measuring uncertainty in class assignment for natural resource maps under fuzzy logic. *Photogrammetric Engineering and Remote Sensing*, 63, 1195–1202.

Zhu, A.X. (1999) A personal construct-based knowledge acquisition process for natural resource mapping. *International Journal of Geographical Information Science*, 13, 119–141.

Zhu, A.-X. & Mackay, D.S. (2001) Effects of spatial detail of soil information on watershed modeling. *Journal of Hydrology*, 248, 54–77.

Zhu, A.-X., Band, L.E., Vertessy, R. & Dutton, B. (1997) Derivation of soil properties using a soil land inference model (SoLIM). *Soil Science Society of America Journal*, 61, 523–533.

Zhu, A.-X., Hudson, B., Burt, J.E. & Lubich, K. (2001) Soil mapping using GIS, expert knowledge, and fuzzy logic. *Soil Science Society of America Journal*, 65, 1463–1472.

Zhu, A.-X., Burt, J.E., Smith, M., Wang, R.X. & Gao, J. (2008) The impact of neighborhood size on terrain derivatives and digital soil mapping. In: Q. Zhou, B. Lees & G. Tang (eds) *Advances in Digital Terrain Analysis*, pp. 333–348. New York, NY: Springer.

Zhu, X., Zhuang, Q., Lu, X. & Song, L. (2014) Spatial scale-dependent land–atmospheric methane exchanges in the northern high latitudes from 1993 to 2004. *Biogeosciences*, 11, 1693–1704.

Zinko, U., Seibert, J., Dynesius, M. & Nilsson, C. (2005) Plant species numbers predicted by a topography based groundwater-flow index. *Ecosystems*, 8, 430–441.

Zulkafli, Z., Buytaert, W., Onof, C., Lavado, W. & Guyot, J.L. (2013) A critical assessment of the JULES land surface model hydrology for humid tropical environments. *Hydrology and Earth Systems Science*, 17, 1113–1132.

# Index



*Environmental Applications of Digital Terrain Modeling*, First Edition. John P. Wilson.

Printed in the USA/Agawam, MA
July 20, 2026

907143.109